E. Coli Infections - Importance of Early Diagnosis and Efficient Treatment

Edited by Luis Rodrigo

Published in London, United Kingdom

IntechOpen

Supporting open minds since 2005

E. Coli Infections - Importance of Early Diagnosis and Efficient Treatment
http://dx.doi.org/10.5772/intechopen.80139
Edited by Luis Rodrigo

Contributors
Marit Nilsen-Hamilton, Soma Banerjee, Baoping Zhang, Paul Onkundi Nyangaresi, Liang Shen, Yanhong Liu, Peng Ji, Xunde Li, Saloua Helali, Adnane Abdelghani, Luis Ángel Fernández, Massiel Cepeda-Molero, Stephanie Schüller, Gad Frankel, Hussein O.M. Al-Dahmoshi, Noor Salman Al-Khafaji, Ukun Soedjanaatmadja, Iman P. Maksum, Safri Ishmayana, Ahmad Nabiel, Mario Galindo-Méndez, Ashok Palaniappan, Sangeetha Muthamilselvan, Nivedita Jaishankar, Akshay Kumar, Nimisha Shiwalkar, Ana Francesca Vommaro Leite, Lais Sanches Maekewa, Silas Jose Braz Filho, Purnadeo Narpaul Persaud, Roopvir Kaur, Asim Kichloo, Juber Dastagir Shaikh, Mohammed H.O. Al-Allak

First published in London, United Kingdom, 2020 by IntechOpen
IntechOpen is the global imprint of INTECHOPEN LIMITED, registered in England and Wales, registration number: 11086078, 5 Princes Gate Court, London, SW7 2QJ, United Kingdom
Printed in Croatia

British Library Cataloguing-in-Publication Data
A catalogue record for this book is available from the British Library

Additional hard and PDF copies can be obtained from orders@intechopen.com

E. Coli Infections - Importance of Early Diagnosis and Efficient Treatment
Edited by Luis Rodrigo
p. cm.
Print ISBN 978-1-83962-523-7
Online ISBN 978-1-83962-524-4
eBook (PDF) ISBN 978-1-83962-525-1

Meet the editor

Luis Rodrigo, MD, is Emeritus Professor of Medicine at the University of Oviedo, Spain. He has been Chief of Gastroenterology Service at HUCA Hospital in Oviedo for more than 40 years. He obtained a PhD in 1975 and has had a long teaching and research career. He has published 590 scientific papers in both English and Spanish and written thirty-five book chapters and edited twenty-four books. Dr. Rodrigo has participated as a main investigator in forty-five clinical trials. He has directed forty doctoral theses and contributed to the formation of 100 specialists in gastroenterology working in hospitals in Spain and abroad. His areas of interest are celiac disease and autoimmune diseases.

Contents

Preface

Food-borne microorganisms are major pathogens that affect food safety and cause human illness worldwide. Bacteria are the causative agents of two-thirds of human food-borne diseases globally, with high burden in developing countries. Meat, dairy products, and eggs are the main avenues by which people are exposed to zoonotic bacteria such as *Staphylococcus aureus*, *Salmonella*, *Campylobacter*, *Listeria monocytogenes*, and *Escherichia coli*. These are the major zoonotic bacterial pathogens responsible for food-borne illness and death associated with consumption of contaminated animal products.

Production of toxins and structural virulent factors are responsible for the pathogenesis of these bacteria. These major zoonotic bacteria cause human infections characterized mainly by gastrointestinal symptoms including nausea, vomiting, diarrhea, abdominal cramps, and other agent-specific symptoms. Some bacteria may cause severe complications. Conventional (culturing), serological, and molecular techniques are important for detecting these common zoonotic bacteria and their toxins in food.

The emergence of multidrug-resistant zoonotic bacteria associated with consumption of contaminated animal products is a great concern for public health. As such, there should be coordinated surveillance and a monitoring system for food-borne zoonotic bacterial pathogens, particularly in developing countries. International trade and the ever-growing flow of goods and people enable animal diseases and zoonotic pathogens to travel worldwide. The risk of reintroducing previously eradicated animal diseases is omnipresent all over the world, as considerable amounts of food products of animal origin from endemic countries are continuously imported legally and illegally.

E. coli is one of the most-studied microorganisms worldwide, but its characteristics are continually changing. Extra-intestinal *E. coli* infections, such as urinary tract infections and neonatal sepsis, represent a huge public health problem. They are caused mainly by specialized extra-intestinal pathogenic *E. coli* (ExPEC) strains that can innocuously colonize human hosts, but can also cause disease upon entering a normally sterile body site. The virulence capability of such strains is determined by a combination of distinctive accessory traits called virulence factors in conjunction with their distinctive phylogenetic background. It is conceivable that by developing interventions against the most successful lineages or their key virulence/colonization factors, the associated burden of disease and healthcare costs could be reduced in the future. On the other hand, one important problem worldwide is the increase of antimicrobial resistance shown by bacteria. As underscored in the last World Health Organization (WHO) global report, within a wide range of infectious agents including *E. coli*, antimicrobial resistance has reached an extremely worrisome situation that "threatens the achievements of modern medicine."

This book includes ten chapters covering the main aspects of infections related to *E. coli*, their pathogenic mechanisms, treatments, and resistance to diverse antibiotics.

I want to thank all authors for their excellent contributions and the editorial team at IntechOpen, especially Mrs. Mia Vulovic for her continuous collaboration and kind support during the book's preparation.

Luis Rodrigo MD
Professor,
Emeritus Full Professor of Medicine,
University of Oviedo,
Oviedo, Spain

Section 1

Importance of *E. coli* Infections and Pathogenic Mechanism

Chapter 1

Fast Detection of Pathogenic *Escherichia coli* from Chicken Meats

Saloua Helali and Adnane Abdelghani

Abstract

Food is a means to sustain and enjoy life, but it is also a medium for microbial contamination, causing disease and death. Fruits, vegetables, meat, and water are the common sources of contamination. *Escherichia coli* is one of the most frequent Pathogenic Bacteria responsible for food poisoning and food-related infections. *E. coli* infection causes severe bloody diarrhea, abdominal cramps, and occasional vomiting. In the present study, electrochemical impedance spectroscopy (EIS), surface plasmon resonance (SPR), and physisorption techniques were evaluated to decrease sample preparation time and to improve the sensitivity and specificity for the detection of low levels of pathogenic *Escherichia coli* in frozen chicken meat. The electrical and optical properties of the immobilized anti-*E. coli* antibody were studied. Moreover, the developed biosensor was used for *E. coli* detection in inoculated frozen chicken meat.

Keywords: physisorption, impedance spectroscopy, SPR imaging, *E. coli*, frozen chicken meat

1. Introduction

According to the World Health Organization (WHO), foodborne illnesses are defined as diseases of infectious or toxic nature caused by consumption of contaminated foods or water. The main causes of foodborne illness are viruses, bacteria, parasites, toxins, metals, and prions where bacteria constitute 66% of the problems [1]. *Campylobacter*, *Salmonella*, *Yersinia enterocolitica*, *Clostridium perfringens*, *Listeria monocytogenes*, *Staphylococcus aureus*, and *E. coli O157:H7* are the major Pathogenic Bacteria causing different food illness [2]. The most common clinical symptoms of foodborne illnesses are diarrhea, vomiting, abdominal cramps, headache, and nausea. In many developing countries, the elderly and children under the age of 5 years are at higher risk of food infection. Among the foodborne pathogens, *E. coli* and *Salmonella* are the most common and frequent pathogens responsible for food poisoning and food-related infections. These two Pathogenic Bacteria can be transferred by poultry meat, red meat, desserts, and egg.

Nowadays, *Escherichia coli* and *Salmonella* are the most important and frequent pathogens responsible for food poisoning and food-related infections in chicken meat [2–6]. *E. coli* is a normal inhabitant of intestinal tracts and can be found in chicken feces, litter, dust, and rodent droppings [7]. Preventing food contamination and human infection from *E. coli* requires continuous control measures at all stages of the food production: from agricultural production to processing, manufacturing, transporting, storing, and preparation of foods in both commercial establishments and the domestic environment. Recently, many analytical applications have been developed for the determination of poultry meat contamination [8–10].

Several microbiological techniques such as conventional culturing, PCR, and ELISA are still considered the oldest and the most accurate approach for bacteria detection. These techniques need traditional sample preparation, though very efficient in extracting the target analyte, which is time-consuming and produces large amount of solvent wastes. Among the various techniques, electrochemical impedance spectroscopy technique and surface plasmon resonance imaging have previously been investigated to study the detection of Pathogenic Bacteria on gold [11]. These two techniques offer several advantages: First, they are label-free and direct detection method for biomolecular interactions because the measurements are based on small electric signal and very large range of frequency (100 mHz–100 kHz) and refractive index changes [12–14]. The analyte does not require any special characteristics (scattering bands) or labels (radioactive or fluorescent) and can be detected directly without the need for multistep detection protocols (sandwich assay). Second, the measurements can be performed in real time, allowing the user to collect kinetic data, as well as thermodynamic data.

In this contribution, an innovative way for sensitive detection of *E.coli* bacteria based on anti-*E. coli* antibody immobilized onto gold surface by physisorption technique is presented. The electrical properties of the immobilization of anti-*E. coli* antibody were studied. Moreover, the developed biosensor was used for *E. coli* detection in inoculated frozen chicken meat.

2. What is *Escherichia coli* (*E. coli*)?

- *Escherichia coli* are gram-negative bacilli of the family *Enterobacteriaceae*.
- *E. coli* is commonly found in the lower intestine of warm-blooded organisms.
- *E. coli* is the most common human and animal pathogens as it is responsible for a broad spectrum of diseases.
- There are many different types (strains) of *E. coli* which cause a number of illnesses. Virulence types of *E. coli* include enterotoxigenic (ETEC), enteroinvasive (EIEC), enteropathogenic (EPEC), and enterohemorrhagic *E. coli* (EHEC) [15].

3. History

In 1982, Riley LW et al. and colleagues were the first to recognize the EHEC serotype O157:H7 as a human pathogen associated with outbreaks of bloody diarrhea in Oregon and Michigan, USA [16]. Since then, *E. coli O157:H7* has become one of the most important foodborne pathogens.

4. Symptoms and mode of transmission

Virulent strains of *Escherichia coli* are responsible for most diarrheal infections, meningitis, septicemia, and urinary tract infections in children worldwide. A person who is infected with *E. coli O157* can pass it on to other people if there is a situation of insufficient hygiene or handwashing. Small children can still pass the infection on for a couple of weeks after they have recovered from any illness.

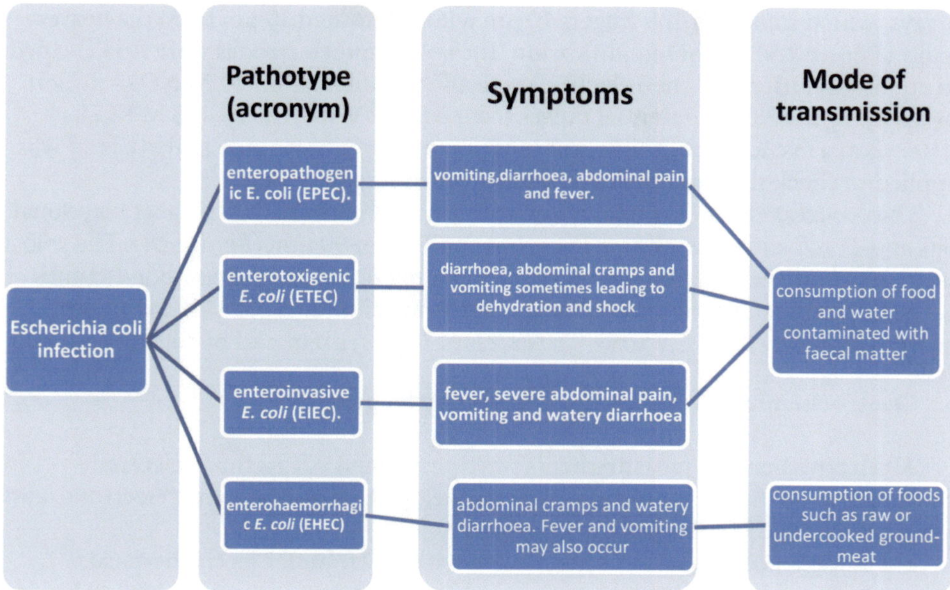

Figure 1.
Escherichia coli *symptoms and mode of transmission.*

On the basis of World Health Organization report, the most important symptoms and mode of transmission of *Escherichia coli* are represented in the diagram (**Figure 1**).

5. Immunosensor conception

5.1 Working electrodes

Interdigitated microelectrodes were provided by the Microelectronics Institute of Barcelona, National Microelectronics Centre (IMB-CNM), Spain. The different steps of the fabrication of the gold interdigitated electrodes were extensively characterized as described in Ref. [17]. The electrode consists in 3 mm × 3 mm square

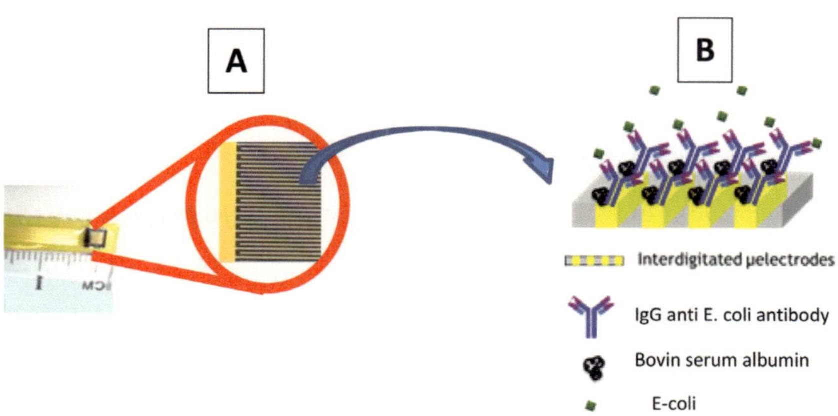

Figure 2.
*(A) The actual planar interdigital electrode impedance sensors. (B) Schematic representation of physisorption of IgG anti-*E. coli *antibody followed by* E. coli *detection.*

arrays, which consist of 108 fingers 10 μm wide, separated 10 μm from the nearest band (**Figure 2A**). Before modification, the gold microelectrodes were first cleaned in ethanol solution and then electrochemically activated in 0.5 M NaNO3 solution by applying a series of potential pulses from 0 to −2 V vs. Ag/AgCl (3 M KCl). After that, a cyclic voltammetry in 1 mM potassium ferrocyanide [$K_4Fe(CN)_6$] was applied to check the degree of activation of the microelectrodes.

The pre-treated working microelectrodes were immersed in 100 μL goat polyclonal IgG anti-*E. coli* antibody solution (5 mg/mL in PBS) for 90 min (**Figure 2B**). The gold substrates were then rinsed with PBS buffer in order to remove the non-bonded antibody; finally the substrate was kept in bovine serum albumin (BSA) 1% for 40 min in order to block any defective areas. The excess of BSA was removed by rinsing with PBS.

5.2 Electrochemical impedance spectroscopy technique

All electrochemical measurements were performed with a three-electrode configuration using a Pt foil counter electrode, an Ag/AgCl reference electrode, and a modified gold μ-electrodes as a work electrode.

The impedance analysis was performed with a CHI604E Electrochemical Instrumentation (CH Instruments, Inc) in the frequency range of 0.1–100 kHz, using a modulation voltage of 10 mV in sterile PBS buffer.

5.3 SPR imaging technique

John Mitchell [18] has been successfully explaining the physical principles of surface plasmon resonance. The SPR is an optoelectronic phenomenon that occurs when a photon of light is incident upon a noble metal surface such as gold or silver. When the wavelength of the photon equals the resonance wavelength of the metal, then the photon couples with the surface and induces the electrons in the metal surface to move as a single electrical entity called a plasmon. This oscillation of electrons sets up an electromagnetic field that exponentially decays out from the metal surface, with significant electrical field strength typically occurring within 300 nm of the surface. When molecules with sufficient mass bind to the surface within the range of the electric field, they perturb the plasmon and change the resonance wavelength. When dealing with a fixed planar surface, this is seen as a shift in the resonance angle of the incoming photons.

In this work, the surface plasmon resonance imaging system was from GWC technologies (USA). The system is based on Charge-Couple Device (CCD) camera which can simultaneously capture all data for all the gold spots and converts the reflectivity changes to pixels data. The sensor surface was an array format with 16 gold spots (each gold spot has a surface of 0.004 cm^2) deposited on glass substrate. An incident beam of excitation wavelength of 850 nm was used. At resonance condition, the variation of the reflected light was due to the refractive index variation of the external dielectric medium or immobilized thin layer. The noise of such system is equal to 0.5 pixel (**Figure 3**).

5.4 Physisorption of polyclonal antibody on interdigitated gold microelectrodes

Physisorption is defined as weak electrostatic interactions including Van Der Waals interactions, dipole-dipole, and London forces. This physical interaction resulting from nonspecific was forming on substrate have energy range from 0.2 to 4 kJ/mol. The binding energy depends on the polarizability and on the number of atoms involved of the molecules. It takes place on all surfaces provided that temperature and pressure conditions are favorable.

Random physisorption is the easiest and fastest strategy for biomolecule immobilization onto substrates. Mainly, physisorption does not depend on multistep, long

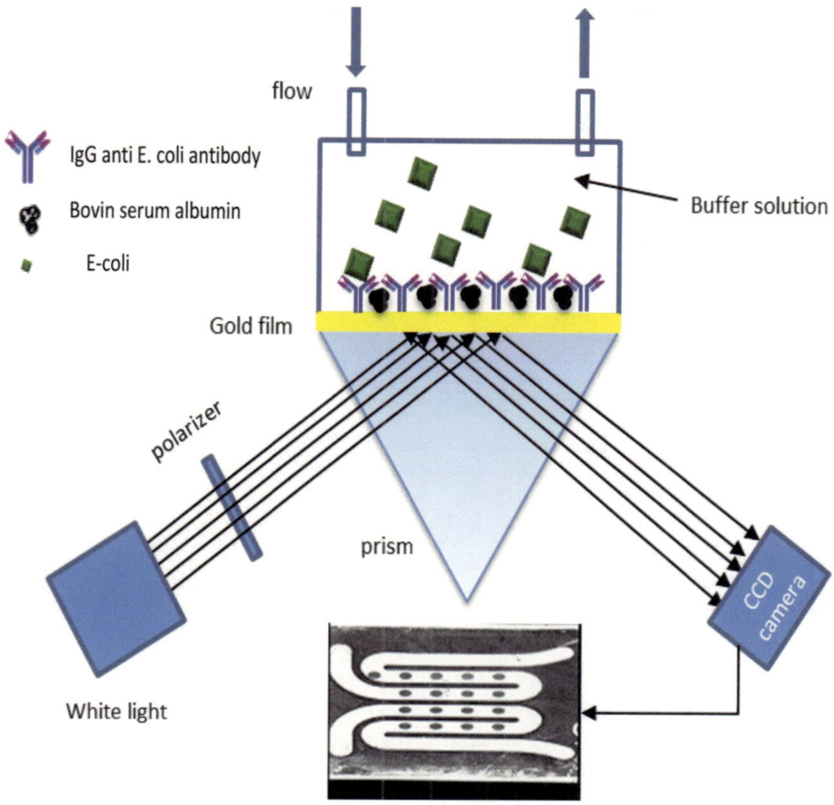

Figure 3.
Schematic representation of the SPR imaging system.

experimental procedures and is easily reversed [19]. In addition, physisorbed phages have been described to promote bacteria-specific capture, infection, and lysis, when monitored by SPR [20, 21]. This work was carried using physisorption functionalization based on its simplicity. First, the gold microelectrodes were modificated with anti-*E. coli* antibody, followed by washing with PBS then physical blocking with BSA. Blocking prevented nonspecific adsorption of unwanted nontarget components during subsequent incubations. Then, in this work, a fast and suitable immunosensor for *E. coli* bacteria detection, using physically adsorbed antibodies, SPR and EIS, is developed.

6. Electrochemical impedance measurement

The rapid and specific detection of Pathogenic Bacteria has become an increasingly demanding field in recent years for ensuring the safety of human health. EIS is a sensitive technique, which monitors the electrical response of the system studied after the application of a periodic small amplitude AC signal [22]. With this aim, the gold microelectrode surface and antibody coverage are of high importance for ensuring high reactivity and stability of the immunosensor.

The typical response of electrochemical impedance spectra of gold, "gold/ Antibody /BSA" interfaces was illustrated in **Figure 4**(I). The curve shows the typical Nyquist plots presented as a combination of the real, Zre, and imaginary, Zim, components originating mainly from the resistance and capacitance of the cell, respectively. The impedance spectra corresponding to each step were fitted with computer-simulated spectra using Randles circuit in **Figure 4**(II) by Zview modeling program (Scriber and Associates, Charlottesville, VA).

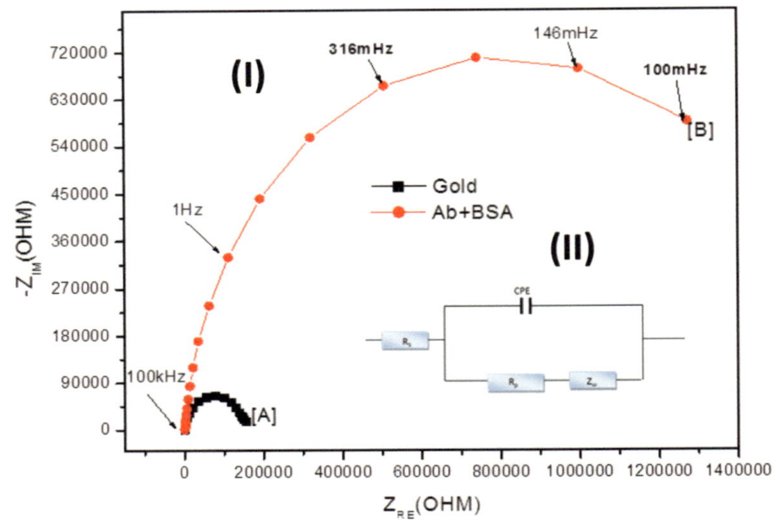

Figure 4.
*(I) [A] Nyquist impedance plots for gold microelectrode and [B] Nyquist impedance plots after physisorption of anti-*E. coli *and BSA on gold microelectrodes. (II) equivalent circuit used to model impedance data.*

This equivalent circuit includes the ohmic resistance of the electrolyte solution, R_s, at 100 kHz; the Warburg impedance, Z_w, from the diffusion; the constant phase element, CPE, which was introduced into the circuit instead of a capacitance in order to depict the nonhomogeneous quality of the deposited layer, respectively [23, 24]; and the charge transfer resistance, R_{ct}. The constant phase element impedance (CPE) was introduced into the circuit instead of a capacitance:

$$Z_{CPE} = \frac{1}{K(j\omega)^n} \tag{1}$$

where $\omega = 2\pi f$ is the angular frequency and K and n are the experimental parameters. When n approaches 1, the CPE acts as an ideal capacitor. The CPE can be viewed as a heuristic method to incorporate the effects of surface heterogeneity both along and through the electrochemical interface. Data was fitted to the Randles circuit shown in **Figure 4**(II). The plot shows the expected increases in the charge transfer resistance after immobilization of anti-*E. coli* antibody from 0.14 to 1.22 MΩ. This increase could be attributed to a rearrangement in the structure of the antibody and showed that the grafted layer becomes more insulating, whereas the gradual decrease in capacitance was related to the positive change in thickness after immobilization of anti-*E. coli* antibody and BSA. This behavior is consistent with the successful immobilization of anti-*E. coli* antibody and BSA molecules.

7. Surface plasmon resonance measurement

In the last few years, surface plasmon resonance was used as a sensitive method and as a label-free detection method for biomolecular interactions.

The immunosensing protocols exposing sensor surface to the PBS buffer as the baseline, followed by injecting anti-*E. coli* antibody, which is allowed to flow over the sensor surface, leading to binding. This binding alters the mass of the surface layer which translates into refractive index variation and resonance angle shift. **Figure 5** shows the resulting response obtained from the injection of 5 mg/mL

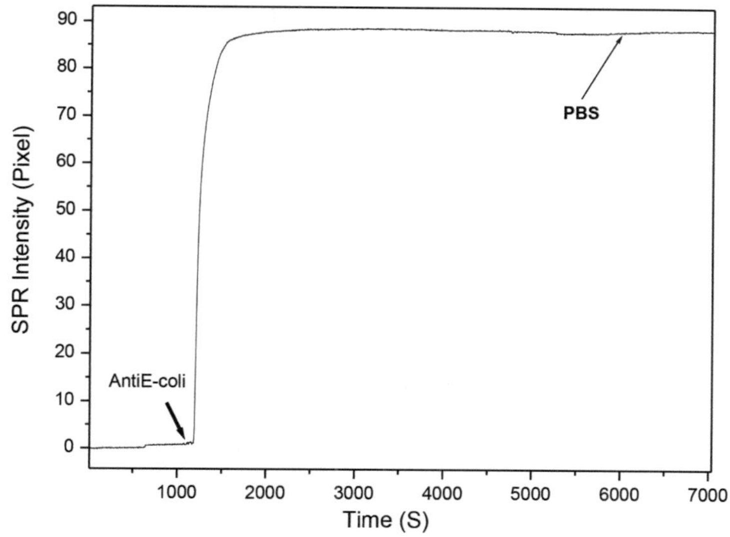

Figure 5.
*SPRi signal versus time for anti-*E. coli*.*

anti-*E. coli* antibody solution. The real-time sensorgram showed a gradual increase in the response with antibody immobilization onto the sensor chip. The highest response obtained was equal to 87 pixels.

8. Detection of *E. coli* bacteria in inoculated frozen chicken meat sample

Although chicken is the most consumed meat in the world and is one of the most important sources of good-quality proteins, it is highly susceptible to microbial contamination and often implicated in foodborne disease. Epidemiological reports suggest that poultry meat is still the primary cause of human food poisoning [25]. According to Osman Albarri (2017) [26], in turkey the highest percentage (93.75%) of *E. coli* was isolated from chicken, while the lowest percentage (56.25%) was isolated from meat. Therefore, ensuring the microbial safety of chicken meat products is an important issue in the context of increasing consumption and production [24]. This moved us to develop rapid, easy, simple, sensitive, and non-time-consuming physical adsorption methods for the detection of *E. coli* bacteria in frozen chicken meat.

With this aim, two samples of fresh chicken meat were kept in freezers at −18°C during 45 days [27]. The first sample (S1) was inoculated with *E. coli* with a concentration of 10^5 CFU/ML in PBS, and the second sample (S2) was kept in PBS buffer (reference). These samples were characterized by EIS as previously described.

Figure 6 shows Nyquist plots for gold microelectrode (curve A), gold electrode with immobilized anti-*E. coli* antibody with BSA blocking layer (curve B), gold electrode with immobilized anti-*E. coli* antibody with BSA blocking layer without inoculated bacteria (sample S2, curve C), and gold electrode with immobilized anti-*E. coli* antibody with BSA blocking layer with inoculated bacteria (sample S1, curve D). Using the same equivalent circuit model described in **Figure 4 (II)**, the best-fit equivalent circuit parameters for each step are given in **Table 1**. The injection of sample S1 and S2 on the microelectrode induce an increase in the charge transfer resistance proving the detection of *E. coli* bacteria in the chicken meat sample.

It is obvious that the chicken was initially contaminated by the *E. coli* bacteria and the concentration of the bacteria is about 10^5 CFU/ML.

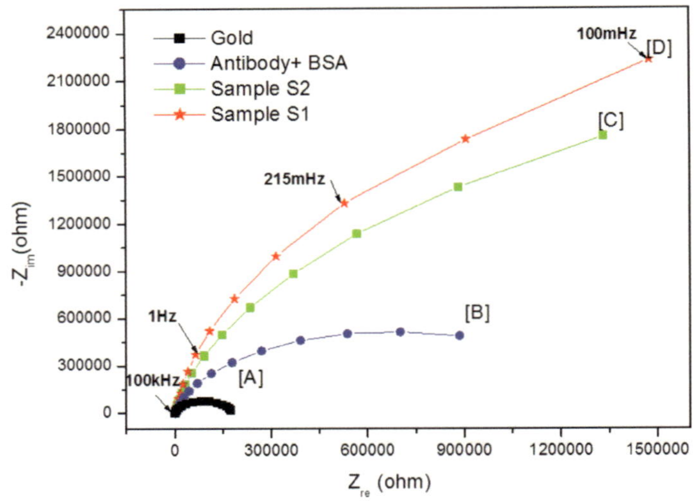

Figure 6.
*[A] Nyquist impedance plots for gold microelectrode, [B] nyquist impedance plots for gold microelectrode with anti-*E. coli *and BSA, and [C and D] nyquist impedance plots for gold microelectrode with anti-*E. coli *and BSA after injection of sample S2 and S1, respectively.*

Parameters	Gold microelectrode	BSA/anti-*E. coli*/ gold	Sample S2	Sample S1
Capacitance, CPE (F)	8.34E-7	7.464E-7	5.76E-7	5.128E-7
n	0.93	0.86	0.88	0.9
Resistance, R_s (Ω)	110	74.74	70.7	71
Resistance, R_{ct} (Ω)	1.43E5	1.225E6	4.68E6	5.01E6
Warburg, Z_W (Ω)	45,364	1452	132	120

Table 1.
The electrical parameters of Randle's circuit.

9. Conclusion

In this work, we describe an approach of detecting of *Escherichia coli* bacteria by Electrochemical Impedance Spectroscopy technique and surface plasmon resonance imaging technique. The physisorption method used for immobilization of anti-*E. coli* into interdigitated microelectrode is rapid and easy. Furthermore, *Escherichia coli* bacteria detection was also possible in frozen chicken meat. This method can be used for real-time detection of meat contamination.

Acknowledgements

This work is funded by the Science for Peace and Security Program of the North Atlantic Treaty Organization (NATO) under grant no. SFP G5571. The authors dedicate this work to the memory of Mr. Naoufel Gaouar, Associate Professor at the National Institute of Applied Science and Technology (INSAT, Tunisia) who died on March 8, 2020.

Author details

Saloua Helali[1,2]* and Adnane Abdelghani[3]

1 Faculty of Science, Department of Physics, University of Tabuk, Tabuk, Kingdom of Saudi Arabia

2 Centre de Recherche et Technologies de L'Energie, Tunisia

3 Carthage University, National Institute of Applied Science and Technology, Tunisia

*Address all correspondence to: s.helali@ut.edu.sa

References

[1] Addis M, Sisay D. A review on major food borne bacterial illnesses. Journal of Tropical Diseases. 2015;**3**:1-7. DOI: 10.4176/2329-891X.1000176

[2] Adu-Gyamfi A, Torgby-Tetteh W, Appiah V. Microbiological quality of chicken sold in accra and determination of D10-value of *E. coli*. Food and Nutrition Sciences. 2012;3:693-698

[3] Borch E, Arinder P. Bacteriological safety issues in beef and ready-to-eat meat products, as well as control measures. Meat Science. 2002;**62**:381-390. DOI: 10.1016/S0309-1740(02)00125-0

[4] Alvarez-Astorga M, Capita R, Alonso-Callega C, Moreno B, Del Camoni Garcia-Fernandez M. Microbiological quality of retail chicken in by-products in Spain. Meat Science. 2002;**62**:45-50. DOI: 10.1016/S0309-1740(01)00225-X

[5] Akbar A, Sitara U, Khan SA, Ali I, Khan MI, Phadungchob T, et al. Presence of Escherichia coli in poultry meat: A potential food safety threat. International Food Research Journal. 2014;**21**:941-945

[6] Rouger A, Tresse O, Zagorec M. Bacterial contaminants of poultry meat: Sources, species, and dynamics. Microorganisms. 2017;**5**:1-16. DOI: 10.3390/microorganisms5030050

[7] Kozačinski L, Hadžiosmanović M, Zdolec N. Microbiological quality of poultry meat on the Croatian market. Veterinarski Arhiv. 2006;**76**:305-313. DOI: hrcak.srce.hr/5139

[8] Kamruzzaman M, Makino Y, Oshita S. Non-invasive analytical technology for the detection of contamination, adulteration, and authenticity of meat, poultry, and fish: A review. Analytica Chimica Acta. 2015;**853**:19-29. DOI: 10.1016/j.aca.2014.08.043

[9] Skóra J, Matusiak K, Wojewódzki P, Nowak A, Sulyok M, Ligocka A, et al. Evaluation of microbiological and chemical contaminants in poultry farms. International Journal of Environmental Research and Public Health. 2016;**13**:1-16. DOI: 10.3390/ijerph13020192

[10] Feng YZ, Elmasry G, Sun DW, Scannell AG, Walsh D, Morcy N. Near-infrared hyperspectral imaging and partial least squares regression for rapid and reagentless determination of *Enterobacteriaceae* on chicken fillets. Food Chemistry. 2013;**138**:1829-1836. DOI: 10.1016/j.foodchem.2012.11.040

[11] Moghtader F, Congur G, Zareie HM, Erdem A, Piskin E. Impedimetric detection of Pathogenic Bacteria with bacteriophages using gold nanorod deposited graphite electrodes. RSC Advances. 2016;**6**:97832

[12] Boozer C, Kim G, Cong S, Guan HW, Londergan T. Looking towards label-free biomolecular interaction analysis in a high-throughput format: A review of new surface plasmon resonance technologies. Current Opinion in Biotechnology. 2006;**17**:400-405. DOI: 10.1016/j.copbio.2006.06.012

[13] Huber W, Mueller F. Biomolecular interaction analysis in drug discovery using surface plasmon resonance technology. Current Pharmaceutical Design. 2006;**12**:3999-4021. DOI: 10.2174/138161206778743600

[14] Yang L, Bashir R. Electrical/electrochemical impedance for rapid detection of foodborne Pathogenic Bacteria. Biotechnology Advances. 2008;**26**:135-150. DOI: 10.1016/j.biotechadv.2007.10.003

[15] Allocati N, Masulli M, Alexeyev MF, Di Ilio C. Escherichia coli in Europe: An overview. International Journal of

Environmental Research and Public Health. 2013;**10**:6235-6254. DOI: 10.3390/ijerph10126235

[16] Lim JY, Yoon JW, Hovde CJ. A brief overview of Escherichia coli O157:H7 and its plasmid O157. Journal of Microbiology and Biotechnology. 2010;**20**:5-14

[17] Laczka O, Baldrich E, Javier del Campo F, Muñoz FX. Immunofunctionalisation of gold transducers for bacterial detection by physisorption. Analytical and Bioanalytical Chemistry. 2008;**391**: 2825-2835. DOI: 10.1007/s00216-008-2199-1

[18] Mitchell J. Small molecule immunosensing using surface plasmon resonance. Sensors. 2010;**10**:7323-7346. DOI: 10.3390/s100807323

[19] Marrakchi M, Campos Sánchez I, Helali S, Mejri N, Soto Camino J, Gonzalez-Martinez MA, et al. A label-free interdigitated microelectrodes immunosensor for pesticide detection. Sensor Letters. 2011;**9**:2203-2206. DOI: 10.1166/sl.2011.1800

[20] Mejri MB, Baccar H, Baldrich E, Del Campo FJ, Helali S, Ktari T, et al. Impedance biosensing using phages for bacteria detection: Generation of dual signals as the clue for in-chip assay confirmation. Biosensors and Bioelectronics. 2010;**26**:1261-1267. DOI: 10.1016/j.bios.2010.06.054

[21] Chammem H, Hafaid I, Meilhac O, Menaa F, Mora L, Abdelghani A. Surface plasmon resonance for C-reactive protein detection in human plasma. Journal of Biomaterials and Nanobiotechnology. 2014;**5**:153-158. DOI: 10.4236/jbnb.2014.53018

[22] Hleli S, Martelet C, Abdelghani A, Burais N, Jaffrezic-Renault N. Atrazine analysis using an impedimetric immunosensor based on mixed biotinylated self-assembled monolayer. Sensors and Actuators B. 2006;**113**:711-717

[23] Prodromidis MI. Impedimetric immunosensors - a review. Electrochimica Acta. 2010;**55**:4227-4233. DOI: 10.1016/j.electacta.2009.01.081

[24] Nahar MK, Zakaria Z, Hashim U. Protein solubility behaviour of fresh and frozen chicken meat in slaughtering and non-slaughtering condition. International Food Research Journal. 2014;**21**:135-137

[25] Kozačinski L, Hadžiosmanović M, Zdolec N. Microbiological quality of poultry meat on the Croatian market. Veterinarski Arhiv. 2006;**76**(4):305-313

[26] Albarri O, Var I, Meral M, Bedir B, Heshmati B, Köksal F. Prevalence of Escherichia coli isolated from meat, chicken and vegetable samples in Turkey. Journal of Biotechnology Science Research. 2017;**4**(3):214-2022

[27] Helali S, Alatawi ASE, Abdelghani A. Pathogenic Escherichia coli biosensor detection on chicken food samples. Journal of Food Safety. 2018;**38**:1-6. DOI: 10.1111/jfs.12510

Chapter 2

Dietary Intervention to Reduce *E. coli* Infectious Diarrhea in Young Pigs

Peng Ji, Xunde Li and Yanhong Liu

Abstract

Postweaning piglets are immediately imposed to remarkable environmental and psychosocial stressors, which adversely affect their intestinal development and health and predispose them to diarrhea. The ratio of postweaning mortality is 6–10% and may rise up to 20% with poor management strategies. Diarrhea per se accounts for 20–30% of cases of mortality in weanling pigs. *E. coli* postweaning diarrhea is one of the most important causes of postweaning diarrhea in pigs. This diarrhea is responsible for huge economic losses due to high mortality and morbidity, weight loss, and cost of medication. Burgeoning evidence suggested feed-based intervention are one of the promising measures to prevent postweaning diarrhea and to enhance overall health of weaned pigs. Although the exact protective mechanisms may vary and are still not completely understood, a number of feed ingredients or feed additives are marketed to assist in boosting intestinal immunity and regulating gut microbiota. The promising results have been demonstrated in several nutrients (i.e., functional amino acids, organic acids, micro minerals, nondigestible carbohydrates, and antimicrobial peptides), non-nutrients (i.e., phytochemicals and probiotics), and many other feed additives. The efficiencies of each candidate may differ based on their exact modes of action, the basal diet formulation, and the health status of pigs.

Keywords: dietary intervention, *E. coli* infectious diarrhea, ETEC, young pigs

1. Introduction

Escherichia coli (*E. coli*), a Gram-negative rod-shaped bacterium, was first discovered in 1885 by Theodor Escherich, who noted that *E. coli* are highly prevalent in the intestinal microflora of healthy individuals and have potential to cause disease when directly inoculated into extraintestinal sites. Diarrheagenic *E. coli* can be further divided into six groups: enterotoxigenic *E. coli* (ETEC), enteropathogenic *E. coli*, enterohemorrhagic *E. coli*, enteroinvasive *E. coli*, diffusely adhering *E. coli*, and enteroaggregative *E. coli* [1]. Different groups of diarrheagenic *E. coli* express different virulence genes, exhibit different adhesion characteristics, and therefore have different mechanisms of pathogenicity. This book chapter only covers the infection caused by ETEC.

ETEC is the major etiological agent causing acute watery diarrhea in postweaning piglets. The duration of diarrheal symptom may be shortened by antibiotic treatment, but ETEC is relative refractory to common antibiotics. A growing evidence suggested some nutritional components (e.g., functional amino acids, nondigestible carbohydrates, etc.) and non-nutrients (e.g., phytochemicals, probiotics, etc.) may provide preventive benefits to control ETEC infection. In general, the compounds listed above are supplemented into animal feed with small amount, also named as feed additives. The exact protective mechanisms are largely unknown and may differ for each compound. However, based on the literature review, these feed additives may alleviate ETEC infection by targeting at least one of the following mechanisms: (1) modification of intestinal microbiota by directly killing pathogens or competitively inhibiting the binding of pathogens and toxins to gut epithelium and (2) regulation or stimulation of host immunity that may include intestinal mucosal immunity and systemic immune defense.

2. *E. coli* infectious diarrhea

E. coli postweaning diarrhea is an important cause of death in weaned pigs. This diarrhea is responsible for economic losses due to mortality, morbidity, decreased growth performance, and cost of medication [2, 3]. ETEC are the most predominant types of pathogenic *E. coli* that cause diarrhea in both preweaning and postweaning piglets [4, 5].

2.1 Clinical signs

Clinical signs of ETEC infection in pigs include reduced appetite, depression, weakness, rapid dehydration, watery diarrhea (light orange-colored feces), anorexia, and shock due to hypovolemia and electrolyte imbalance [6, 7]. Cyanotic discoloration may appear on the tip of the nose, the ears, and the abdomen. The rectal temperature of infected pigs is generally normal. Pigs may spontaneously recover within 1 week if the infection is mild. However, severe infection may cause death within 12 hours, even without the symptoms of diarrhea. Dehydration of the carcass and distension of the small intestine by colorless mucoid fluid are most common necropsy characteristics for ETEC-infected pigs. During ETEC infection, bacteria normally line the epithelial cells of the intestine rather than invade the mucosa; gross and histological lesions, therefore, may not be directly caused by the bacteria. However, physiological changes in the intestine may be caused by the toxins released by ETEC [8, 9].

2.2 The pathogenesis of *E. coli* infection

There are two major virulence factors involved in the pathogenesis of ETEC infection, including the expression of fimbriae that enable the attachment of bacteria to the small intestinal epithelial cells, and the production of toxins by the colonized ETEC [10–12]. In addition, other structural components from *E. coli*, such as capsular polysaccharides, cell wall lipopolysaccharides (LPS), and iron-binding proteins, may also be involved in the pathogenesis of ETEC [13]. The endotoxins produced by ETEC could induce intestinal physiological changes, which are leading to the disrupted water and fluid absorption and ion secretion, finally causing dehydration and acidosis. The bacterial structural components could also initial a cascade of immune stimulation, resulting in intestinal inflammation and systemic inflammation [14, 15].

2.2.1 *Fimbriae of* E. coli

Fimbriae are proteinaceous appendages located at the outer membrane of the bacterial cells. They are straight or kinky shapes. The major role of fimbriae is to facilitate the adhesion and colonization of ETEC at the small intestinal mucosa [16–18]. The adhesion of bacteria is extremely important for ETEC infection. It will stabilize the location of bacteria in the intestinal lumen, which allow the pathogens with better access to luminal nutrition, facilitate the secretion and delivery of endotoxins through epithelium, and help the bacteria penetrate into the tissue if needed [16–18]. Diarrheagenic ETEC may express many kinds of fimbria, including F4 (K88), F5 (K99), F6 (987p), F18, etc.; F4 (K88) and F18 ETEC are the most common pathogenic ETEC in young pigs.

F4 fimbriae are typically identified in ETEC isolated from pre- and postweaning pigs. F4 ETEC tend to colonize throughout the whole segments of the small intestine in pigs [19]. F4 fimbriae are encoded by the *fae* operon, which comprises genes coding for several regulatory proteins, distal tip protein, minor subunits, and a major subunit, FaeG, that enable F4+ ETEC binding to specific receptors on intestinal brush border cells [2, 3, 20]. There are three naturally occurring serological variants of F4 fimbriae, including F4ab, F4ac, and F4ad. They are interchangeable by changing a residue stretch in the FaeG protein. However, F4ac variant is the most common F4 fimbriae variant expressed in porcine pathogenic ETEC in the United States [2, 3]. The adhesion receptors of F4 fimbriae appear to be glycoconjugates, including glycoproteins and glycolipids, which have been identified from the brush borders of epithelial cells, intestinal membranes, and mucosa [21, 22]. It is interesting to note that F4ad adhesin appears to preferentially bind to glycolipids, whereas F4ab and F4ac adhesins preferentially bind to glycoproteins [22–24].

F18 fimbriae are associated with *E. coli* strains isolated from postweaning diarrhea and edema disease in pigs. These fimbriae are long flexible appendages that show a characteristic zigzag pattern [16]. Based on morphological, serological, functional, and genetic characteristics, two antigenic variants of F18 fimbriae were determined and designated: F18ab and F18ac [25]. F18ab-positive strains are usually isolated from cases of edema disease, whereas F18ac-positive strains are associated with cases of postweaning diarrhea [16, 26]. F18 fimbriae are composed of protein subunits (FedA) with molecular weights of approximately 15.1 kDa [27]. Five structural genes (*fed*A, *fed*B, *fed*C, *fed*E, and *fed*F) encoded on a plasmid have been identified [28]. Among these genes, *fed*E and *fed*F genes are essential for F18 adhesion and fimbrial length [29]. However, receptors for F18 fimbriae actually increase with age and have not been detected in newborn pigs [30]. This may in part explain the reason why ETEC strains carrying F18 are more prevalent in weaned pigs.

2.2.2 *Toxin effects*

After adhering to the small intestinal surface, ETEC induce enteric infectious disease and diarrhea through release of enterotoxins, which stimulate copious secretion by the small intestinal mucosa. The enterotoxins include heat-labile toxin, heat-stable toxin, LPS, and Shiga toxins.

Heat-labile toxins. Heat-labile toxin mainly accumulates in the periplasmic space, with limited amount appears on the surface of the bacteria. Heat-labile toxin consists of a single A subunit and five B subunits. The binding of B subunits to the monosialotetrahexosylganglioside (GM1) ganglioside on the cell surface facilitates the translocation of a fragment of A domain into the cell, which then activates the adenylate cyclase system and increases the expression of cyclic adenosine monophosphate (cAMP) [12]. Several physiological changes are mediated by the increased cAMP. First, cAMP stimulates the phosphorylation of the cystic fibrosis transmembrane conductance regulator (CFTR), inducing chloride secretion from the apical region of enterocytes [31]. Second, cAMP stimulates the activation of an apical chloride channel and a basolateral Na/K/2Cl cotransporter, which results in the release of prostaglandin E2 and vasoactive intestinal peptide and loosening of tight junctions [32, 33]. These activities all contribute to increased chloride secretion, reduced sodium absorption, and a concomitant massive loss of water into the intestinal lumen. The effect of heat-labile toxin is irreversible [12].

2.2.2.1 Heat-stable toxins

Heat-stable toxins in porcine isolates are further classified as STa and STb [32]. STa is a small, non-immunogenic protein with a molecular weight of approximately 2 kDa [34]. The major receptor for STa is a particular transmembrane form of guanylate cyclase (GC-C) [35]. Therefore, STa can stimulate the GC-C system, leading to excessive levels of cyclic guanosine monophosphate (cGMP) in enterocytes. Signals resulting from cGMP accumulation induce the activation of CFTR, elevated secretion of Cl- and water in crypt cells, but reduced Na^+ and Cl^- absorption from cells at the tips of villi [11]. Heat-stable toxin a mainly induces small intestinal fluid secretion in newborn but not in weaned pigs [4].

Heat-stable toxin b is a 48-amino acid protein with a molecular weight of approximately 5.1 kDa [36]. STb is antigenically and genetically unrelated to STa and is poorly immunogenic [37]. Production of STb is restricted to porcine ETEC [13]. The mechanisms of action and molecular characteristics of STb are still less known than heat-labile toxins and STa. STb does not stimulate an increase in intracellular levels of either cAMP or cGMP, either Na^+ or Cl^- [38], but stimulates the secretion of HCO_3– from intestinal epithelial cells [10, 39]. Heat-stable toxin b causes mild histological damage in the intestinal epithelium, including loss of villous epithelial cells and villous atrophy. This damage may be responsible for impaired absorption of fluids [40]. Another proposed mechanism of action is that STb could increase the level of prostaglandins by opening a G-protein-linked receptor-operated calcium channel in the plasma membrane and elevating intracellular Ca^{++} [41, 42]. STb can induce small intestinal fluid secretion in newborn and weaned pigs [4].

2.2.2.2 LPS

Lipopolysaccharide is the major surface component of the outer membrane of most Gram-negative bacteria, including ETEC [43]. LPS consists of three distinct regions, lipid A, core oligosaccharides, and the O-antigen polysaccharide with the structural variability from low to high. Lipid A is the primary immunostimulatory component of LPS and is highly recognized by numerous cellular signaling pathways in the innate immunity. The receptors that respond to LPS are mainly located on the cells in the innate immune system, such as macrophages and endothelial cells [44]. Therefore, LPS not only contributes to the physiological membrane functions but also plays an important role in the pathogenesis of Gram-negative bacterial infection [44]. The pathogenic impacts of LPS are mainly through stimulating immune cells of the host, resulting in the release of large amounts of cytokines. The CD14, expressed on monocytes or macrophages, are highly involved in this process. Briefly, the soluble CD14 (sCD14) and LPS-binding protein facilitate the interaction of LPS and the membrane CD14 (mCD14). After binding to mCD14 in the cell surface, LPS is recognized by TLR4-MD-2 complex, which transduces intracellular LPS signals via several signal pathways [45], ultimately resulting in the activation of NFκB and subsequently the release of inflammatory cytokines [15].

2.2.2.3 Shiga toxin

Some ETEC strains that cause postweaning diarrhea possess additional genes that encode Shiga toxin, allowing them to cause edema disease as well [4]. Similar to heat-labile toxin, Shiga toxin is also a protein toxin, consisting of one A subunit and five B subunits. However, Shiga toxin has completely different mechanisms infecting cells, in comparison with heat-labile toxins. Briefly, Shiga toxin first binds to the cells that possess the glycolipid receptors, globotriaosylceramide (Gb)3 or Gb4 [46, 47]. After binding, Shiga toxin is transported to the Golgi apparatus through endocytosis. The Golgi apparatus further transports Shiga toxin to the endoplasmic reticulum, where the subunit A is cleaved by trypsin and is separated into A1 and A2 subunits. The A1 subunit is released into the cytosol and subsequently impacts ribosomes [14]. Shiga toxin can inhibit protein synthesis and induce synthesis of cytokines, including IL-1, IL-6, IL-8, and TNF-α [47, 48]. In addition, Shiga toxin also induces DNA degradation and release of the cellular contents that facilitate proteolytic attack on neighboring cells, contribute to cell apoptosis, and have a toxic effect in the whole organism [14].

3. Dietary intervention on *E. coli* infectious diarrhea

Nutrients are compounds in feed ingredients that are essential to animal maintenance and production, by providing animals with energy, the building components for repair and growth, and the substances to regulate biological processes. Nutrients are generally grouped into six major classes: water, carbohydrates, proteins, lipids, minerals, and vitamins. With the exception of carbohydrates, all five classes of nutrients are indispensable and have to be provided through animal feed. In addition, a group of specific nutrients, such as functional amino acids, nondigestible carbohydrates, short-chain fatty acids (SCFA), and several micro minerals, has beneficial effects on animal health and performance beyond their nutritional contributions. Recently, a novel concept, non-nutrients, is illuminated to describe a group of dietary compounds, which has no nutrient contribution to animals, but have physiological activities beyond the nutritional pyramid, formulation practices, and feeding methods that similarly alter physiological condition. Emerging evidence suggested that several non-nutrient feed additives (i.e., plant extracts, probiotics, enzymes, etc.) improved animal health through modulating microbial ecology in the digestive tract and/or enhancing immune responses of animals to enteric infections.

3.1 Functional amino acids

Amino acids are defined as organic substances that contain both amino and carboxyl groups. Amino acids are classified according to their molecular weights, chemical structures, the composition of nitrogen and sulfur, and physiological functions. The 20 common proteinogenic amino acids shared by all animal species are further categorized into indispensable, semi-dispensable (conditional essential), and dispensable amino acids, dependent on their dietary essentiality. Functional amino acids are defined with a group of amino acids that are traditionally classified as dispensable amino acids, but with extra biological functions [49]. The well-investigated functional amino acids are the arginine family, which includes arginine, glutamine, glutamate, aspartate, proline, etc. The basic functions of these amino acids have been well summarized by Wu et al. [49, 50], which include but not limited to (1) providing substrates for the synthesis of tissue protein; (2) impacting hormone synthesis and secretion; (3) regulating endothelial function, vasodilation, and blood flow; (4) affecting nutrient metabolism; and (5) maintaining acid-base balance and whole-body homeostasis.

Large amounts of literature have reported the impacts of dietary supplementation of functional amino acids on health and performance of newly weaned pigs. For example, supplementation of 0.2 to 1% L-arginine could enhance growth performance and alleviate the negative effects of different insults or challenges in young pigs [51–54]. Administration of proline was shown to improve mucosal proliferation, intestinal morphology, as well as intestinal tight junction of weaned pigs [55]. Dietary supplementation of glutamine or dipeptides that are composed of glutamine has shown positive impacts on intestinal integrity, enzyme activities, and growth performance of weaned pigs [56–58]. Several mechanisms are highly involved in the benefits of the arginine family on intestinal health of weaned pigs, which could prevent the intestinal dysfunction caused by *E. coli* infectious diarrhea in weaned pigs. First, these amino acids could provide major fuel for small intestinal epithelial cell proliferation and provide energy required for intestinal ATP-dependent metabolic processes [59]. Second, catabolism of these amino acids provides precursors or substrates for the synthesis of nitric oxide, polyamines, and creatine, which are important regulators in blood flow, intestinal integrity and secretion, and epithelial cell repair and migration [60–62]. Third, glutamine is also a major substrate for glutathione synthesis, which is an important endogenous antioxidant in cells regulating the homeostasis of free radicals [63]. Fourth, these amino acids could enhance intestinal secretory IgA production via regulating the intestinal microbiota and immunity [53, 64].

3.2 Fatty acids

Dietary fat and lipids are extremely important for animal health and production. They have three major fundamental roles in swine nutrition by providing energy, compound lipids, and steroids to animals. Triglycerides and free fatty acids are the primary forms of metabolic energy storage and transport in the animal body. Short-chain fatty acids and medium-chain fatty acids (MCFA) have recently attracted increased research attention as potential candidates to reduce enteric infectious disease in animal production due to their potential antimicrobial activities [65–67].

SCFA are fatty acids with a chain of less than six carbon atoms, which are primarily produced by hindgut fermentation of dietary fiber. The most abundant SCFA in the gastrointestinal tract are acetic (C2), propionic (C3), and butyric acid (C4). They are the major fuel source for colonocytes and are essential for maintaining the normal metabolism of colon mucosa, including colonocyte growth and proliferation [68, 69]. Butyric acid has received particular attention and has been widely investigated to enhance disease resistance of weaned pigs. Addition of this acid directly to a swine diet may be limited because of its pungent odor and unpalatable flavor. Thus, the salt form (sodium or calcium) or glyceryl form (monobutyrin or tributyrin) of butyric acid has been adopted in animal feed industry. One major advantage of glyceryl forms in comparison with salt forms is that they stay intact in the stomach and are slowly released as butyrate and/or monobutyrin in the small intestine where pancreatic lipase appears [70]. Many research have confirmed the positive protective effects of sodium butyrate or tributyrin on intestinal health of weaned pigs, as they reduced diarrhea, enhanced gut integrity, and improved overall immunity of newly weaned pigs [71–74]. The mechanisms resulting in improved disease resistance of weaned pigs are highly associated with the antimicrobial activities of SCFA; however, other mechanisms may be also involved. Butyric acid could penetrate into epithelial cells either by simple diffusion or monocarboxylate transporter [75]. Butyric acid could also bind to G-protein-coupled receptor expressed in epithelial cells or immune cells. The binding will mediate a cascade of immune regulation and regulate large amount of gene expression [76–78].

MCFA are saturated fatty acids containing 6 to 12 carbon atoms, including caproic (C6), caprylic (C8), capric (C10), and lauric acids (C12). MCFA are naturally occurred lipids that are enriched in animal milk fat and in the oil fraction of various plants, such as coconuts, palm kernels, and *Cuphea* seeds. Similar to SCFA, MCFA have unpleasant smell; therefore, they are commonly used in their glyceryl forms in animal feed. MCFA may have particular nutritional and metabolic effects on young animals due to their rapid digestion, passive absorption, and obligatory oxidation [79, 80]. Although the evidence for the favorable energetic attributes of MCFA is strong, the results of in vivo studies using weaned pigs have been inconsistent [81]. However, the beneficial effects of MCFA on gut health of weaned pigs have been suggested, as they could influence intestinal morphology and physiology, gut microbiome, and intestinal immunity [80]. More research is necessary in the future to explore the influences of MCFA on disease resistance of weaned pigs.

3.3 Micro minerals

Minerals required in smaller quantities are called micro minerals or trace minerals, which include Zn, Cu, Mn, Fe, Se, and others. Micro minerals have confirmed physiological roles and are needed for normal bodily functions of pigs. However, unlike most other minerals, Cu and Zn have antimicrobial properties, and they are, therefore, often added to animal feed in quantities greater than the amount of nutritional requirements.

Zinc serves as a component or an activator of several metalloenzymes and is involved in many intracellular and intercellular signaling pathways. Zinc also plays important roles in skin and wound healing and in regulating immune system [82]. Zinc deficiency in weaned pigs leads to growth retardation, loss of appetite, skeletal abnormalities, and parakeratosis if Zn concentration in feed is much lower than the requirement of 80–100 mg/kg for nursery pigs [83–85]. However, pharmacological levels (2000–3000 mg/kg) of inorganic Zn in the form of ZnO have been widely adopted to control postweaning diarrhea and enhance feed intake and overall growth performance [86–89]. The benefits of pharmacological Zn on disease resistance of weaned pigs are likely related to several mechanisms: enhancement of intestinal integrity [90], regeneration of injured intestinal mucosa, stability of intestinal microbiota diversity [91], reduction of intestinal permeability [92], and modulation of intestinal immunity [93]. However, more recent research indicate that feeding pharmacological ZnO could reduce the digestibility of Ca and P, reduce the effectiveness of microbial phytase in pig diet, and increase the abundance of multiresistant bacteria in weaned pigs [94–96]. Inclusion of pharmacological levels of ZnO has recently been banned in the European Union due to increased Zn pollution from pigs fed with high Zn diets. Meantime, animal feed industry and nutritionists are actively working together to search alternatives that could replace pharmacological ZnO. For instance, low dose of organic Zn sources (i.e., 125 mg/kg of Zn-methionine) has been confirmed to have beneficial effects that are equivalent to addition of pharmacological ZnO due to their greater bioavailability [97, 98].

Copper is also an essential component of several metalloenzymes including cytochrome oxidase and lysyl oxidase. Copper is highly involved in oxidation-reduction reactions, transport of oxygen and electrons, antioxidant system, and many other metabolic functions, including cellular respiration, tissue pigmentation, hemoglobin formation, and connective tissue development [82]. In general, neonatal pigs only require 5–6 mg/kg of Cu for normal metabolism [85, 99], and Cu requirement decreases as animal gets older. Cu deficiency may lead to critical dysfunctions and hypocuprosis in pigs [100]. Pigs may also suffer from microcytic anemia and bone abnormalities [101, 102]. Addition of pharmacological levels of Cu (125–500 mg/kg) in pig diets has been a common practice to reduce postweaning diarrhea and improve growth performance [103–106]. The beneficial effects of pharmacological Cu have been attributed to its bacteriostatic and bactericidal properties [107, 108]. Similar to zinc, many Cu forms could be used in animal feed, including copper sulfate ($CuSO_4$), copper chloride ($CuCl_2$), tribasic copper chloride ($Cu_2(OH)_3Cl$), and copper citrate. Copper sulfate and copper chloride are the most common supplementing forms. Tribasic copper chloride has been suggested to have similar bioavailability but less negative impacts on phosphorus digestibility and intestinal microbiota than copper sulfate [109–111]. Chelated Cu, such as Cu citrate, has greater availability than inorganic Cu sources, which may be used in animal feed as low dose, resulting in reduced Cu excretion [112].

3.4 Prebiotics and probiotics

Prebiotics are a category of nutritional compounds that may not share similar structures but have the ability to improve the growth of beneficial microorganism in the gastrointestinal tract. Gibson et al. [113] offered a definition of prebiotics, which contains three key aspects: resistance to digestion, fermentation by the large intestinal microbiota, and a selective effect on the microbiota associated with health-promoting effects. Most well-studied prebiotics are nondigestible oligosaccharides or polysaccharides [114]. For instance, inulin-type prebiotics are a group of nondigestible carbohydrates that mainly comprise fructose, including inulin, oligofructose, and fructo-oligosaccharides. They are commonly used in the pig industry and human foods [115]. Galactooligosaccharides that exist in human milk have been reported to have prebiotic effects by enhancing colonic health of breast-fed infants [116]. Many other naturally occurring prebiotics have been reported as well, including polydextrose, trans-galactooligosaccharides, xylo-oligosaccharides, lactulose, pyrodextrins, and isomalto-oligosaccharides. However, a few other nondigestible carbohydrates are not categorized as prebiotics (e.g., mannan-oligosaccharides, β-glucan etc.), but manifest health-promoting functions [117]. For example, a growing evidence demonstrates that β-glucans, either produced by bacteria or extracted from different sources (i.e., cereal, algae, and fungi), could boost host immunity, therefore enhancing disease resistance of human and animals [118–120].

Probiotics, also known as direct-fed microbials, are live microorganisms and, when administered in adequate amounts, confer a health benefit on the host [121]. Probiotics are categorized into three main groups, including *Bacillus*, lactic acid-producing bacteria, and yeast [122]. Based on the Food and Drug Administration instruction, the term probiotics is used for human microbial products, whereas the term direct-fed microbials is used for the US feed industry. However, "probiotics" are interchangeably used with human and animal feed worldwide. *Bacillus*-based probiotics are spore-forming, which makes them thermostable and able to survive at low pH. *Bacillus*-based probiotics have been identified as potent producers of extracellular fiber-degrading enzymes, which may aid nutrient digestion and utilization [123]. Lactic acid-producing bacteria are not spore-forming; therefore, their survival during feed processing is of concern [124]. Lactic acid-producing bacteria dominate the gastrointestinal tract of the nursing pig [125], which helps reduce the pH in the gut by producing lactic acid through fermentation, inhibiting enteric pathogens [126], and improving host immunity [124, 127]. However, after weaning of pigs, the concentration of lactic acid-producing bacteria diminishes; therefore, supplementation of weaned pig diets with lactic acid-producing probiotics may be beneficial [122]. Yeast include a broad range of products that may be available in pig feed, including whole live yeast cells, heat-treated yeast cells, ground yeast cells, purified yeast cell cultures, and yeast extracts. The efficacy of yeast-based products varies depending on their forms. Yeast or yeast-based product supplementation may boost feed intake and overall growth performance, augment mucosal immunity, promote intestinal development, adsorb mycotoxins, reduce postweaning diarrhea, and modulate gut microbiota in weaned pigs [128–131].

The most notable effect of prebiotics and probiotics is their modification of intestinal microbiota. They may control or prevent pathogenic bacterial infection by specifically stimulating the growth of beneficial microorganisms in the intestine. The beneficial microorganisms may include but not limited to *Bifidobacteria* and *Lactobacilli*, which have confirmed benefit to suppress the growth of pathogenic microorganisms, such as *E. coli*, through the potential mechanisms described below. For example, the desired bacteria produce SCFA and lactic acid, which may indirectly and specifically kill or inhibit the growth of pathogens [132]. The production of acids may reduce the pH of the intestinal environment, which is unsupportive of the growth of several pathogens [133]. The desired bacteria may produce antimicrobial compounds such as bacteriocins or antibiotics [134]. The desired bacteria compete the available nutrients against pathogens [135].

Many research articles have been published on the impacts of prebiotics and probiotics on infectious diseases in young pigs. For instance, supplementation of 8% inulin reduce the incidence and severity of postweaning diarrhea, probably by increasing SCFA production in the cecum and proximal colon [136]. The addition of fructo-oligosaccharide prevented the mortality and morbidity of weaned pigs infected with K88 ETEC [137]. Supplementation of β-glucan originated from different sources (yeast or algae) could enhance the resistance of pigs against K88 or F18 ETEC infection [120, 138]. The α-D-mannans from yeast could bind to mannose-specific receptors that are present on many bacteria such as *E. coli* and *Salmonella* spp., which prevents adhesion of these pathogens to the mannose-rich glycoproteins lining the intestinal lumen [128]. Indeed, pigs supplemented with live yeast or a yeast fermentation product had reduced disease-related stress, diarrhea scores, duration of diarrhea, and shedding of *E. coli* and enhanced intestinal integrity in pigs challenged with ETEC [139–141]. Supplementation of *Bacillus subtilis* also enhanced disease resistance and growth performance and reduced diarrhea of weaned pigs infected with F18 ETEC [142].

3.5 Phytochemicals

Phytochemicals are secondary plant metabolites that are either naturally obtained from plant materials or directly synthetized. Phytochemicals are used in solid powder form, as crude extracts, or as concentrated extracts. The extracts are further classified as essential oils or oleoresins based on the extraction methods. Essential oils are volatile lipophilic substances obtained by cold extraction or distillation, whereas oleoresins are derived by nonaqueous solvents [143]. A few examples of well-known phytochemicals are curcumin, flavonoids, phenolic acids, isoflavones, carotenoids, etc. The major bioactive compounds in phytochemicals are polyphenols, terpenoids, alkaloids, or sulfur-containing compounds.

However, the composition and concentration of bioactive compounds in different phytochemicals may vary a lot, completely depending on the types of plant, the parts of plant, geographical origins, growing conditions, harvesting seasons, processing techniques, and storage conditions [144]. The in vitro biological properties of many phytochemicals have been well investigated, including antimicrobial, antioxidant, anti-inflammatory, and antiviral effects [145–148]. Therefore, phytochemicals have been largely applied in food processing, cosmetics, and other areas related to human nutrition and health.

Various phytochemicals have been reported to exhibit a broad spectrum of antimicrobial activities against Gram-negative and Gram-positive bacteria [149–151]. The potential mechanisms of action of antimicrobial activities of phytochemicals are described below. Many phytochemicals are lipophilic, which could damage bacterial membrane, eventually causing the leakage of intracellular materials and cell death [152–156]. In addition, the phenolic compounds possess strong antibacterial properties by inhibiting virulence factors, such as enzymes and toxins [157–159]. Lastly, certain bioactive components may also prevent the development of virulent structure (i.e., flagella) in bacteria, therefore inhibiting ETEC adhesion and toxin binding [160, 161].

Our previously published research reported that dietary supplementation of 10 mg/kg of capsicum oleoresin, garlic botanical, or turmeric oleoresin reduced the frequency of diarrhea and enhanced disease resistance of pigs infected with F18 ETEC [162]. The active components in these phytochemicals are capsaicin, propyl thiosulfonates, and curcuminoides, respectively. The results of gene expression profiles in ileal mucosa indicated that supplementation of these phytochemicals modified the expression of genes related to mucin production, cell membrane integrity, and antigen processing and presentations in ETEC-infected pigs [163]. In addition to the enhanced intestinal mucosal health, pigs fed with those phytochemicals had less recruitment of macrophages and neutrophils in the ileum [162]. These observations also suggest that the weaned pigs supplemented with those phytochemicals actually had less gut inflammation than infected control. The gene expression profile analysis by microarray also confirmed the reduced gut inflammation by feeding those phytochemicals to weaned pigs [163]. The phytochemicals discussed above can be naturally obtained from seasonings that are commonly used in kitchen. Many other phytochemicals have been thoroughly investigated to against ETEC infection as well. For example, the anti-diarrheal activity of back or green tea extract has been revealed, because the reduced net fluid and electrolyte losses were observed when F4 ETEC-infected jejunal segments were perfused with black or green tea extract [164]. The administration of cranberry extract (1 g/L) in drinking water also reduced the diarrhea of F18 ETEC-infected piglets [165].

3.6 Antimicrobial peptides

Antimicrobial peptides, also called host defense peptides, are polypeptides that are naturally occurring molecules in various organisms from prokaryotes to mammals. Antimicrobial peptides can be synthesized as recombinant molecules, such as recombinant lactoferrin, or can be isolated from bacteria, insects, vertebrates, or plants, such as bovine lactoferrin and plant defensins [166, 167]. Most of antimicrobial peptides are cationic (positively charged) and amphiphilic (hydrophobic and hydrophilic). Antimicrobial peptides were firstly discovered in the 1980s. They can be classified into different groups based on the different amino acid components, structures, and biological function. The antimicrobial peptides derived from mammals are mainly classified into two families, defensins or cathelicidins. Defensins are further subgrouped into α-, β-, and θ-defensins according to the spacing patterns of their cysteine residues [168]. Defensins are more abundant in epithelial cells and phagocytic cells, whereas cathelicidins are highly expressed in mammalian neutrophils [169].

Antimicrobial peptides possess a strong and wide-spectrum activity against Gram-negative and Gram-positive bacteria, as well as against parasites, fungi, and viruses [170]. One potential advantage of antimicrobial peptides is that they may kill pathogenic bacteria that are resistant to specific medically important antibiotics [168, 171]. Most antimicrobial peptides are small, positively charged, and amphipathic molecules that allow them to actively interact with bacterial membranes through different models, such as barrel-stave model, carpet model, or toroidal pore model [172, 173]. These properties will also allow them to disrupt cell membrane structure, penetrate into cells, regulate intracellular signaling pathways, and ultimately cause bacterial cell death. Many research findings have demonstrated that antimicrobial peptide treatment could inhibit protein and nucleic acid synthesis, suppress bacterial cell wall synthesis, as well as inhibit enzyme activities in bacteria [174]. In addition to their antibacterial properties, antimicrobial peptides may also act as epithelial "preservatives" or immunomodulators to protect host against enteric infectious agents [166, 175, 176].

The protective effects of antimicrobial peptides on infectious diarrhea and intestinal integrity have been reported in weaned pigs [177, 178]. For example, feeding 0.4% of a mixture of bovine lactoferrin, plant defensins, and active yeast increased intestinal integrity and reduced gut permeability of weaned pigs [178]. Addition of cecropin AD reduced incidence of diarrhea and enhanced intestinal *Lactobacilli* counts in *E. coli*-challenged piglets [177]. The regulation of gut microbiota may also attribute to the potential benefits of antimicrobial peptides. Supplementation of recombinant lactoferrin or lactoferramoin-lactoferricin reduced the total viable counts of *E. coli* and *Salmonella* but enriched the abundance of *Lactobacillus* and *Bifidobacterium* in the colon of weaned pigs [179, 180].

3.7 Lysozyme

Lysozyme, also known as muramidase or N-acetylmuramide glycanhydrolase, is an antimicrobial enzyme that is naturally present in body fluids of all mammalian species [181–183]. Lysozyme could catalyze the hydrolysis of its natural substrate peptidoglycan that is the major component of bacterial cell wall. The hydrolysis of peptidoglycan eventually results in cell lysis. Gram-positive bacteria have cell walls composed of thick layers of peptidoglycan; they are, therefore, more sensitive to the enzymatic degradation of lysozyme [184]. However, a growing evidence supports that lysozyme also displays bactericidal activity against a variety of Gram-negative strains through non-enzymatic mechanisms [184–186]. For instance, lysozyme has been found to act synergistically with antimicrobial peptides, lactoferrin, in killing Gram-negative bacteria, such as *E. coli* [181, 187]. In addition to its antimicrobial activity, lysozyme also exhibited anti-inflammatory property that was mediated through inhibiting neutrophil migration and shown the ability to modulate intestinal microbiota [188].

Consumption of lysozyme-rich milk significantly enhanced the relative abundance of *Bifidobacteriaceae* and *Lactobacillaceae* in feces of weaned pigs [189]. Those bacterial families are known for their health-promoting functions in lower gastrointestinal tract of human and pigs. Interestingly, consumption of lysozyme-rich milk reduced the relative abundance of bacteria (*Mycobacteriaceae*, *Streptococcaceae*, *Campylobacterales*) that are associated with diseases in pig feces [189]. In another trial from the same research group, feeding of lysozyme milk reduced the incidence of diarrhea and reduced total bacteria translocation into the mesenteric lymph nodes by 83% in ETEC (O149:F4 strain)-infected pigs [190]. Feeding lysozyme-rich milk also tended to reduce fecal *Enterobacteriaceae* family, in which many prevalent enteric pathogens such as *E. coli* and *Salmonella* belong to [190]. Similar results were also reported in one animal trial focusing on human lysozyme-rich milk [191]. In this experiment, neonatal pigs were used and infected with F4 ETEC. Consumption of human lysozyme-rich milk (approximately 1300 mg/L lysozyme) increased survival rate, reduced diarrhea, and facilitated the recovery of infected pigs [191]. Lysozyme treatment also increased the relative abundance of *Lactobacillus* in feces and enhanced intestinal integrity and mucosa immunity of these neonatal pigs [191].

4. Conclusions

Accumulating evidence has confirmed the importance of nutritional interventions, including modified feeding strategies and nutrient supplements, in the control of diarrheal diseases, and preventing enteric infection as the use of antibiotics will be progressively restricted in many countries. Interest is particularly growing

Strain[1]	Dietary supplements	Outcome	Reference
K88	Milk from human lysozyme transgenic goats	Reduced diarrhea, reduced bacterial translocation in mesenteric lymph nodes	Brundige et al. [193]; Cooper et al. [194]; Garas et al. [190]
K88	Chito-oligosaccharide	Reduced diarrhea	Liu et al. [195]
K88	Combination of raw potato starch and probiotic *E. coli* strains	Reduced diarrhea, enhanced gut microbial diversity	Krause et al. [196]
K88, F18	Probiotics: *Pediococcus acidilactici*, *Saccharomyces cerevisiae boulardii*, *Bacillus subtilis*	Reduced ETEC attachment to ileal mucosa, upregulated inflammatory responses in the gut	Kim et al. [142]; Daudelin et al. [197]
K88	*Saccharomyces cerevisiae* fermented products	Enhanced appetite and ileal digesta bacteria richness, reduced ETEC adhering to the mucosa and colonic ammonia	Kiarie et al. [139, 140]
K88	Probiotics: *Lactobacillus plantarum* CJLP243	Enhanced growth performance, reduced diarrhea, reduced gut inflammation, enhanced gut barrier function	Lee et al. [198]; Yang et al. [199]
K88	Phytogenics	Enhanced growth performance	Devi et al. [200]
K88	Nucleotides	Enhanced growth performance and nutrient digestibility, reduced diarrhea	Li et al. [201]
F18	Clays (smectite, zeolite, kaolinite)	Reduced diarrhea, enhanced gut integrity	Song et al. [202]; Almeida et al. [203]
F18	Phytochemicals (capsicum oleoresin, garlic botanical, turmeric oleoresin)	Reduced diarrhea, enhanced gut morphology, decreased systemic and gut mucosal inflammation	Liu et al. [162, 163]
K88, F18	β-glucan	Enhanced gut barrier function, reduced systemic inflammation	Stuyven et al. [138], Kim et al. [120]

[1]*Modified from Liu and Ji [192].*

Table 1.
Dietary interventions on enterotoxigenic Escherichia coli *infection of weaned pigs.*

in the use of probiotics and/or prebiotics to increase the populations of target microbes in the digestive tract, thereby improving gut health and performance of animals. Phytochemicals can be an additional tool that producers use to keep pigs healthy and reduce the negative impacts of disease. Dietary supplementation of certain phytochemicals may enhance disease resistance of pigs by improving gut mucosal integrity and optimizing immune response. There are much more candidates of feed additives/nutritional interventions, which may be effective in regulating intestinal environments and immunity and alleviating postweaning enteric infection (**Table 1**) [192]. It is very important to keep in mind that the efficiencies of each candidate may differ on the basis of their modes of action, the basal diet formulation, and the health status of pigs.

Conflict of interest

The authors declare no conflict of interest.

Author details

Peng Ji[1], Xunde Li[2] and Yanhong Liu[3]*

1 Department of Nutrition, University of California, Davis, CA, USA

2 School of Veterinary Medicine, University of California, Davis, CA, USA

3 Department of Animal Science, University of California, Davis, CA, USA

*Address all correspondence to: yahliu@ucdavis.edu

References

[1] Torres AG, Zhou X, Kaper JB. Adherence of diarrheagenic *Escherichia coli* strains to epithelial cells. Infection and Immunity. 2005;**73**(1):18-29

[2] Fairbrother JM, Nadeau E, Gyles CL. *Escherichia coli* in postweaning diarrhea in pigs: An update on bacterial types, pathogenesis, and prevention strategies. Animal Health Research Reviews. 2005;**6**(1):17-39

[3] Nagy B, Fekete PZ. Enterotoxigenic *Escherichia coli* in veterinary medicine. International Journal of Medical Microbiology. 2005;**295**(6-7):443-454

[4] Nagy B, Fekete PZ. Enterotoxigenic *Escherichia coli* (ETEC) in farm animals. Veterinary Research. 1999;**30**(2-3):259-284

[5] Hampson DJ. Postweaning *Escherichia coli* diarrhea in pigs. In: Gyles CL, editor. *Escherichia coli* in Domestic Animals and Humans. Guildford: CAB International; 1994. pp. 171-191

[6] Sarmiento JI, Casey TA, Moon HW. Postweaning diarrhea in swine: Experimental model of enterotoxigenic *Escherichia coli* infection. American Journal of Veterinary Research. 1988;**49**(7):1154-1159

[7] Nollet H, Deprez P, Van Driessche E, Muylle E. Protection of just weaned pigs against infection with F18+ *Escherichia coli* by non-immune plasma powder. Veterinary Microbiology. 1999;**65**(1):37-45

[8] Faubert C, Drolet R. Hemorrhagic gastroenteritis caused by *Escherichia coli* in piglets: Clinical, pathological and microbiological findings. The Canadian Veterinary Journal = La revue veterinaire canadienne. 1992;**33**(4):251-256

[9] Berberov EM, Zhou Y, Francis DH, Scott MA, Kachman SD, Moxley RA. Relative importance of heat-labile enterotoxin in the causation of severe diarrheal disease in the gnotobiotic piglet model by a strain of enterotoxigenic *Escherichia coli* that produces multiple enterotoxins. Infection and Immunity. 2004;**72**(7):3914-3924

[10] Weikel CS, Guerrant RL. STb enterotoxin of *Escherichia coli*: Cyclic nucleotide-independent secretion. Ciba Foundation Symposium. 1985;**112**:94-115

[11] Forte LR, Thorne PK, Eber SL, Krause WJ, Freeman RH, Francis SH, et al. Stimulation of intestinal Cl– transport by heat-stable enterotoxin: Activation of cAMP-dependent protein kinase by cGMP. The American Journal of Physiology. 1992;**263**(3 Pt 1): C607-C615

[12] O'Brien ADRKH. Protein toxins of *Escherichia coli* and Salmonella. In: Neidhart FC, Curtiss R, Ingraham JL, Lin ECC, Low KB, Magasanik B, Reznikoff WS, Riley M, et al., editors. *Escherichia coli* and Salmonella: Cellular and Molecular Biology. Washington, DC: ASM Press; 1996. pp. 2788-2802

[13] Gyles CL. *Escherichia coli*. In: Gyles CL, Charles O. Thoen editor. Pathogenesis of Bacterial Infections in Animals. Ames, IA: Iowa State University Press; 1993. p. 164-187

[14] Sandvig K. Shiga toxins. Toxicon : Official Journal of the International Society on Toxinology. 2001;**39**(11):1629-1635

[15] Bannerman DD, Goldblum SE. Mechanisms of bacterial lipopolysaccharide-induced endothelial

apoptosis. American Journal of Physiology Lung Cellular and Molecular Physiology. 2003;**284**(6):L899-L914

[16] Nagy B, Whipp SC, Imberechts H, Bertschinger HU, Dean-Nystrom EA, Casey TA, et al. Biological relationship between F18ab and F18ac fimbriae of enterotoxigenic and verotoxigenic *Escherichia coli* from weaned pigs with oedema disease or diarrhoea. Microbial Pathogenesis. 1997;**22**(1):1-11

[17] Isaacson RE, Fusco PC, Brinton CC, Moon HW. In vitro adhesion of *Escherichia coli* to porcine small intestinal epithelial cells: Pili as adhesive factors. Infection and Immunity. 1978;**21**(2):392-397

[18] Morris JA, Thorns C, Scott AC, Sojka WJ, Wells GA. Adhesion in vitro and in vivo associated with an adhesive antigen (F41) produced by a K99 mutant of the reference strain *Escherichia coli* B41. Infection and Immunity. 1982;**36**(3):1146-1153

[19] Nagy B, Moon HW, Isaacson RE. Colonization of porcine small intestine by *Escherichia coli*: Ileal colonization and adhesion by pig enteropathogens that lack K88 antigen and by some acapsular mutants. Infection and Immunity. 1976;**13**(4):1214-1220

[20] Van den Broeck W, Cox E, Oudega B, Goddeeris BM. The F4 fimbrial antigen of *Escherichia coli* and its receptors. Veterinary Microbiology. 2000;**71**(3-4):223-244

[21] Blomberg L, Krivan HC, Cohen PS, Conway PL. Piglet ileal mucus contains protein and glycolipid (galactosylceramide) receptors specific for *Escherichia coli* K88 fimbriae. Infection and Immunity. 1993;**61**(6):2526-2531

[22] Grange PA, Erickson AK, Levery SB, Francis DH. Identification of an intestinal neutral glycosphingolipid as a phenotype-specific receptor for the K88ad fimbrial adhesin of *Escherichia coli*. Infection and Immunity. 1999;**67**(1):165-172

[23] Erickson AK, Baker DR, Bosworth BT, Casey TA, Benfield DA, Francis DH. Characterization of porcine intestinal receptors for the K88ac fimbrial adhesin of *Escherichia coli* as mucin-type sialoglycoproteins. Infection and Immunity. 1994; **62**(12):5404-5410

[24] Jin LZ, Zhao X. Intestinal receptors for adhesive fimbriae of enterotoxigenic *Escherichia coli* (ETEC) K88 in swine—A review. Applied Microbiology and Biotechnology. 2000;**54**(3):311-318

[25] Rippinger P, Bertschinger HU, Imberechts H, Nagy B, Sorg I, Stamm M, et al. Designations F18ab and F18ac for the related fimbrial types F107, 2134P and 8813 of *Escherichia coli* isolated from porcine postweaning diarrhoea and from oedema disease. Veterinary Microbiology. 1995;**45**(4):281-295

[26] Wittig W, Klie H, Gallien P, Lehmann S, Timm M, Tschape H. Prevalence of the fimbrial antigens F18 and K88 and of enterotoxins and verotoxins among *Escherichia coli* isolated from weaned pigs. Zentralblatt fur Bakteriologie: International Journal of Medical Microbiology. 1995;**283**(1):95-104

[27] Imberechts H, De Greve H, Lintermans P. The pathogenesis of edema disease in pigs. A review. Veterinary Microbiology. 1992;**31**(2-3):221-233

[28] Smeds A, Hemmann K, Jakava-Viljanen M, Pelkonen S, Imberechts H, Palva A. Characterization of the adhesin of *Escherichia coli* F18 fimbriae. Infection and Immunity. 2001;**69**(12):7941-7945

[29] Imberechts H, Wild P, Charlier G, De Greve H, Lintermans P, Pohl P. Characterization of F18 fimbrial genes fedE and fedF involved in adhesion and length of enterotoxemic *Escherichia coli* strain 107/86. Microbial Pathogenesis. 1996;**21**(3):183-192

[30] Nagy B, Casey TA, Whipp SC, Moon HW. Susceptibility of porcine intestine to pilus-mediated adhesion by some isolates of piliated enterotoxigenic *Escherichia coli* increases with age. Infection and Immunity. 1992;**60**(4):1285-1294

[31] Thiagarajah JR, Verkman AS. CFTR pharmacology and its role in intestinal fluid secretion. Current Opinion in Pharmacology. 2003;**3**(6):594-599

[32] Nataro JP, Kaper JB. Diarrheagenic *Escherichia coli*. Clinical Microbiology Reviews. 1998;**11**(1):142-201

[33] De Haan L, Hirst TR. Cholera toxin: A paradigm for multi-functional engagement of cellular mechanisms (review). Molecular Membrane Biology. 2004;**21**(2):77-92

[34] Lallier R, Bernard F, Gendreau M, Lazure C, Seidah NG, Chretien M, et al. Isolation and purification of *Escherichia coli* heat-stable enterotoxin of porcine origin. Analytical Biochemistry. 1982;**127**(2):267-275

[35] Schulz S, Green CK, Yuen PS, Garbers DL. Guanylyl cyclase is a heat-stable enterotoxin receptor. Cell. 1990;**63**(5):941-948

[36] Arriaga YL, Harville BA, Dreyfus LA. Contribution of individual disulfide bonds to biological action of *Escherichia coli* heat-stable enterotoxin B. Infection and Immunity. 1995;**63**(12):4715-4720

[37] Dubreuil JD. *Escherichia coli* STb enterotoxin. Microbiology (Reading, England). 1997;**143**(Pt 6):1783-1795

[38] Weikel CS, Nellans HN, Guerrant RL. In vivo and in vitro effects of a novel enterotoxin, STb, produced by *Escherichia coli*. The Journal of Infectious Diseases. 1986;**153**(5):893-901

[39] Argenzio RA, Liacos J, Berschneider HM, Whipp SC, Robertson DC. Effect of heat-stable enterotoxin of *Escherichia coli* and theophylline on ion transport in porcine small intestine. Canadian Journal of Comparative Medicine: Revue Canadienne de Medecine Comparee. 1984;**48**(1):14-22

[40] Whipp SC, Kokue E, Morgan RW, Rose R, Moon HW. Functional significance of histologic alterations induced by *Escherichia coli* pig-specific, mouse-negative, heat-stable enterotoxin (STb). Veterinary Research Communications. 1987;**11**(1):41-55

[41] Dreyfus LA, Harville B, Howard DE, Shaban R, Beatty DM, Morris SJ. Calcium influx mediated by the *Escherichia coli* heat-stable enterotoxin B (STB). Proceedings of the National Academy of Sciences of the United States of America. 1993;**90**(8):3202-3206

[42] Harville BA, Dreyfus LA. Involvement of 5-hydroxytryptamine and prostaglandin E2 in the intestinal secretory action of *Escherichia coli* heat-stable enterotoxin B. Infection and Immunity. 1995;**63**(3):745-750

[43] Alexander C, Rietschel ET. Bacterial lipopolysaccharides and innate immunity. Journal of Endotoxin Research. 2001;**7**(3):167-202

[44] Raetz CR, Whitfield C. Lipopolysaccharide endotoxins. Annual Review of Biochemistry. 2002;**71**:635-700

[45] Wiese A, Brandenburg K, Ulmer AJ, Seydel U, Muller-Loennies S. The

dual role of lipopolysaccharide as effector and target molecule. Biological Chemistry. 1999;**380**(7-8):767-784

[46] Sandvig K, van Deurs B. Endocytosis, intracellular transport, and cytotoxic action of Shiga toxin and ricin. Physiological Reviews. 1996;**76**(4):949-966

[47] Paton JC, Paton AW. Pathogenesis and diagnosis of Shiga toxin-producing *Escherichia coli* infections. Clinical Microbiology Reviews. 1998;**11**(3):450-479

[48] Hughes AK, Stricklett PK, Kohan DE. Shiga toxin-1 regulation of cytokine production by human glomerular epithelial cells. Nephron. 2001;**88**(1):14-23

[49] Wu G. Amino acids: Metabolism, functions, and nutrition. Amino Acids. 2009;**37**(1):1-17

[50] Wu G, Bazer FW, Davis TA, Jaeger LA, Johnson GA, Kim SW, et al. Important roles for the arginine family of amino acids in swine nutrition and production. Livestock Science. 2007;**112**(1):8-22

[51] Kim SW, Wu G. Dietary arginine supplementation enhances the growth of milk-fed young pigs. The Journal of Nutrition. 2004;**134**(3):625-630

[52] Hernandez A, Hansen CF, Mullan BP, Pluske JR. L-arginine supplementation of milk liquid or dry diets fed to pigs after weaning has a positive effect on production in the first three weeks after weaning at 21 days of age. Animal Feed Science and Technology. 2009;**154**(1):102-111

[53] Wu X, Ruan Z, Gao Y, Yin Y, Zhou X, Wang L, et al. Dietary supplementation with L-arginine or N-carbamylglutamate enhances intestinal growth and heat shock protein-70 expression in weanling pigs fed a corn- and soybean meal-based diet. Amino Acids. 2010;**39**(3):831-839

[54] Wu L, Liao P, He L, Feng Z, Ren W, Yin J, et al. Dietary L-arginine supplementation protects weanling pigs from deoxynivalenol-induced toxicity. Toxins. 2015;**7**(4):1341-1354

[55] Wang J, Li GR, Tan BE, Xiong X, Kong XF, Xiao DF, et al. Oral administration of putrescine and proline during the suckling period improves epithelial restitution after early weaning in piglets. Journal of Animal Science. 2015;**93**(4):1679-1688

[56] Wu G, Meier SA, Knabe DA. Dietary glutamine supplementation prevents jejunal atrophy in weaned pigs. The Journal of Nutrition. 1996;**126**(10):2578-2584

[57] Li Y, Li J, Jiang J, Li N, Wang X, Wang Z, et al. Glycyl-glutamine-supplemented long-term total parenteral nutrition selectively improves structure and function in heterotopic small-bowel autotransplantation in the pig. Transplant International: Official Journal of the European Society for Organ Transplantation. 2003;**16**(12):866-871

[58] Jiang ZY, Sun LH, Lin YC, Ma XY, Zheng CT, Zhou GL, et al. Effects of dietary glycyl-glutamine on growth performance, small intestinal integrity, and immune responses of weaning piglets challenged with lipopolysaccharide. Journal of Animal Science. 2009;**87**(12):4050-4056

[59] Burrin DG, Reeds PJ. Alternative fuels in the gastrointestinal tract. Current Opinion in Gastroenterology. 1997;**13**(2):165-170

[60] Zhan Z, Ou D, Piao X, Kim SW, Liu Y, Wang J. Dietary arginine supplementation affects microvascular

development in the small intestine of early-weaned pigs. The Journal of Nutrition. 2008;**138**(7):1304-1309

[61] Eklou-Lawson M, Bernard F, Neveux N, Chaumontet C, Bos C, Davila-Gay AM, et al. Colonic luminal ammonia and portal blood L-glutamine and L-arginine concentrations: A possible link between colon mucosa and liver ureagenesis. Amino Acids. 2009;**37**(4):751-760

[62] Tan B, Xiao H, Xiong X, Wang J, Li G, Yin Y, et al. L-arginine improves DNA synthesis in LPS-challenged enterocytes. Frontiers in Bioscience (Landmark Edition). 2015;**20**:989-1003

[63] Wang J, Chen L, Li P, Li X, Zhou H, Wang F, et al. Gene expression is altered in piglet small intestine by weaning and dietary glutamine supplementation. The Journal of Nutrition. 2008;**138**(6):1025-1032

[64] Wu M, Xiao H, Liu G, Chen S, Tan B, Ren W, et al. Glutamine promotes intestinal SIgA secretion through intestinal microbiota and IL-13. Molecular Nutrition & Food Research. 2016;**60**(7):1637-1648

[65] Ruzin A, Novick RP. Equivalence of lauric acid and glycerol monolaurate as inhibitors of signal transduction in *Staphylococcus aureus*. Journal of Bacteriology. 2000;**182**(9):2668-2671

[66] Mallo JJ, Balfagon A, Gracia MI, Honrubia P, Puyalto M. Evaluation of different protections of butyric acid aiming for release in the last part of the gastrointestinal tract of piglets. Journal of Animal Science. 2012;**90**(Suppl 4): 227-229

[67] Kovanda L, Zhang W, Wei X, Luo J, Wu X, Atwill ER, et al. In vitro antimicrobial activities of organic acids and their derivatives on several species of Gram-negative and Gram-positive bacteria. Molecules (Basel, Switzerland). 2019;**24**(20):3770-3783

[68] Rossi R, Pastorelli G, Cannata S, Corino C. Recent advances in the use of fatty acids as supplements in pig diets: A review. Animal Feed Science and Technology. 2010;**162**(1):1-11

[69] Bedford A, Gong J. Implications of butyrate and its derivatives for gut health and animal production. Animal Nutrition (Zhongguo Xu Mu Shou Yi Xue Hui). 2018;**4**(2):151-159

[70] Xiong X, Tan B, Song M, Ji P, Kim K, Yin Y, et al. Nutritional intervention for the intestinal development and health of weaned pigs. Frontiers in Veterinary Science. 2019;**6**:46

[71] Lu JJ, Zou XT, Wang YM. Effects of sodium butyrate on the growth performance, intestinal microflora and morphology of weanling pigs. Journal of Animal and Feed Sciences. 2008;**17**(4):568-578

[72] Fang CL, Sun H, Wu J, Niu HH, Feng J. Effects of sodium butyrate on growth performance, haematological and immunological characteristics of weanling piglets. Journal of animal physiology and animal nutrition. 2014;**98**(4):680-685

[73] Hou Y, Wang L, Yi D, Ding B, Chen X, Wang Q, et al. Dietary supplementation with tributyrin alleviates intestinal injury in piglets challenged with intrarectal administration of acetic acid. The British Journal of Nutrition. 2014;**111**(10):1748-1758

[74] Huang C, Song P, Fan P, Hou C, Thacker P, Ma X. Dietary sodium butyrate decreases postweaning diarrhea by modulating intestinal permeability and changing the bacterial communities in weaned piglets. The Journal of Nutrition. 2015;**145**(12):2774-2780

[75] Cuff M, Dyer J, Jones M, Shirazi-Beechey S. The human colonic monocarboxylate transporter isoform 1: Its potential importance to colonic tissue homeostasis. Gastroenterology. 2005;**128**(3):676-686

[76] Thangaraju M, Cresci GA, Liu K, Ananth S, Gnanaprakasam JP, Browning DD, et al. GPR109A is a G-protein-coupled receptor for the bacterial fermentation product butyrate and functions as a tumor suppressor in colon. Cancer Research. 2009;**69**(7):2826-2832

[77] Fontenelle B, Gilbert KM. n-butyrate anergized effector CD4+ T cells independent of regulatory T cell generation or activity. Scandinavian Journal of Immunology. 2012;**76**(5):457-463

[78] Zeng X, Sunkara LT, Jiang W, Bible M, Carter S, Ma X, et al. Induction of porcine host defense peptide gene expression by short-chain fatty acids and their analogs. PLoS One. 2013;**8**(8):e72922

[79] Odle J. New insights into the utilization of medium-chain triglycerides by the neonate: Observations from a piglet model. The Journal of Nutrition. 1997;**127**(6): 1061-1067

[80] Zentek J, Buchheit-Renko S, Ferrara F, Vahjen W, Van Kessel AG, Pieper R. Nutritional and physiological role of medium-chain triglycerides and medium-chain fatty acids in piglets. Animal Health Research Reviews. 2011;**12**(1):83-93

[81] Hanczakowska E. The use of medium-chain fatty acids in piglet feeding—A review. Annals of Animal Science. 2017;**17**(4):967

[82] McDowell LR. Minerals in Animal and Human Nutrition. New York, NY: Academic Press; 1992

[83] Ku PK, Ullery DE, Miller ER. Zinc deficiency and tissue nucleic acid and protein concentration. In: Mills CF, editor. Trace Element Metabolism in Animals. Edinburgh, UK: E. & S. Livingstone; 1970. pp. 158-164

[84] Fernandez-Madrid F, Prasad AS, Oberleas D. Effect of zinc deficiency on collagen metabolism. The Journal of Laboratory and Clinical Medicine. 1971;**78**(5):853

[85] NRC. Nutrient Requirements of Swine. 11th revised ed. Washington, DC: Natl. Acad. Press; 2012

[86] Hahn JD, Baker DH. Growth and plasma zinc responses of young pigs fed pharmacologic levels of zinc. Journal of Animal Science. 1993;**71**(11):3020-3024

[87] Hill GM, Mahan DC, Carter SD, Cromwell GL, Ewan RC, Harrold RL, et al. Effect of pharmacological concentrations of zinc oxide with or without the inclusion of an antibacterial agent on nursery pig performance. Journal of Animal Science. 2001;**79**(4):934-941

[88] Case CL, Carlson MS. Effect of feeding organic and inorganic sources of additional zinc on growth performance and zinc balance in nursery pigs. Journal of Animal Science. 2002;**80**(7):1917-1924

[89] Starke IC, Pieper R, Neumann K, Zentek J, Vahjen W. The impact of high dietary zinc oxide on the development of the intestinal microbiota in weaned piglets. FEMS Microbiology Ecology. 2014;**87**(2):416-427

[90] Pearce SC, Sanz Fernandez MV, Torrison J, Wilson ME, Baumgard LH, Gabler NK. Dietary organic zinc attenuates heat stress-induced changes in pig intestinal integrity and metabolism. Journal of Animal Science. 2015;**93**(10):4702-4713

[91] Katouli M, Melin L, Jensen-Waern M, Wallgren P, Mollby R. The effect of zinc oxide supplementation on the stability of the intestinal flora with special reference to composition of coliforms in weaned pigs. Journal of Applied Microbiology. 1999;**87**(4):564-573

[92] Zhang B, Guo Y. Supplemental zinc reduced intestinal permeability by enhancing occludin and zonula occludens protein-1 (ZO-1) expression in weaning piglets. The British Journal of Nutrition. 2009;**102**(5):687-693

[93] van Heugten E, Spears JW, Kegley EB, Ward JD, Qureshi MA. Effects of organic forms of zinc on growth performance, tissue zinc distribution, and immune response of weanling pigs. Journal of Animal Science. 2003;**81**(8):2063-2071

[94] Walk CL, Srinongkote S, Wilcock P. Influence of a microbial phytase and zinc oxide on young pig growth performance and serum minerals. Journal of Animal Science. 2013;**91**(1):286-291

[95] Walk CL, Wilcock P, Magowan E. Evaluation of the effects of pharmacological zinc oxide and phosphorus source on weaned piglet growth performance, plasma minerals and mineral digestibility. Animal: An International Journal of Animal Bioscience. 2015;**9**(7):1145-1152

[96] Ciesinski L, Guenther S, Pieper R, Kalisch M, Bednorz C, Wieler LH. High dietary zinc feeding promotes persistence of multi-resistant *E. coli* in the swine gut. PloS One. 2018;**13**(1):e0191660

[97] Ward TLAG, Louis GF, Pollman DS. Zinc-methionine improves growth performance of starter pigs. Jornal of Animal Science. 1996;**74**(Suppl 1):182

[98] Mavromichalis I, Webel DM, Parr EN, Baker DH. Growth-promoting efficacy of pharmacological doses of tetrabasic zinc chloride in diets for nursery pigs. Canadian Journal of Animal Science. 2001;**81**(3):387-391

[99] Underwood EJ. Trace Elements in Human and Animal Nutrition. 4th ed. New York: Academic Press; 1977

[100] Suttle NF. Mineral Nutrition of Livestock. 4th ed. Oxon, UK: CABI Publishing; 2010

[101] Baxter JH, Van Wyk JJ, Follis RH Jr. A bone disorder associated with copper deficiency. II. Histological and chemical studies on the bones. Bulletin of the Johns Hopkins Hospital. 1953;**93**(1):25-39

[102] Hart EB, Steenbock H, Waddell J, Elvehjem CA. Iron in nutrition. VII. Copper as a supplement to iron for hemoglobin building in the rat. 1928. The Journal of Biological Chemistry. 2002;**277**(34):e22

[103] Zhao J, Harper AF, Estienne MJ, Webb KE Jr, McElroy AP, Denbow DM. Growth performance and intestinal morphology responses in early weaned pigs to supplementation of antibiotic-free diets with an organic copper complex and spray-dried plasma protein in sanitary and nonsanitary environments. Journal of Animal Science. 2007;**85**(5):1302-1310

[104] Perez VG, Waguespack AM, Bidner TD, Southern LL, Fakler TM, Ward TL, et al. Additivity of effects from dietary copper and zinc on growth performance and fecal microbiota of pigs after weaning. Journal of Animal Science. 2011;**89**(2):414-425

[105] Ma YL, Zanton GI, Zhao J, Wedekind K, Escobar J, Vazquez-Anon M. Multitrial analysis of the effects of copper level and source on performance in nursery pigs. Journal of Animal Science. 2015;**93**(2):606-614

[106] Kloubert V, Blaabjerg K, Dalgaard TS, Poulsen HD, Rink L, Wessels I. Influence of zinc supplementation on immune parameters in weaned pigs. Journal of Trace Elements in Medicine and Biology: Organ of the Society for Minerals and Trace Elements (GMS). 2018;**49**:231-240

[107] Stahly TS, Cromwell GL, Monegue HJ. Effects of the dietary inclusion of copper and(or) antibiotics on the performance of weanling pigs. Journal of Animal Science. 1980;**51**(6):1347-1351

[108] Hojberg O, Canibe N, Poulsen HD, Hedemann MS, Jensen BB. Influence of dietary zinc oxide and copper sulfate on the gastrointestinal ecosystem in newly weaned piglets. Applied and Environmental Microbiology. 2005;**71**(5):2267-2277

[109] Cromwell GL, Lindemann MD, Monegue HJ, Hall DD, Orr DE Jr. Tribasic copper chloride and copper sulfate as copper sources for weanling pigs. Journal of Animal Science. 1998;**76**(1):118-123

[110] Banks KM, Thompson KL, Rush JK, Applegate TJ. Effects of copper source on phosphorus retention in broiler chicks and laying hens. Poultry Science. 2004;**83**(6):990-996

[111] Pang Y, Applegate TJ. Effects of copper source and concentration on in vitro phytate phosphorus hydrolysis by phytase. Journal of Agricultural and Food Chemistry. 2006;**54**(5):1792-1796

[112] Armstrong TA, Cook DR, Ward MM, Williams CM, Spears JW. Effect of dietary copper source (cupric citrate and cupric sulfate) and concentration on growth performance and fecal copper excretion in weanling pigs. Journal of Animal Science. 2004;**82**(4):1234-1240

[113] Gibson GR, Probert HM, Loo JV, Rastall RA, Roberfroid MB. Dietary modulation of the human colonic microbiota: Updating the concept of prebiotics. Nutrition Research Reviews. 2004;**17**(2):259-275

[114] Macfarlane S, Macfarlane GT, Cummings JH. Review article: Prebiotics in the gastrointestinal tract. Alimentary Pharmacology & Therapeutics. 2006;**24**(5):701-714

[115] Kelly G. Inulin-type prebiotics—A review: Part 1. Alternative Medicine Review: A Journal of Clinical Therapeutic. 2008;**13**(4):315-329

[116] Boehm G, Jelinek J, Stahl B, van Laere K, Knol J, Fanaro S, et al. Prebiotics in infant formulas. Journal of Clinical Gastroenterology. 2004;**38**(6 Suppl):S76-S79

[117] Zeković DB, Kwiatkowski S, Vrvić MM, Jakovljević D, Moran CA. Natural and modified (1→3)-β-D-glucans in health promotion and disease alleviation. Critical Reviews in Biotechnology. 2005;**25**(4):205-230

[118] Samuelsen AB, Schrezenmeir J, Knutsen SH. Effects of orally administered yeast-derived beta-glucans: A review. Molecular Nutrition & Food Research. 2014;**58**(1):183-193

[119] Soltanian S, Stuyven E, Cox E, Sorgeloos P, Bossier P. Beta-glucans as immunostimulant in vertebrates and invertebrates. Critical Reviews in Microbiology. 2009;**35**(2):109-138

[120] Kim K, Ehrlich A, Perng V, Chase JA, Raybould H, Li X, et al. Algae-derived β-glucan enhanced gut health and immune responses of weaned pigs experimentally infected with a pathogenic *E. coli*. Animal Feed Science and Technology. 2019;**248**:114-125

[121] World Heath Organization, Food and Agriculture Organization of

the United Nations, United Nations University. Protein and amino acid requirements in human nutrition. 2007. Report of a joint FAO/WHO/UNU expert consultation (WHO Technical Report Series 935)

[122] Stein HH, Kil DY. Reduced use of antibiotic growth promoters in diets fed to weanling pigs: Dietary tools, part 2. Animal Biotechnology. 2006;**17**(2):217-231

[123] Ferrari E, Jarnagin A, Schmidt BF. Commercial production of extracellular enzymes. In: Sonenshein AL, Hoch JA, Losick R, editors. *Bacillus subtilis* and Other Gram-Positive Bacteria. Washington, DC: American Society for Microbiology; 1993. pp. 917-937

[124] de Lange CFM, Pluske J, Gong J, Nyachoti CM. Strategic use of feed ingredients and feed additives to stimulate gut health and development in young pigs. Livestock Science. 2010;**134**(1):124-134

[125] Richards JD, Gong J, de Lange CFM. The gastrointestinal microbiota and its role in monogastric nutrition and health with an emphasis on pigs: Current understanding, possible modulations, and new technologies for ecological studies. Canadian Journal of Animal Science. 2005;**85**(4):421-435

[126] Vandenbergh PA. Lactic acid bacteria, their metabolic products and interference with microbial growth. FEMS Microbiology Reviews. 1993;**12**(1):221-237

[127] Niers LE, Timmerman HM, Rijkers GT, van Bleek GM, van Uden NO, Knol EF, et al. Identification of strong interleukin-10 inducing lactic acid bacteria which down-regulate T helper type 2 cytokines. Clinical and Experimental Allergy: Journal of the British Society for Allergy and Clinical Immunology. 2005;**35**(11):1481-1489

[128] Kogan G, Kocher A. Role of yeast cell wall polysaccharides in pig nutrition and health protection. Livestock Science. 2007;**109**(1):161-165

[129] Sauer N, Eklund M, Roth S, Rink F, Jezierny D, Bauer E, et al. Short-term effect of dietary yeast nucleotide supplementation on small intestinal enzyme activities, bacterial populations and metabolites and ileal nutrient digestibilities in newly weaned pigs. Journal of Animal Physiology and Animal Nutrition. 2012;**96**(4):700-708

[130] Shen YB, Piao XS, Kim SW, Wang L, Liu P, Yoon I, et al. Effects of yeast culture supplementation on growth performance, intestinal health, and immune response of nursery pigs. Journal of Animal Science. 2009;**87**(8):2614-2624

[131] Jiang Z, Wei S, Wang Z, Zhu C, Hu S, Zheng C, et al. Effects of different forms of yeast *Saccharomyces cerevisiae* on growth performance, intestinal development, and systemic immunity in early-weaned piglets. Journal of Animal Science and Biotechnology. 2015;**6**:47

[132] Toure R, Kheadr E, Lacroix C, Moroni O, Fliss I. Production of antibacterial substances by bifidobacterial isolates from infant stool active against listeria monocytogenes. Journal of Applied Microbiology. 2003;**95**(5):1058-1069

[133] Gibson GR, Wang X. Regulatory effects of bifidobacteria on the growth of other colonic bacteria. The Journal of Applied Bacteriology. 1994;**77**(4):412-420

[134] Corr SC, Li Y, Riedel CU, O'Toole PW, Hill C, Gahan CG. Bacteriocin production as a mechanism for the antiinfective activity of

Lactobacillus salivarius UCC118. Proceedings of the National Academy of Sciences of the United States of America. 2007;**104**(18):7617-7621

[135] Gibson GR, McCartney AL, Rastall RA. Prebiotics and resistance to gastrointestinal infections. The British Journal of Nutrition. 2005;**93**(Suppl 1): S31-S34

[136] Halas D, Hansen CF, Hampson DJ, Mullan BP, Wilson RH, Pluske JR. Effect of dietary supplementation with inulin and/or benzoic acid on the incidence and severity of post-weaning diarrhoea in weaner pigs after experimental challenge with enterotoxigenic *Escherichia coli*. Archives of Animal Nutrition. 2009;**63**(4):267-280

[137] Bunce TJ, Howard MD, Kerley MS, Alee GL, Pace LW. Protective effect of fructooligosaccharide (FOS) in prevention of mortality and morbidity from infectious *E. coli* K:88 challenge. Journal of Animal Science. 1995;**63**(Suppl 1):69

[138] Stuyven E, Cox E, Vancaeneghem S, Arnouts S, Deprez P, Goddeeris BM. Effect of beta-glucans on an ETEC infection in piglets. Veterinary Immunology and Immunopathology. 2009;**128**(1-3):60-66

[139] Kiarie E, Bhandari S, Scott M, Krause DO, Nyachoti CM. Growth performance and gastrointestinal microbial ecology responses of piglets receiving Saccharomyces cerevisiae fermentation products after an oral challenge with *Escherichia coli* (K88). Journal of Animal Science. 2011;**89**(4):1062-1078

[140] Kiarie E, Scott M, Krause DO, Khazanehei H, Khafipour E, Nyachoti CM. Interactions of *Saccharomyces cerevisiae* fermentation product and in-feed antibiotic on gastrointestinal and immunological responses in piglets challenged with *Escherichia coli* K88+. Journal of Animal Science. 2012;**90**(Suppl 4):1-3

[141] Trckova M, Faldyna M, Alexa P, Sramkova Zajacova Z, Gopfert E, Kumprechtova D, et al. The effects of live yeast *Saccharomyces cerevisiae* on postweaning diarrhea, immune response, and growth performance in weaned piglets. Journal of Animal Science. 2014;**92**(2):767-774

[142] Kim K, He Y, Xiong X, Ehrlich A, Li X, Raybould H, et al. Dietary supplementation of Bacillus subtilis influenced intestinal health of weaned pigs experimentally infected with a pathogenic *E. coli*. Journal of Animal Science and Biotechnology. 2019;**10**(1):52

[143] Gadde U, Kim WH, Oh ST, Lillehoj HS. Alternatives to antibiotics for maximizing growth performance and feed efficiency in poultry: A review. Animal Health Research Reviews. 2017;**18**(1):26-45

[144] Windisch W, Schedle K, Plitzner C, Kroismayr A. Use of phytogenic products as feed additives for swine and poultry. Journal of Animal Science. 2008;**86**(14 Suppl):E140-E148

[145] Baydar NG, Özkan G, Sağdiç O. Total phenolic contents and antibacterial activities of grape (*Vitis vinifera* L.) extracts. Food Control. 2004;**15**(5):335-339

[146] Sokmen M, Serkedjieva J, Daferera D, Gulluce M, Polissiou M, Tepe B, et al. In vitro antioxidant, antimicrobial, and antiviral activities of the essential oil and various extracts from herbal parts and callus cultures of Origanum acutidens. Journal of Agricultural and Food Chemistry. 2004;**52**(11):3309-3312

[147] Dundar E, Olgun EG, Isiksoy S, Kurkcuoglu M, Baser KH, Bal C.

The effects of intra-rectal and intra-peritoneal application of *Origanum onites* L. essential oil on 2,4,6-trinitrobenzenesulfonic acid-induced colitis in the rat. Experimental and Toxicologic Pathology: Official Journal of the Gesellschaft fur Toxikologische Pathologie. 2008;**59**(6):399-408

[148] Liu Y, Song M, Che TM, Bravo D, Pettigrew JE. Anti-inflammatory effects of several plant extracts on porcine alveolar macrophages in vitro. Journal of Animal Science. 2012;**90**(8):2774-2783

[149] Hammer KA, Carson CF, Riley TV. Antimicrobial activity of essential oils and other plant extracts. Journal of Applied Microbiology. 1999;**86**(6):985-990

[150] Dorman HJ, Deans SG. Antimicrobial agents from plants: Antibacterial activity of plant volatile oils. Journal of Applied Microbiology. 2000;**88**(2):308-316

[151] Wong SY, Grant IR, Friedman M, Elliott CT, Situ C. Antibacterial activities of naturally occurring compounds against *Mycobacterium avium* subsp. paratuberculosis. Applied and Environmental Microbiology. 2008;**74**(19):5986-5990

[152] Carson CF, Mee BJ, Riley TV. Mechanism of action of *Melaleuca alternifolia* (tea tree) oil on *Staphylococcus aureus* determined by time-kill, lysis, leakage, and salt tolerance assays and electron microscopy. Antimicrobial Agents and Chemotherapy. 2002;**46**(6):1914-1920

[153] Burt S. Essential oils: Their antibacterial properties and potential applications in foods—A review. International Journal of Food Microbiology. 2004;**94**(3):223-253

[154] Xu J, Zhou F, Ji BP, Pei RS, Xu N. The antibacterial mechanism of carvacrol and thymol against *Escherichia coli*. Letters in Applied Microbiology. 2008;**47**(3):174-179

[155] Trombetta D, Castelli F, Sarpietro MG, Venuti V, Cristani M, Daniele C, et al. Mechanisms of Antibacterial Action of Three Monoterpenes. 2005;**49**(6):2474-2478

[156] Sikkema J, de Bont JA, Poolman B. Mechanisms of membrane toxicity of hydrocarbons. Microbiological Reviews. 1995;**59**(2):201-222

[157] Farag RS, Daw ZY, Hewedi FM, El-Baroty GSA. Antimicrobial activity of some Egyptian spice essential oils. Journal of Food Protection. 1989;**52**(9):665-667

[158] Lambert RJ, Skandamis PN, Coote PJ, Nychas GJ. A study of the minimum inhibitory concentration and mode of action of oregano essential oil, thymol and carvacrol. Journal of Applied Microbiology. 2001;**91**(3):453-462

[159] Ankri S, Mirelman D. Antimicrobial properties of allicin from garlic. Microbes and Infection. 1999;**1**(2):125-129

[160] Burt SA, van der Zee R, Koets AP, de Graaff AM, van Knapen F, Gaastra W, et al. Carvacrol induces heat shock protein 60 and inhibits synthesis of flagellin in *Escherichia coli* O157:H7. Applied and Environmental Microbiology. 2007;**73**(14):4484-4490

[161] Verhelst R, Schroyen M, Buys N, Niewold T. The effects of plant polyphenols on enterotoxigenic *Escherichia coli* adhesion and toxin binding. Livestock Science. 2010;**133**(1):101-103

[162] Liu Y, Song M, Che TM, Almeida JA, Lee JJ, Bravo D, et al.

Dietary plant extracts alleviate diarrhea and alter immune responses of weaned pigs experimentally infected with a pathogenic *Escherichia coli*. Journal of Animal Science. 2013;**91**(11):5294-5306

[163] Liu Y, Song M, Che TM, Lee JJ, Bravo D, Maddox CW, et al. Dietary plant extracts modulate gene expression profiles in ileal mucosa of weaned pigs after an *Escherichia coli* infection. Journal of Animal Science. 2014;**92**(5):2050-2062

[164] Bruins MJ, Cermak R, Kiers JL, van der Meulen J, van Amelsvoort JM, van Klinken BJ. In vivo and in vitro effects of tea extracts on enterotoxigenic *Escherichia coli*-induced intestinal fluid loss in animal models. Journal of Pediatric Gastroenterology and Nutrition. 2006;**43**(4):459-469

[165] Coddens A, Loos M, Vanrompay D, Remon JP, Cox E. Cranberry extract inhibits in vitro adhesion of F4 and F18(+)*Escherichia coli* to pig intestinal epithelium and reduces in vivo excretion of pigs orally challenged with F18(+) verotoxigenic *E. coli*. Veterinary Microbiology. 2017;**202**:64-71

[166] Zasloff M. Antimicrobial peptides of multicellular organisms. Nature. 2002;**415**(6870):389-395

[167] Xiao H, Shao F, Wu M, Ren W, Xiong X, Tan B, et al. The application of antimicrobial peptides as growth and health promoters for swine. Journal of Animal Science and Biotechnology. 2015;**6**(1):19

[168] Wang S, Zeng X, Yang Q, Qiao S. Antimicrobial peptides as potential alternatives to antibiotics in food animal industry. International Journal of Molecular Sciences. 2016;**17**(5):603-614

[169] Selsted ME, Ouellette AJ. Mammalian defensins in the antimicrobial immune response. Nature Immunology. 2005;**6**(6):551-557

[170] Hancock RE. Cationic peptides: Effectors in innate immunity and novel antimicrobials. The Lancet Infectious Diseases. 2001;**1**(3):156-164

[171] Zhang L, Parente J, Harris SM, Woods DE, Hancock RE, Falla TJ. Antimicrobial peptide therapeutics for cystic fibrosis. Antimicrobial Agents and Chemotherapy. 2005;**49**(7):2921-2927

[172] Shai Y. Mechanism of the binding, insertion and destabilization of phospholipid bilayer membranes by alpha-helical antimicrobial and cell non-selective membrane-lytic peptides. Biochimica et Biophysica Acta. 1999;**1462**(1-2):55-70

[173] Yang L, Harroun TA, Weiss TM, Ding L, Huang HW. Barrel-stave model or toroidal model? A case study on melittin pores. Biophysical Journal. 2001;**81**(3):1475-1485

[174] Brogden KA. Antimicrobial peptides: Pore formers or metabolic inhibitors in bacteria? Nature Reviews Microbiology. 2005;**3**(3):238-250

[175] Mansour SC, Pena OM, Hancock RE. Host defense peptides: front-line immunomodulators. Trends in Immunology. 2014;**35**(9):443-450

[176] Ren ZH, Yuan W, Deng HD, Deng JL, Dan QX, Jin HT, et al. Effects of antibacterial peptide on cellular immunity in weaned piglets. Journal of Animal Science. 2015;**93**(1):127-134

[177] Wu S, Zhang F, Huang Z, Liu H, Xie C, Zhang J, et al. Effects of the antimicrobial peptide cecropin AD on performance and intestinal health in weaned piglets challenged with *Escherichia coli*. Peptides. 2012;**35**(2):225-230

[178] Xiao H, Tan BE, Wu MM, Yin YL, Li TJ, Yuan DX, et al. Effects of composite antimicrobial peptides

in weanling piglets challenged with deoxynivalenol: II. Intestinal morphology and function. Journal of Animal Science. 2013;**91**(10):4750-4756

[179] Wang Y-Z, Shan T-Z, Xu Z-R, Feng J, Wang Z-Q. Effects of the lactoferrin (LF) on the growth performance, intestinal microflora and morphology of weanling pigs. Animal Feed Science and Technology. 2007;**135**(3):263-272

[180] Tang Z, Yin Y, Zhang Y, Huang R, Sun Z, Li T, et al. Effects of dietary supplementation with an expressed fusion peptide bovine lactoferricin-lactoferrampin on performance, immune function and intestinal mucosal morphology in piglets weaned at age 21 d. The British Journal of Nutrition. 2009;**101**(7):998-1005

[181] Lonnerdal B. Biochemistry and physiological function of human milk proteins. The American journal of clinical nutrition. 1985;**42**(6):1299-1317

[182] Wiesner J, Vilcinskas A. Antimicrobial peptides: The ancient arm of the human immune system. Virulence. 2010;**1**(5):440-464

[183] Masschalck B, Michiels CW. Antimicrobial properties of lysozyme in relation to foodborne vegetative bacteria. Critical Reviews in Microbiology. 2003;**29**(3):191-214

[184] Ibrahim HR, Thomas U, Pellegrini A. A helix-loop-helix peptide at the upper lip of the active site cleft of lysozyme confers potent antimicrobial activity with membrane permeabilization action. The Journal of Biological Chemistry. 2001;**276**(47):43767-43774

[185] Laible NJ, Germaine GR. Bactericidal activity of human lysozyme, muramidase-inactive lysozyme, and cationic polypeptides against *Streptococcus sanguis* and *Streptococcus faecalis*: Inhibition by chitin oligosaccharides. Infection and Immunity. 1985;**48**(3):720-728

[186] Ibrahim HR, Matsuzaki T, Aoki T. Genetic evidence that antibacterial activity of lysozyme is independent of its catalytic function. FEBS letters. 2001;**506**(1):27-32

[187] Ellison RT, Giehl TJ. Killing of gram-negative bacteria by lactoferrin and lysozyme. Journal of Clinical Investigation. 1991;**88**(4):1080-1091

[188] Gordon LI, Douglas SD, Kay NE, Yamada O, Osserman EF, Jacob HS. Modulation of neutrophil function by lysozyme. potential negative feedback system of inflammation. Journal of Clinical Investigation. 1979;**64**(1):226-232

[189] Maga EA, Desai PT, Weimer BC, Dao N, Kültz D, Murray JD. Consumption of lysozyme-rich milk can alter microbial fecal populations. Applied and Environmental Microbiology. 2012;**78**(17):6153-6160

[190] Garas LC, Cooper CA, Dawson MW, Wang JL, Murray JD, Maga EA. Young pigs consuming lysozyme transgenic goat milk are protected from clinical symptoms of enterotoxigenic *Escherichia coli* infection. The Journal of Nutrition. 2017;**147**(11):2050-2059

[191] Huang G, Li X, Lu D, Liu S, Suo X, Li Q, et al. Lysozyme improves gut performance and protects against enterotoxigenic *Escherichia coli* infection in neonatal piglets. Veterinary Research. 2018;**49**(1):20

[192] Liu Y, Ji P. Dietary factors in prevention of pediatric *Escherichia coli* infection: A model using domestic piglets. ILAR Journal. 2018;**59**(3):338-351

[193] Brundige DR, Maga EA, Klasing KC, Murray JD. Lysozyme

transgenic goats' milk influences gastrointestinal morphology in young pigs. The Journal of Nutrition. 2008;**138**(5):921-926

[194] Cooper CA, Garas Klobas LC, Maga EA, Murray JD. Consuming transgenic goats' milk containing the antimicrobial protein lysozyme helps resolve diarrhea in young pigs. PLoS One. 2013;**8**(3):e58409

[195] Liu P, Piao XS, Thacker PA, Zeng ZK, Li PF, Wang D, et al. Chito-oligosaccharide reduces diarrhea incidence and attenuates the immune response of weaned pigs challenged with *Escherichia coli* K88. Journal of Animal Science. 2010;**88**(12):3871-3879

[196] Krause DO, Bhandari SK, House JD, Nyachoti CM. Response of nursery pigs to a synbiotic preparation of starch and an anti-*Escherichia coli* K88 probiotic. Applied and Environmental Microbiology. 2010;**76**(24):8192-8200

[197] Daudelin JF, Lessard M, Beaudoin F, Nadeau E, Bissonnette N, Boutin Y, et al. Administration of probiotics influences F4 (K88)-positive enterotoxigenic *Escherichia coli* attachment and intestinal cytokine expression in weaned pigs. Veterinary Research. 2011;**42**:69

[198] Lee JS, Awji EG, Lee SJ, Tassew DD, Park YB, Park KS, et al. Effect of lactobacillus plantarum CJLP243 on the growth performance and cytokine response of weaning pigs challenged with enterotoxigenic *Escherichia coli*. Journal of Animal Science. 2012;**90**(11):3709-3717

[199] Yang KM, Jiang ZY, Zheng CT, Wang L, Yang XF. Effect of lactobacillus plantarum on diarrhea and intestinal barrier function of young piglets challenged with enterotoxigenic *Escherichia coli* K88. Journal of Animal Science. 2014;**92**(4):1496-1503

[200] Devi SM, Lee SI, Kim IH. Effect of phytogenics on growth performance, fecal score, blood profiles, fecal noxious gas emission, digestibility, and intestinal morphology of weanling pigs challenged with *Escherichia coli* K88. Polish Journal of Veterinary Sciences. 2015;**18**(3):557-564

[201] Li H, Zhao P, Lei Y, Li T, Kim I. Response to an *Escherichia coli* K88 oral challenge and productivity of weanling pigs receiving a dietary nucleotides supplement. Journal of Animal Science and Biotechnology. 2015;**6**:49

[202] Song M, Liu Y, Soares JA, Che TM, Osuna O, Maddox CW, et al. Dietary clays alleviate diarrhea of weaned pigs. Journal of Animal Science. 2012;**90**(1):345-360

[203] Almeida JA, Liu Y, Song M, Lee JJ, Gaskins HR, Maddox CW, et al. *Escherichia coli* challenge and one type of smectite alter intestinal barrier of pigs. Journal of Animal Science and Biotechnology. 2013;**4**(1):52

Chapter 3

Safety Aspect of Recombinant Protein Produced by *Escherichia coli*: Toxin Evaluation with Strain and Genomic Approach

Iman Permana Maksum, Ahmad Nabiel, Safri Ishmayana and Ukun M.S. Soedjanaatmadja

Abstract

Escherichia coli is a Gram-negative bacteria which is well known for its pathogenic properties that can cause serious food poisoning, mostly indicated by diarrhea or other severe symptoms. Despite of its well-known properties due to its ability to produce toxin, most of *E. coli* strains are harmless and even beneficial especially in recombinant protein production. This bacterium is suitable for protein recombinant host since it has rapid growth, high expression rate, and well-known genome. Various proteins have been produced using *E. coli* expression systems, with therapeutic protein for medical application being the most notably produced. Apart from that, our group succeeded in producing beta galactosidase from a wild type *E. coli* strain B130. Furthermore, recombinant human serum albumin was successfully produced using *E. coli* strains BL21 (DE3). However, studies on *E. coli* toxin contamination in recombinant protein productions, strains, and genomic comprehension are indispensable, particularly in therapeutic protein. Therefore, this chapter will discuss the safety aspects of recombinant therapeutic proteins in terms of toxin contamination by strain and genomic approaches.

Keywords: *E. coli*, genomic maps, recombinant therapeutic proteins, toxin

1. Introduction

Escherichia coli is a member of *Enterobacteria* family which can be found in gastrointestinal tracts [1–3]. In general, it is well known to cause broad diseases, including gastrointestinal problems. Aside the fact that *E. coli* was normal to be found in colon, a number of its strains were discovered with the ability to produce toxins. Shiga toxin *E. coli* (STEC) and enterotoxin *E. coli* (ETEC) were groups of *E. coli* strains that have the ability to produce toxin that may cause several diseases, such as diarrhea [1, 4, 5].

Although *E. coli* may cause numerous gastrointestinal diseases; in fact, strains that are responsible for pathogenic properties were relatively minor in numbers. Furthermore, *E. coli* was considered harmless and even useful as a host for producing recombinant proteins. Even this bacteria becomes favorite host chosen in industrial and medical applications since it has rapid growth, well-characterized gene, and its ability to grow under aerobic and anaerobic system, and facilitates to form high cell density culture (HCDC) [6–8].

The discussion about advantage in producing recombinant proteins and worries of toxins of *E. coli* is like talking about two opposite sides of a coin. This will certainly raise a question "Is it safe to produce recombinant protein in *E. coli*? Will it be toxin-free contamination?" Therefore, this chapter will discuss the safety aspects of recombinant protein produced by *E. coli* against toxins using genomic and strains approach.

2. Toxin produced by *Escherichia coli*

Several pathogenic *E. coli* strains are known to be responsible for broad diseases, from mild to complicated cases. It is varying from mild diarrhea, hemorrhagic colitis, to hemolytic uremic syndrome. Among the pathogenic strains, STEC is an example of common strains which occupy high number in *E. coli* serotypes that produce toxin called Shiga toxins (Stx) [1, 2, 5, 9–11]. While STEC is a common pathogenic example, it belongs to a larger group named enterohemorrhagic *E. coli* (EHEC); also, there still exist numerous pathogenic *E. coli* and cause different diseases and complications. Pathogenic *E. coli* were classified in **Table 1** along with its diseases they caused and virulence factors [1].

Considering the number of pathogenic *E. coli*, it is useful to classify the toxins' properties and structure. It will be convenience to determine whether the toxins belong to organic compound or peptide-based structure; therefore, we could analyze contamination probabilities in terms of producing recombinant protein. Most of the virulence factors stated in **Table 1** were protein attached in bacterial membrane with the role of adhesion or recognition to host cell [12]. Meanwhile, shiga toxin, heat-stable and heat-labile toxin, and other cytotoxins were protein released by pathogenic *E. coli*. These toxins have specific receptors to induce invagination to the host cell, while their virulence mechanism also differs depending on the nature of each toxin and their molecular target [4].

STEC serotypes vary and differ in number of incidences, although the O157:H7 is a serotype considered to be responsible of numerous outbreaks. Shiga toxin occupies AB5 structure (see **Figure 1**), the catalytic subunit A (StxA) and homopentamer of subunit B (Stx B) as recognition site to globotetraosylceramide (Gb3/Gb4), which are present in the host cell surface, which leads to invagination of the toxin. STEC can produce either Stx1 (Stx1 and Stx1c), Stx2 variant (Stx2, Stx2c, Stx2d, Stx2e, and Stx2f) or range combination of both variants [4, 12]. Once invagination succeeds, catalytic subunit A would disrupt cell metabolism by inhibiting elongation factor-dependent aminoacyl tRNA binding (see detailed mechanism in [4]). The highly specific RNA *N*-glycosidase activity cleaves adenine base in eukaryotic ribosomal RNA, precisely at 28S subunit on the α-sarcin loop located in position 4324 [4].

Meanwhile, heat-labile (LT) and heat-stable (ST) toxin belong to ETEC groups. Nevertheless, LT enterotoxin shares similar structures to Stx which occupy AB5 conformation. Subunit A acts as a toxin by binding to its receptor, guanylyl cyclase C (GC-C). The interaction will activate guanine nucleotide protein Gsα by ADP-ribosylation, which trigger stimulation of secretion by cAMP-dependent mechanism. Elevated numbers of cAMP cause CTFR channel to secrete water and ions, thus generating diarrhea [3]. By contrast, ST structure is relatively simple. The STa class was made up with 18–19 cysteine-rich amino acids, while STb has 48 amino acids. ST virulence acts by triggering secretion of water and ions by triggering signaling cascade through guanylyl cyclase C (GC-C) in intestine [13, 14]. The structure of both ST and LT is shown in **Figure 2**.

Pathogenic	Diseases	Virulence factors	Reference
Enteric *E. coli*			
Enteropathogenic *E. coli* (EPEC)	Diarrhea in children	Bfp, intimin, LEE	[1]
Enterohaemorrhagic *E. coli* (EHEC)	Hemorrhagic colitis, hemolytic uremic syndrome	Shiga toxins, intimin, Bfp	
Enterotoxigenic *E. coli* (ETEC)	Traveler's diarrhea	Heat-labile and heat-stable toxins, CFA	
Enteroaggregative *E. coli* (EAEC)	Diarrhea in children	AAFs, cytotoxins	
Enteroinvasive *E. coli* (EIEC)	Shigellosis-like	Shiga toxin, hemolysin, cellular invasion, Ipa	
Diffusely adherent *E. coli* (DAEC)	Acute diarrhea in children	Daa, AIDA	
Adherent invasive *E. coli* (AIEC)	Crohn's disease-associated	Type 1 fimbriae, cellular invasion	
Extraintestinal *E. coli*			
Uropathogenic *E. coli* (UPEC)	Lower urinary tract infection and systemic infection	Type 1 and P fimbriae, AAFs, hemolysin	
Neonatal meningitis *E. coli* (NMEC)	Neonatal meningitis	S fimbriae, K1 capsule	
Avian pathogenic *E. coli* (APEC)	Food-borne source disease	Type 1 and P fimbriae, K1 capsule	

AAF, aggregative adhere fimbria; AIDA, adhesion involved in diffuse adherence; Bfp, bundle-forming pili; CFA, colonization factor antigen; Daa, diffuse adhesin; Ipa, invasion plasmid antigen; LEE, locus for enterocyte effacement.

Table 1.
Classification of pathogenic E. coli.

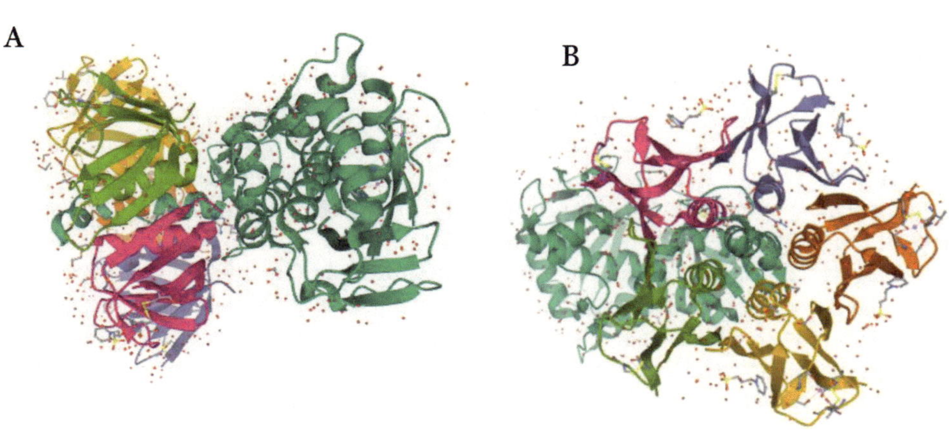

Figure 1.
Crystal structure of shiga-like toxin in E. coli *(PDB: 1R4P). (A) Transversal view of Stx subunits, StxA represented as teal-colored chain, while StxB presented in rainbow color; (B) axial view of homopentamer StxB [3].*

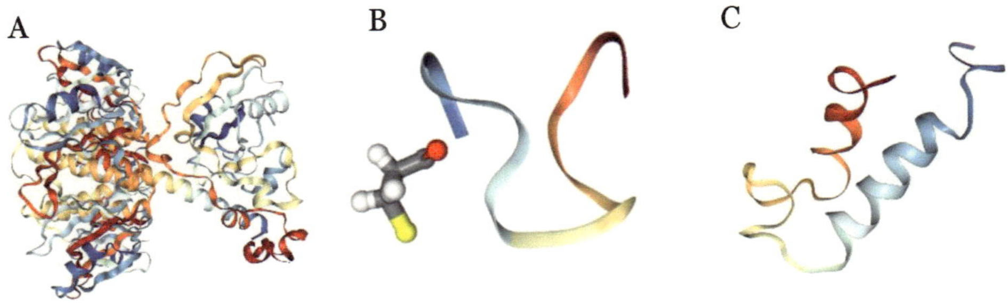

Figure 2.
Enterotoxin secreted by ETEC. (A) Heat-labile enterotoxin (PDB: 1tii) [16]; (B) heat-stable enterotoxin, STa class (1etn) [17]; (C) heat-stable enterotoxin, STb class (1ehs) [18].

The fact both STEC and ETEC toxins (Stx and LT, and ST, respectively) are peptide based elucidates its origin that were genetically listed in their DNA. These toxins were made under central dogma of protein synthesis. Therefore, analysis through genomic approach on recombinant *E. coli* host is possible to be conducted.

3. *E. coli* as host for recombinant protein expression

The production of recombinant proteins in microbial systems was started in 1970 and continued to boom in 1980 with the production of insulin. There is no doubt that this method has revolutionized and widened the field of biochemistry [19]. The ability to express large quantity of protein with less effort, relative to manual synthesis, allows industrial processes to produce in commercial scale. However, several considerations should be discussed before executing the production such as, appropriate vector, location of the protein of interest (whether as soluble fraction or inclusion bodies), optimum condition (pH, medium, temperature, aerobic/anaerobic system), genetic design for convenience of purification, and at the top of it, microbial selection [7, 8, 20].

E. coli become preferred microbes in terms of recombinant protein host among researchers and industrial use. The simplicity of its expression system, compared to other higher level organism, and large quantity of well-characterized genomic database offer advantages in constructing the vector to be used [20]. A plenty number of research regarding *E. coli* also become an advantage to give amount of consideration of various expression conditions. Nevertheless, *E. coli* expression system has limited post-translational modification, which means that some proteins that require modification, such as alkylation or glycosylation, may not be perfectly expressed in *E. coli*. However, several strains of *E. coli* have the ability to perform specific post-translational modification [19, 21]. Therefore, we provide a simple summary on recombinant proteins produced by *E. coli* along with strains and expression strategies in **Table 2**.

Among recombinant proteins mentioned in **Table 2**, hEGF and hPT-2 are examples of therapeutic protein. Regarding its use in medical interests, therapeutic proteins produced in *E. coli* have to be safe for administration into human bodies; therefore, purification steps and any contaminants present become a huge concern in producing recombinant protein. Idetifying location of protein target is a prominent fundamental to determine source of contamination and to predict any possible contamination. Understanding the protein location also helps with the purification strategies needed to separate contaminants, specifically toxins, with the result that highly pure proteins were recovered. Choi et al. [29] through **Figure 3** classify locations of protein expressed in *E. coli* and its general purification steps needed.

Recombinant proteins	Strain	Expression strategies	Ref.
α-Cyclodextrin glycosyltransferase	BL21 (DE3)	Extracellular, using OmpA constructed on pET-20b(+) vector	[22]
P64k	*E. coli* K12 GC366	Soluble intracellular, using pM-152 vector	[23]
SpA-βgal (*Staphylococcus* protein A-*E. coli β*-galactosidase)	*E. coli* RR1 LacZ	Soluble intracellular as fusion protein	[24]
Human prethrombin-2 (hPT-2)	*E. coli* TOP10F'	Fusion protein with intein, CBD-*Ssp* DnaB-hPT-2	[25]
Human prethrombine-2 (hPT-2)	*E. coli* BL21 (DE3) ArcticExpress	Soluble intracellular using pTWIN1	[26]
Human epidermal growth factor (hEGF)	*E. coli* BL21 (DE3)	Extracellular using TorA signal peptide constructed on pD881 vector	[6]
Human epidermal growth factor (hEGF)	*E. coli* BL21 (DE3)	Extracellular, using OmpA constructed in pD881 vector	[27]
Human serum albumin	*E. coli* BL21 (DE3)	Extracellular, using TorA constructed on pD881 vector	[28]
Pyrococcus furiosus α-amylase (PFA)	*E. coli* DH5α	Soluble protein, designed with co-expression of prefoldin, chaperonin HSP60, and sHSP	[29]
Hepcidin	*E. coli* BL21 (DE3)	Inclusion bodies, fused with His-tag	[30]
Keratinase	*E. coli* AD494(DE3) pLys	Intracellular expression using His-tag	[31]
Pig liver esterase (γ-PLE)	*E. coli* origami (DE3)	Soluble intracellular, using Strep-tag and combination of His- and Trx-tag	[32]

Table 2.
Summary of recombinant protein using E. coli *expression system and its expression strategy.*

The distinction of protein location is affected by either the nature of the expression system or the protein construction design. Both extracellular and intracellular strategies on expressing protein give its own advantages and disadvantages. Extracellular expression offers simple purification, improved folding, and soluble products. This strategy can be achieved using signal peptide, co-expression with phospholipase, or co-expression with chaperon [33, 34]. In contrast, intracellular expression prefers inclusion bodies formation. While inclusion bodies give easy separation and prevent protease degradation, it has complex purification steps and refolding process is compulsory. Fusion partners, such as intein, often added in gene construction in intracellular works to provide efficient strategy in purification steps [21, 33].

Based on protein location, toxin contamination can be investigated. Both Stx and LT-ST toxins are secreted by *E. coli*, increasing the risk of contamination when the protein of interest is produced extracellularly. Even so, since extracellular protein exists in soluble state, purification might not be impossible. Whereas intracellular expression may put more concern at contamination risk since toxins might be clumped together in the form of inclusion bodies. This case may put more consideration in solubilization and purification process. However, these allegations are only an assessment of risk factors with the assumption that toxins are produced in *E. coli*, which is used for recombinant protein expression.

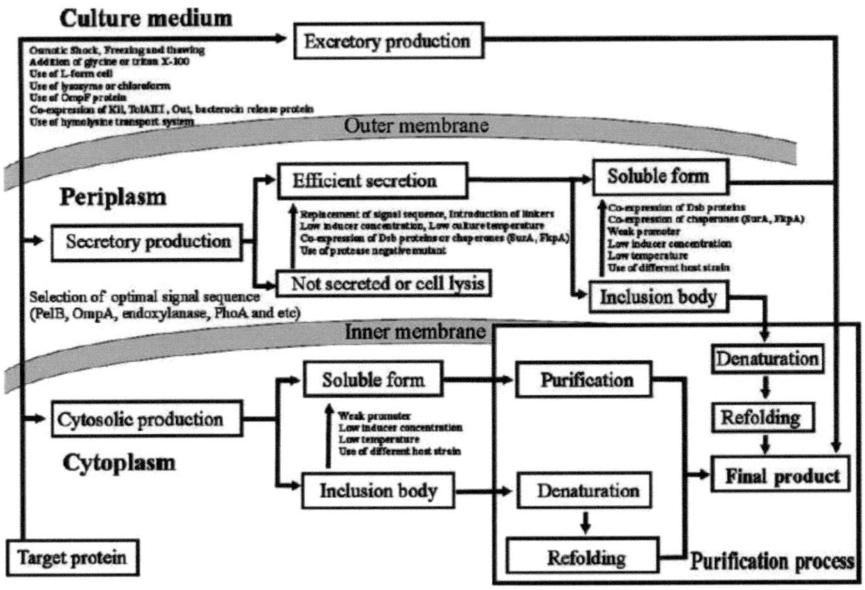

Figure 3.
Scheme of protein expressed in E. coli *and its following purification steps [33].*

4. Safety aspects of recombinant protein production against toxin

Using comprehensive understanding of toxin origin, specifically Shiga toxin and enterotoxin, it is clear that these toxins were peptide based and generated by certain gene in STEC and ETEC. The gene *stx* was responsible for producing the Stx toxin using central dogma of *E. coli*, reciprocally to ST and LT encoding gene. Moreover, *E. coli* strains that are commonly used for recombinant protein work are also known. Therefore, it is possible to examine the safety aspect of recombinant protein against toxin through genetic alignment between common *E. coli* strains in recombinant work and toxin genes. Here, *E. coli* BL21 (DE3) (ACC: NC_012892) and K-12 MG1655 (ACC: U00096.3) were used as representative. While toxin genes used are Stx (ACC: AY143336.1), LT (ACC: JQ031712), and ST (ACC: P22542.1).

In term of the existence of *stx* gene, common recombinant host strains are absence of the *stx* gene. Therefore, since the strains were clearly different, it is considerably safe to use *E. coli* as recombinant host without neglecting other contaminants.

5. Expression and characterization of *HSA* gene in *E. coli* BL21 (DE3)

This step started with growing *E. coli* BL21 (DE3) [pD881-torA-HSA] transformant as starter culture at 200 rpm, 37°C for 16–18 hours. Then starter culture was moved as much as 1% into 25 mL Luria-Bertani medium containing kanamycin as selection marker. *E. coli* BL21 (DE3) cell culture was grown until OD_{600nm} reached 0.8 for induction. Before induction was performed, 1 mL sample from culture was separated as protein fraction before induction (t_0). Induction was initiated by adding L-rhamnose into the expression medium to bring the final concentration to 4 mM. To obtain protein fraction in cytoplasm, sonication method was used. Lysate from six *E. coli* BL21 (DE3) [pD881-torA-HSA] transformant colonies showed that HSA was expressed in cytoplasm, it was characterized with the presence of ±67.0 kDa and in the SDS-PAGE electrophoresis [28]. The result of expression is presented in **Figure 4**.

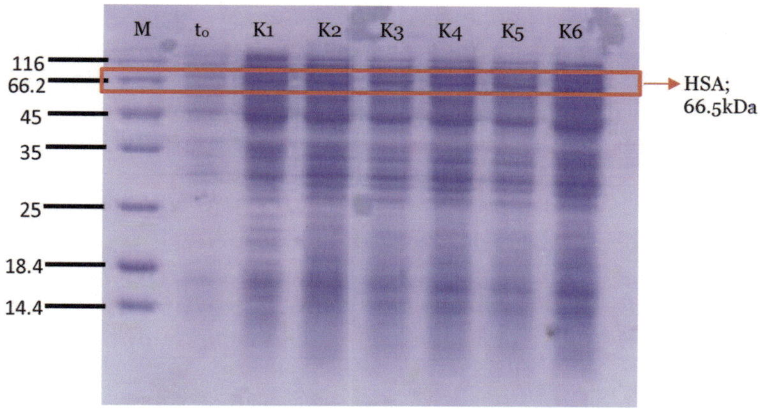

Figure 4.
SDS-PAGE electropherogram of HSA soluble fraction in cytoplasm after induction using 4 mM L-rhamnose from E. coli *BL21 (DE3) [pD881-torA-HSA] colonies. (M) Protein marker; (t_o) fraction before L-rhamnose induction; (K_i) colony transformant 1–6 after induction L-rhamnose induction [28].*

6. Recombinant hEGF expression in *E. coli* BL21 (DE3)

E. coli BL21 (DE3) [pD881-PelB] that has been characterized was grown in 5 mL LB medium containing kanamycin for about 16–18 hours at 37°C with 200 rpm shaking. Then, 1000 μL *E. coli* BL21 (DE3) [pD881-PelB] culture was transferred into 100 mL LB medium containing kanamycin in shaken flask. Native *E. coli* BL21 (DE3) was also grown and received the same treatment as the transformant. The culture was incubated until OD_{600nm} reached 0.7. *E. coli* BL21 (DE3) [pD881-PelB] culture was taken as a sample (before L-Rhamnose induction) and transferred into a microtube and centrifuged at 3000 g, 4°C for 20 minutes. L-rhamnose was added into *E. coli* BL21 (DE3) [pD881-PelB] and native *E. coli* BL21 (DE3) culture to give final concentration of 2 mM. Incubation was continued at 37°C for 20 hours with 200 rpm shaking. Culture of the *E. coli* BL21 (DE3) [pD881-PelB] was taken as a sample (after L-rhamnose induction) and transferred into 2 microtubes then was centrifuged at 3000 g, 4°C for 20 minutes [35].

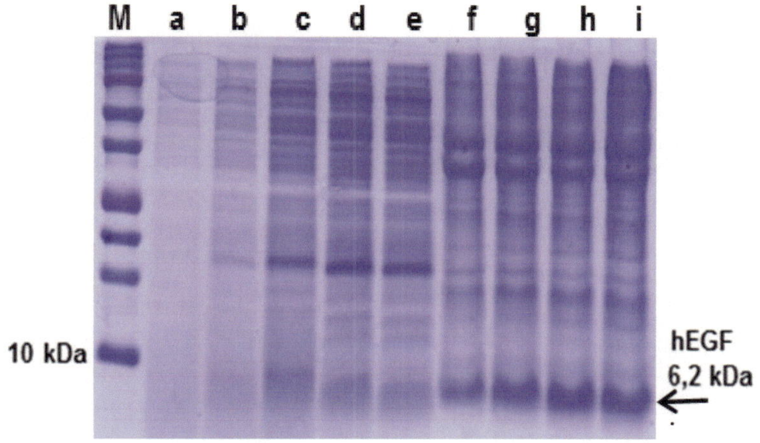

Figure 5.
Electropherogram of proteins expressed by E. coli *BL21 (DE3) [pD881-PelB] with varying L-rhamnose concentration. Lane M is the protein markers. Lane (a-e) are soluble fractions of the medium with 40 μM, 1 mM, 2 mM, 4 mM, and 6 mM L-rhamnose as inducer, respectively. While lane (f-i) are inclusion bodies obtained from cell with 1 mM, 2 mM, 4 mM and 6 mM L-rhamnose as inducer. Samples were collected after 20 hours of induction [35].*

Based on SDS-PAGE electrophoresis of the protein produced by the *E. coli* cell at varying concentration of L-rhamnose as inducer (**Figure 5**), it can be concluded that the best concentration of L-rhamnose that induces the production of the protein of interest was 4 mM because it produces more target protein, either in the insoluble fraction of the medium or in the form of inclusion bodies at $t_{(20)}$. The results also indicate that not all rhEGF translocated into the periplasm were secreted to medium. The hEGF was expressed in *E. coli* BL21 (DE3) with molecular weight of 6.2 kDa. The result of expression is presented in **Figure 5** [35].

Apart from that, our group succeeded in producing beta-galactosidase from a wild type *E. coli* strain B130, with high purity. Kinetical parameter (K_m and V_{max}) of the enzyme were 2.417×10^{-4} mol and 4.664×10^{-4} mol.minute^{-1}, respectively [36].

7. Conclusions

E. coli is renowned by its pathogenic properties, specifically in causing gastrointestinal disease. While in contrast, the same species also being helpful in expressing recombinant protein. Thus, contrary properties leave questions in terms of safety in expressing recombinant protein. Pathogenic *E. coli* strains were identified and classified in accordance with the disease caused. While most of pathogenic group gain its virulence by their membrane protein, some of it secretes toxins, like Stx from STEC or LT and ST from ETEC group. This toxin-secreting *E. coli* were important to understand contamination risk in recombinant protein. All three toxins were considered as peptide-based structure, in which production relies on respective genes. Alignment of toxin genes to commonly used *E. coli* in recombinant work makes a way to investigate toxin presence in recombinant-host *E. coli*. The BL21 (DE3) and K-12 MG1655 strains used as representative in alignment process, which generate non-overlapping alignment. This clears up the risk of toxin contamination on recombinant protein since the absence of toxin gene in these strains. Therefore, expressing recombinant protein, especially therapeutic protein, in *E. coli* was considered to be safe against toxin.

Acknowledgements

The authors would like to acknowledge the Executive Board of Universitas Padjadjaran, Dean of the Faculty of Mathematics and Natural Sciences, Head of the Department of Chemistry, Head of the Laboratory, and the entire staff of the Laboratory of Biochemistry.

Funding sources

Academic Leadership Grant from the Directorate of Research, Public Services, and Innovations, Universitas Padjadjaran.

DOI: http://dx.doi.org/10.5772/intechopen.92031

Author details

Iman Permana Maksum, Ahmad Nabiel, Safri Ishmayana and
Ukun M.S. Soedjanaatmadja*
Laboratory of Biochemistry, Department of Chemistry, Faculty of Mathematics and Natural Sciences, Universitas Padjadjaran, Sumedang, Indonesia

*Address all correspondence to: ukun_28@yahoo.com; ukun@unpad.ac.id

References

[1] Allocati N, Masulli M, Alexeyev MF, Di Ilio C. *Escherichia coli* in Europe: An overview. International Journal of Environmental Research and Public Health. 2013;**10**(12):6235-6254

[2] Beutin L, Geier D, Zimmermann S, Aleksic S, Gillespie HA, Whittam TS. Epidemiological relatedness and clonal types of natural populations of *Escherichia coli* strains producing Shiga toxins in separate populations of cattle and sheep. Applied and Environmental Microbiology. 1997;**63**(6):2175-2180

[3] Fraser ME, Fujinaga M, Cherney MM, Melton-Celsa AR, Twiddy EM, O'Brien AD, et al. Structure of Shiga toxin type 2 (Stx2) from *Escherichia coli* O157:H7. The Journal of Biological Chemistry. 2004;**279**(26):27511-27517. [cited 01 March 2020]. Available from: https://www.rcsb.org/structure/1R4P

[4] Johannes L, Römer W. Shiga toxins from cell biology to biomedical applications. Nature Reviews Microbiology [Internet]. 2010;**8**(2): 105-116. DOI: 10.1038/nrmicro2279

[5] Hunt JM. Shiga toxin-producing *Escherichia coli* (STEC). The Journal of Clinical and Laboratory Medicine. 2010;**30**(1):21-45. DOI: 10.1016/j.cll.2009.11.001

[6] Maksum IP, Utama E, Sriwidodo ST. Extracellular secretion of recombinant human epidermal growth factor by using trimethylamine N-oxide reductase a (TORA) signal peptide in *Escherichia coli* BL21 (DE3). Journal of Pharmaceutical Sciences and Research. 2017;**9**(6):1007-1016

[7] Chae YK, Kim SH, Markley JL. Relationship between recombinant protein expression and host metabolome as determined by two-dimensional NMR spectroscopy. PLoS One. 2017;**12**(5):e0177233

[8] Rosano GL, Ceccarelli EA. Recombinant protein expression in *Escherichia coli*: Advances and challenges. Frontiers in Microbiology. 2014;**5**(APR):1-17

[9] Yang X, Bai X, Zhang J, Sun H, Fu S, Fan R, et al. *Escherichia coli* strains producing a novel Shiga toxin 2 subtype circulate in China. International Journal of Medical Microbiology [Internet]. 2020;**310**(1):151377. DOI: 10.1016/j.ijmm.2019.151377

[10] Melton-celsa AR. Shiga toxin (Stx) classification, structure, and function. 2014:1-13

[11] Reis MHL, Matos DP, Pestana De Castro AF, Toledo MR, Trabulsi LR. Relationship among enterotoxigenic phenotypes, serotypes, and sources of strains in enterotoxigenic *Escherichia coli*. Infection and Immunity. 1980;**28**(1):24-27

[12] Farrokh C, Jordan K, Auvray F, Glass K, Oppegaard H, Raynaud S, et al. Review of Shiga-toxin-producing *Escherichia coli* (STEC) and their significance in dairy production. International Journal of Food Microbiology [Internet]. 2013;**162**(2): 190-212. DOI: 10.1016/j.ijfoodmicro.2012.08.008

[13] Weiglmeier PR, Rösch P, Berkner H. Cure and curse: *E. coli* heat-stable enterotoxin and its receptor guanylyl cyclase C. 2010;**2**(9): 2213-2229

[14] Taxt A, Aasland R, Sommerfelt H, Nataro J, Puntervoll P. Heat-stable enterotoxin of enterotoxigenic *Escherichia coli* as a vaccine target. Infection and Immunity. 2010;**78**(5):1824-1831

[15] Ozaki H, Sato T, Kubota H, Hata Y, Katsube Y, Shimonishi Y. Molecular structure of the toxic domain of heat-stable enterotoxin produced by a pathogenic strain of *Escherichia coli*. The Journal of Biological Chemistry. 1991;**266**:5934-5941

[16] van den Akker F, Sarfaty S, Twiddy EM, Connell TD, Holmes R, Hol WG. Crystal structure of a new heat-labile enterotoxin, LT-IIb. Structure; 1996;**4**(6):665-678

[17] Ozaki H, Sato T, Kubota H, Hata Y, Katsube Y, Shimonishi Y. Molecular structure of the toxin domain of heat-stable enterotoxin produced by a pathogenic strain of *Escherichia coli*. A putative binding site for a binding protein on rat intestinal epithelial cell membranes. The Journal of Biological Chemistry. 1996;**266**(9):5934-5941

[18] Sukumar M, Rizo J, Wall M, Dreyfus LA, Kupersztoch YM, Gierasch LM. The structure of *Escherichia coli* heat-stable enterotoxin b by nuclear magnetic resonance and circular dichroism. Protein Science. 1995;**4**:1718-1729 [cited 01 March 2020] Available from: https://www.rcsb.org/structure/1ehs

[19] Demain AL, Vaishnav P. Production of recombinant proteins by microbes and higher organisms. Biotechnology Advances [Internet]. 2009;**27**(3):297-306. DOI: 10.1016/j.biotechadv.2009.01.008

[20] Kim S, Jeong H, Kim EY, Kim JF, Lee SY, Yoon SH. Genomic and transcriptomic landscape of *Escherichia coli* BL21(DE3). Nucleic Acids Research. 2017;**45**(9):5285-5293

[21] Maksum IP. Sriwidodo Y, Ekspresi Protein Rekombinan secara Intraselular pada Inang *Escherichia coli*. Sumedang: Alqaprint Jatinangor; 2019. p. 92

[22] Cheng J, Wu D, Chen S, Chen J, Wu J. High-level extracellular production of α-cyclodextrin glycosyltransferase with recombinant *Escherichia coli* BL21 (DE3). Journal of Agricultural and Food Chemistry. 2011;**59**(8):3797-3802

[23] Pérez ER, Suárez GJ, Diaz NE, Rodríguez SR, Menéndez CE, Balaguer DH, et al. Scaling-up fermentation of *Escherichia coli* for production of recombinant P64k protein from *Neisseria meningitidis*. Electronic Journal of Biotechnology [Internet]. 2018;**33**:29-35. DOI: 10.1016/j.ejbt.2018.03.004

[24] Strandberg L, Veide A, Enfors SO. Production of the hybrid protein staphylococcal protein A/*Escherichia coli* β-galactosidase with *E. coli*. Journal of Biotechnology. 1987;**6**(3):225-238

[25] Silaban S, Gaffar S, Simorangkir M, Maksum IP, Subroto T. Construction and optimization of prethrombin-2 human genes in *E. coli* for the production of active thrombin. Journal of Physics Conference Series. 2019;**1374**(1):6

[26] Silaban S, Gaffar S, Simorangkir M, Maksum IP, Subroto T. Effect of IPTG Concentration on Recombinant Human Prethrombin-2 Expression in *Escherichia coli* BL21(DE3). Arctic Express. 2019

[27] Indriyani A, Anggraeni NI, Sriwidodo S, Maksum IP. Optimization extracellular secretion of recombinant human epidermal growth factor (hEGF) in *Escherichia coli* BL21 (DE3) pD881-OmpA-hEGF by using response surface method (RSM). International Journal of Research in Pharmaceutical Sciences. 2019;**10**(3):1824-1831

[28] Maksum IP, Lestari A, Fauzia RP, Rachman SD, Soedjanaatmadja UMS. *Escherichia coli* BL21(DE3) expression system using TorA signal peptide for recombinant human albumin (rHA) secretion. International Journal of

Research in Pharmaceutical Sciences. 2019;**10**(4):3319-3324

[29] Peng S, Chu Z, Lu J, Li D, Wang Y, Yang S, et al. Co-expression of chaperones from *P. furiosus* enhanced the soluble expression of the recombinant hyperthermophilic α -amylase in *E. coli*. Cell Stress and Chaperones. 2016;**21**(3):477-484

[30] Zhang H, Yuan Q, Zhu Y, Ma R. Expression and preparation of recombinant hepcidin in *Escherichia coli*. Protein Expression and Purification. 2005;**41**:409-416

[31] Lin H-H, Yin L-J, Jiang S-T. Cloning, expression, and purification of *Pseudomonas aeruginosa* keratinase in *Escherichia coli* AD494 (DE3) pLysS expression system. Journal of Agricultural and Food Chemistry. 2009;**494**:3506-3511

[32] Brüsehaber E, Schwiebs A, Schmidt M, Bottcher D, Bornscheuer UT. Production of pig liver esterase in batch fermentation of *E. coli*. Origami. Applied Microbiology and Biotechnology. 2010;**86**(5): 1337-1344

[33] Choi JH, Keum KC, Lee SY. Production of recombinant proteins by high cell density culture of *Escherichia coli*. Chemical Engineering Science. 2006;**61**(3):876-885

[34] Maksum IP. Sriwidodo, Indriyani A. Sistem Ekspresi Protein Rekombinan dalam *Eschercia coli* secara Ekstraselular. Sumedang: Alqaprint Jatinangor; 2018. p. 207

[35] Sriwidodo MIP, Riswanto N, Rostinawati T, Subroto T. Extracellular secretion recombinant of human epidermal growth factor (hEGF) using pectate lyase B (PelB) signal peptide in *Escherichia coli* BL21(DE3). International Journal of Research in Pharmaceutical Sciences. 2016;**8**(1):1-8

[36] Ningsih E, Padmadisastra T, Soedjanaatmadja UMS. Isolasi Pemurnian dan Karakterisasi Enzim beta-galaktosidase dari *Escherichia coli* B-130. Bandung: Laporan Tugas Akhir Penelitian. Universitas Padjadjaran; 1994

Section 2

Diagnosis and Treatment

Chapter 4

Aptamers for Infectious Disease Diagnosis

Soma Banerjee and Marit Nilsen-Hamilton

Abstract

Aptamers are in vitro-selected, nucleic acids with unique abilities to bind strongly and specifically to their selective targets (ligands) based on their three-dimensional structures. Target binding is generally associated with a change in aptamer structure, which provides a means of linking many output signals to the binding event. Being synthetic, aptamers are less expensive compared to antibodies. Aptamers are also more easily modified chemically or their sequence changed to optimize properties such target specificity, storability and stability. In this chapter we will discuss the potential benefits of applying aptamers to diagnostics with a focus on infectious disease and the unique challenges posed by aptamers for their successful incorporation into reliable aptasensors.

Keywords: aptamers, SELEX, aptasensors, portable diagnostic tools, electrochemical impedance spectroscopy

1. Introduction

Aptamers, first disclosed in 1990 by three groups [1–3], are ssDNA or RNA molecules capable of binding strongly and specifically to their target (ligand) molecules. Their target binding specificities and affinities are based on their sequence-specific 3D structures. Such properties of aptamers make them analogues of antibodies with unique advantages. For example, aptamers are relatively small (diam. ~2 nm) compared to antibodies (diam. ~15 nm), which allows them to bind targets that are inaccessible to the larger antibodies. Like antibodies, their properties are defined by the ionic conditions and pH in which they are placed. However, being shorter polymers, aptamers are generally more sensitive than antibodies to their physical and chemical environment.

In contrast to the time-consuming and expensive production and screening procedures for antibodies, aptamers can be produced faster and more cost effectively by a procedure known as Systemic Evolution of Ligands by Exponential Enrichment (SELEX). Once an aptamer sequence has been identified, its further production is by chemical synthesis, for which variation is negligible compared with the batch to batch variation of antibodies generated in animals or by cell culture. Their synthetic production makes aptamers accessible for selective chemical modifications to enhance their binding specificity or to increase their resistance to degradation. With such advantages over antibodies, aptamers have emerged as new generation molecular recognition elements [4]. In the current chapter, we focus on their impact in diagnosis of infectious disease agents. The reader is referred to other reviews of the application of aptamers to therapy, biosensing and molecular probing [5–10].

2. Aptamers in diagnostics

Fast and accurate diagnosis is a key factor for the treatment of infectious disease. Molecular recognition by aptamers can be highly discriminatory such that they can distinguish between two closely related molecules, including conformational isomers [11] or highly related proteins [12–14]. One-to-one comparison between aptamers and antibodies as recognition elements for the same target molecule, demonstrated that aptamers can equal antibodies in their sensitivity and selectivity [15]. As an added advantage, aptamers showed tolerance to repeated regeneration and recycling after ligand binding.

The small sizes and homogeneous structures of aptamers allow them to be immobilized in a dense and well-oriented manner. The higher density of aptamer packing compared with antibodies increases the binding capacity of the sensors and extends their linear range of detection of analyte [9, 15]. With these aspects of aptamers considered, their application as recognition elements in analytical devices offers a multitude of advantages and brings a new dimension to diagnostics.

Aptamers are compatible as the recognition element with many sensor platforms, including quartz crystal microbalances (QCM), surface plasmon resonance (SPR), diamond field effect transistors (FET), electrochemical impedance spectroscopy (EIS), colorimetric and fluorescence-based optical detection. Of these, electrochemical impedance spectroscopy (EIS) has gained popularity as it offers rapid, low-cost, label-free detection with high signal to noise ratios and sensitive detection of target molecules when employing aptamers [16, 17]. EIS is more sensitive than other electrochemical techniques [18] and is one of the best techniques to analyze the properties of electrochemical systems [19]. EIS is a technique used to study the electrochemical response to the application of periodic small amplitude ac signal at different frequencies [20, 21]. It is useful to monitor the changes in the electrochemical properties of the system due to biorecognition events at the surface of modified electrodes. For example, the electrodes can be modified with aptamers to detect the presence of a target analyte. EIS produces high quality data by directly converting a biological event into an electrical signal. Moreover, EIS-based sensors are small and portable and can be employed outside of well-equipped laboratory. EIS is an attractive technique for biosensor development as it provides the advantages of real time monitoring and label-free detection and is compatible with flexible electronics, disposable sensors and wearable devices. Inkjet printing can be applied to produce aptamer-based EIS biosensors for their automated mass production with uniform aptamer deposition [22–24].

In some instances aptamers can also be used for therapeutics or for combined diagnostics and therapeutics (theranostics). Examples include (1) various aptamers to human immunodeficiency virus (HIV) proteins that either prevent virus entry or replication [7, 25–29], (2) the S-$PS_{8.4}$ aptamer that recognizes *Salmonella enterica* and inhibits invasion of the bacteria into human monocytes [30], (3) two DNA aptamers A9 and B4 against the H9N2 avian influenza virus that inhibit viral infection [31], and (4) a DNA aptamer against MUC-1 that was applied in a nanocomposite for fluorescent imaging and demonstrated to inhibit the proliferation of colorectal (HT-29) and breast (MCF-7) cancer cells [32].

Although antibodies dominate the global market of diagnostics and therapeutics, several biotechnology companies have actively started exploring aptamers for diagnostics. The first to develop aptamer-based diagnostic arrays, SomaLogic employed SOMAmers (slow off rate modified aptamers) to detect many protein biomarkers for disease diagnosis [33, 34]. The combination of aptamers as the recognition elements, with long shelf lives at room temperature, inkjet printing methods for immobilizing them and EIS as the method of detection will enable the development of low cost, label free, and rapid response diagnostic devices that could be in disposable or wearable forms as well as in more conventional instrument formats.

2.1 Aptamers that detect biomarkers of microbial infections

Early detection of infectious disease is of primary importance for its management. The conventional methods of diagnosis, which include microbiological methods (isolation, growth and microscopy of pathogens), polymerase chain reaction (PCR) and immunological methods [35, 36] suffer from turn-around times of 24 h or longer. This is especially a problem when the patient(s) are located in remote areas from which samples must be sent to centralized laboratories for their analysis. It is also difficult to grow some microbes in culture, which limits their detection.

Viral diseases are generally detected by serology, immunology or PCR amplification of DNA/RNA fragments corresponding to the pathogen's genome. Although it is sensitive and specific, the performance of PCR requires appropriate instruments, specialized reagents and experienced personnel. Immunological methods, which are widely used for diagnosis, employ antibodies specific to a protein or carbohydrate moiety that is unique to the target pathogen. Some popular immunological methods are the agglutination, ELISA and western blot assays. These well-established and time-tested assays are the work-horses of modern clinical laboratories. They are generally reliable and are likely to be the mainstay of the clinical technical repertoire well into the future. But, these assays limit the ability of communities to respond rapidly to microbial and viral outbreaks because they require laboratory equipment that is not readily portable and trained personnel to perform them. ELISA and western blots also depend on provided antibodies, which require cold storage to prevent their denaturation. Infectious disease outbreaks often start in regions that are distant from clinical laboratories. Therefore, the challenge for future diagnostics is to develop portable devices that require little expertise to perform. Current technology development is moving in this direction with portable PCR machines [37] and lateral flow immunology tests [38]. Whereas the former still requires trained personal to operate, the latter can often be readily used and interpreted by an untrained individual.

Diagnostics based on aptamers stand out as promising options for rapid, cost effective and specific detection of pathogens using devices that can be operated by minimally trained personnel. Many aptamers have been reported that recognize specific viruses and bacteria. Some were selected against recombinant proteins from the target microbe or virus. Others were selected against the intact microbe or virus. For example, RNA aptamer S-PS$_{8.4}$, which specifically recognizes the type IVB pili of *Salmonella enterica*, was isolated using a recombinant pilin structural protein as the selection target [39] and incorporated into a potentiometric biosensor as a recognition element for *S. enterica* [40]. This aptasensor could detect a single CFU of target *S. enterica* and was specific for *S. enterica*, not recognizing either *Escherichia coli* or *Lactobacillus casei*. Using a cell-SELEX approach, two 62 nt DNA aptamers, SA17 and SA61, were selected against intact *Staphylococcus aureus*. As for S-PS$_{8.4}$, these aptamers bound their *S. aureus* target with high affinities and specificities [41]. Many aptamers have been selected with specificities for particular microbe targets including or *Campylobacter jejuni* [42, 43], *L. monocytogenes* [44–48], *Vibrio parahemolyticus* [49], *Shigella dysenteriae* [50], *Streptococcus pyogenes* [51], *Francisella tularensis* [52], *Pseudomonas aeruginosa* [53] and the spores of anthrax *Bacillus anthracis* [54, 55]. Parasites are also good aptamer targets with aptamers identified that recognize Trypanosoma spp., Plasmodium spp., Leishmania spp., *Entamoeba histolytica*, and *Cryptosporidium parvuum* [56]. For the diagnosis of invasive fungal infections, DNA aptamers have been screened against (1 → 3)-β-D-glucans from cell wall of *Candida albicans*. Two selected DNA aptamers (AU1 and AD1) showed high binding affinities in the range of 100 nM and did not bind to the same domain of (1 → 3)-β-D-glucans. The application of these aptamers in a double-aptamer sandwich enzyme-linked oligonucleotide assay (ELONA) resulted in an assay sensitivity and specificity of the detection of ~92% [57]. For viruses, there are aptamers that recognize HIV intracellular proteins [7, 13, 58–60], the HIV envelope glycoprotein [28, 61], hepatitis C [62–64], influenza virus [65–67], herpes simplex virus 1 [68], dengue virus [69], zika virus [70] and ebola virus [71]. In another study, a device was reported for the multiplexed detection of the envelope proteins of Zika and chikungunya viruses. The detection takes place in a microfluidic channel containing microsized pillars with attached aptamers. These pillars increase the surface sensing area, thereby enabling the attachment of more aptamers and increasing the overall sensitivity of the sensor envelope proteins. The working principle of this device depends on the formation of a protein-mediated sandwich with an aptamer-functionalized gold nanoparticle (AuNP) and an unattached aptamer. The signal is obtained upon introduction of silver reagents into the channel, which is selectively deposited on the AuNP surface, providing a gray contrast in the testing zone. This colorimetric aptasensor is reported to detect clinically relevant concentrations of Zika and chikungunya envelope proteins in phosphate-buffered saline (1 pM) and calf blood (100 pM) with high specificity [72].

Many of the aptamers discussed have been employed as recognition elements in diagnostic tools of a wide variety of types with electrochemical sensors being a popular platform. Examples include a potentiometric carbon-nanotube system to detect trypanosomes in blood [73], a voltametric aptasensor for ultrasensitive detection of *Mycobacterium tuberculosis* (MTB) virulence factor antigen ESAT-6 [74] and an EIS aptasensor for influenza virus [67].

Along with fast read-out, another advantage of the electrochemical approach is high sensitivity. The aptamer-based detection threshold is sometimes lower than for RT-PCR as demonstrated for influenza virus [67]. In another example, aptamer conjugates with gold nanoparticles were sensitive enough to detect a single *Staphylococcus aureus* cell [41]. An aptasensor has been reported that detects attomolar concentrations of the variable surface glycoprotein from African trypanosomes as analyte in blood [73]. From these and other examples it was found that immobilization of aptamers on biosensor surfaces increases target binding affinity [75–78]. The increased affinity is most likely due to two effects of immobilization: (1) immobilization creates a multivalent surface that decreases the rate at which the aptamer ligand can leave the surface (off-rate), which is the denominator in calculating the association constant ($K_a = k_{on}/k_{off}$), and (2) molecular crowding promotes aptamer folding to produce the appropriate ligand-binding structure [79–81].

Due to their low concentrations in the blood, infectious disease markers can be difficult to detect [82, 83]. Here, aptamers can play a different role of concentrating the target prior to their quantification. For this purpose, magnetic beads coated with aptamers specific for *Trypanosoma cruzi* were used to capture these parasites from the blood in which they are present at very low concentrations [84].

The stage is set for the development of commercial diagnostics for infectious disease agents. Some have already come to market such as OTA-Sense and Aflasense developed by Neoventures Biotechnology Inc. for detection of toxins in food samples, AptoCyto and AptoPrep developed by Aptsci Inc. for isolation of biomarker positive cells, SOMAscan from SomaLogic for diagnosis of several diseases, CibusDx a food pathogen Diagnostic platform developed by the

USA-based start-up (Pronucleotein, Inc.), OLIGOBIND®Thrombin activity assay by Sekisui Diagnostics [85]. Aptasensors demonstrate remarkably short detection times compared with the conventional methods of ELISA and PCR. The advantage of a faster detection time is of utmost importance for identifying rapidly developing and epidemiologically dangerous diseases, such as influenza, Ebola and SARS (Severe Acute Respiratory Syndrome). The rapid detection capabilities of aptasensors and their ready portability will broaden their scope of acceptance in the field of diagnosis.

3. Aptasensors

Some of the most attractive features of aptamer-based sensors (aptasensors) are their stability to storage at ambient temperatures and their reusability. Moreover, their small size and versatility allow aptamers to be immobilized at high densities, which facilitate their multiplexing in miniaturized systems. Several signaling modes have been coupled to aptamer-based sensors [86, 87]. Some popular outputs for detection include fluorescence [88], chemiluminescence [55], electrochemical [89], field effects (FET) [90], surface plasmon resonance (SPR) [91], changes in resonating frequency of quartz crystal sensor (QCM) [92], surface acoustic waves (SAW) [93], mechanical (microcantilevers) [94]. In this chapter, we will focus mainly on aptamer-based biosensors with fluorescent or electrochemical outputs.

3.1 Fluorescent aptasensors

Aptasensors with fluorescence outputs are designed to take advantage of the flexibility of aptamers, which results in their frequently adopting alternate conformations in the presence or absence of their target molecules. For these sensors, aptamers are modified in key positions with fluorescent dyes that interact in Förster resonance energy transfer (FRET). Upon aptamer binding to its target, the associated structural change alters the distance between the fluorescent dyes and thus the efficiency of energy transfer. The signal change, manifested as an increase (signal-on mode) or decrease (signal-off mode) in fluorescence, is proportional to the extent of target binding. A representative "signal-on" fluorescent aptamer holds a fluorophore, usually at one end, which is quenched by a molecule that is attached to a proximate location in the unoccupied aptamer. Target binding separates fluorophore from quencher allowing recovery of the fluorescent signal, which provides a quantitative measure of the target concentration [95]. FRET can also be used in "signal-off" sensor designs in which the conformational change of the aptamer on target binding brings the donor and quencher into closer proximity with a resulting fluorescence quenching. Sensors based on the "signal-off" mode are usually less sensitive than those based on the "signal-on" mode, but they can help to improve target detection by low-affinity aptamers [88].

An alternative means of signaling an aptamer binding event using fluorescence is with an oligonucleotide (attenuator) that is complementary to a portion of the aptamer and remains bound to the aptamer in the absence of target molecule. The length of the attenuator and its placement on the aptamer must create a condition that prevents aptamer folding to the actively binding conformation, but the affinity of the aptamer for the attenuator should be less than for the target molecule. With these requirements fulfilled, the target molecule can successfully compete with the attenuator to bind the aptamer and release the attenuator. Such a design can be used for "signal-on" reporting if the target and aptamer are labeled with fluorescent dyes that are optimally placed to interact in FRET while the aptamer and attenuator are hybridized. This format can also be used for a "signal-off" system in which a single fluorophore is attached to the attenuator. When the aptamer binds the target molecule, the released attenuator adsorbs to surrounding gold nanoparticles (AuNPs), which quench the fluorescence [96]. Another aptasensor design used upconversion nanoparticles (UCNPs) as donors and AuNPs as acceptors for rapid, ultrasensitive and specific detection of bacteria (e.g., *E. coli* ATCC 8739) [97]. FRET-based aptasensors provide an efficient method for detecting pathogens and their released toxins in one step [98–100].

The concept of using a material to quench the fluorophore was applied to create a paper-based MoS_2 nanosheet-mediated FRET aptasensor for rapid malaria diagnosis [96]. This format uses paper test strips impregnated with fluorescently-labeled aptamers and MoS_2 nanosheets. The MoS_2 quenches the fluorescence until the aptamers are released when they bind their targets. These aptasensors are facile, inexpensive and therefore attractive for point-of-care diagnosis, especially in low-resource areas. Similar "low-tech" FRET-based aptasensors have also been found to be ideal for spacecraft, such as for diagnosing microgravity-induced bone loss in outer space by monitoring urinary C-telopeptide [4, 101]. In such scenarios, where both space and lab resources are limited, handheld fluorometers such as the commercially available QuantiFluor™ (Promega Corp.) or other such portable fluorometers will provide much needed opportunities for point-of-care diagnostics. These applications benefit from the greater stability to ambient temperatures for storage of aptamers compared with antibodies.

3.2 Electrochemical aptasensors

Upon binding to their target molecules, aptamers fold their supple, single-stranded chains into distinct three-dimensional (3D) structures. This structural change can be employed for initiating electron-transfer when the aptamers are labeled with a redox-active moiety and immobilized on a conducting support. Several electrochemical aptasensors have been developed based on this strategy [87], which can also be classified into "signal-on" and "signal-off" aptasensors. For example, an electrochemical thrombin aptasensor was constructed by immobilizing a thrombin aptamer (TBA) labeled with redox-active methylene blue (MB) on an electrode [102]. After binding thrombin, the TBA adopts a G-quadruplex structure, which moves MB away from the electrode. This "signal-off" sensing format has the disadvantage of a decreasing signal with increasing target molecule. An example of the preferred "signal-on" format includes the TBA, which is immobilized on a gold electrode and tagged with a terminal electroactive ferrocene redox label [103]. In the absence of thrombin a low signal is produced because many of the conformations adopted by the aptamer do not bring the ferrocene close to the electrode. Upon binding thrombin, the TBA adopts a G-quadruplex conformation, bringing the ferrocene to the electrode to allow electron-transfer and a positive signal in the presence of target molecule.

Electrochemical signals can be amplified when catalytic events are part of the signaling mechanism. For example, an electrochemical aptasensor was developed to rapidly diagnose tuberculosis (TB) by detecting the *Mycobacterium tuberculosis* antigen, MPT64, in serum samples [104]. MPT64 exists in serum as a disulfide linked homo-multimer. With multiple target sites on the same multimeric particle, MPT64 can be detected by a sandwich assay with the same aptamer on each side of the sandwich. In this study, coil-like fullerene (C60)-doped polyaniline (C60-PAn) nanohybrids were used as redox nanoprobes and catalysts to initiate the oxidation of ascorbic acid. When linked to the MPT64 aptamer, these nanohybrids were brought close to a gold surface (also decorated with MPT64 aptamers) in a sandwich joined by MPT64 multimers. In this configuration the electrons released by the oxidation of ascorbic acid were transferred to the gold electrode. This simple yet elegant approach for TB diagnosis showed selectivity to target antigen over several other serum proteins, a wide linear range of detection from 0.02 to 1000 pg/mL and a detection limit of 20 fg/mL MPT64. The delayed diagnosis and misdiagnosis of patients with MTB infection is the leading cause behind the spread and high mortality rate of TB [105]. Therefore, the possibility of rapid and accurate detection of MTB by these aptasensors is of great significance for the early diagnosis and treatment of TB.

Electrochemical impedance spectroscopy (EIS), an electrochemical label-free detection method, can be an extremely sensitive method for target recognition at the electrode/electrolyte interface [106]. Here, aptamers are immobilized on a gold (Au) electrode and the remaining gold surface filled in by a self-assembled monolayer such as mercaptohexanol (6-MCH). This approach was used with a DNA aptamer as molecular recognition element for malaria detection, for which the response range of 1 pM–10 nM covered the diagnostically relevant concentration range of Plasmodium lactate dehydrogenase protein from the falciparum parasite species (PfLDH) [107, 108]. The aptasensor functioned well with a sample matrix of 10% human serum and could be regenerated for reuse by washing with 6 M urea.

Electrochemical aptasensors have been fabricated to be sufficiently small to insert into a vein for continuous, real-time measurement of specific molecular targets in situ in the living body. The limited surface area of these small devices leads to low faradaic currents and poor signal-to-noise ratios when deployed in the complex, fluctuating environments found in vivo. To circumvent this problem, an electrochemical roughening approach was developed to enhance the signal-to-noise ratios by increasing the microscopic surface area of gold electrodes, thereby allowing more redox reporter-modified aptamers to be packed onto the surface. These high surface area electrochemical aptasensors of less than 200 μm in diameter were used in a proof-of concept study to measure continuous drug pharmacokinetic profiles over a 3 h period in live rats [109].

Colorimetric detection is gaining popularity in the diagnostic field considering its low cost and the minimal training needed to identify and interpret the visible signal. A colorimetric approach was used to develop a diagnostic device for tuberculosis with aptamers that bind to antibodies against the MPT64 protein secreted by *Mycobacterium tuberculosis*. When adsorbed to Fe_3O_4 magnetic nanoparticles (MNPs) the aptamers decrease the ability of the particles to reduce oxygen to H_2O_2 [110]. Upon exposure of the MNP-aptamer suspension to anti-MPT64 antibodies, the aptamers preferentially bind to the antibodies, thereby increasing the available surface area of the MNPs with the resulting higher rates of H_2O_2 production. Inclusion of 2,2′-azino-bis(3-ethylbenzo-thiazoline-6-sulfonic acid) (ABTS), which is oxidized by H_2O_2 and converted to a colored product, signals the presence of anti-MPT64 antibodies [111]. Another format for a colorimetric aptasensor is a paper-based microfluidic chip. For this format, aptamers against bacteria considered as nosocomial and antibiotic-resistant were immobilized by ultraviolet crosslinking on a nitrocellulose membrane housed within the chip. Incubation with bacteria, washing and then the addition of biotinylated aptamers allowed the use of HRP-linked streptavidin to create a blue color based on the oxidation of tetramethyl benzidine (TMB) by the H_2O_2 product of HRP, which was trapped on the surface by way of its linkage to the streptavidin bound to the biotinylated aptamers attached to the target bacteria. This dual-aptamer microfluidic chip possesses many advantages such as rapid output (35 min), small size, higher specificity, and the capability to detect multiple pathogens simultaneously, which are ideal for point-of-care bacterial diagnostics [112]. A similar sandwich type aptasensor has been reported for the early diagnosis of periodontitis, in which chronic inflammation is caused by many factors including pathogenic bacteria. Periodontitis is one of the major causes of tooth loss in adults. Aptamers were targeted against a potential biomarker of this disease, odontogenic ameloblast-associated protein (ODAM). The lateral flow strip format used a cognate pair of aptamers that recognize different sites on ODAM. One aptamer (2^0 aptamer) was attached to gold nanoparticles that were mixed with the sample to capture the biomarker. The second aptamer (1^0 aptamer) was present in a line on the strip to capture nanoparticles with attached biomarker. A control line with DNA complementary to the 2^0 aptamer captured the particles that did not have attached biomarkers. The biosensor had a detection limit of 0.24 and 1.63 nM in buffer and saliva samples, respectively [113].

4. Considerations for further development of aptasensor applications in diagnostics

Of the many aptamers that have been selected, very few have been applied as recognition elements in sensors and fewer have reached the stage of commercial availability. In this section we will consider some of the reasons that aptasensors, with all their promise, have been slow to come to the market in diagnostic devices.

One reason for the limited breadth of application of aptamers to diagnostics is that many of the biosensor platforms are new to the concept of incorporating aptamers as sensors. Therefore, the research focus has been on developing sensor platforms that are compatible with aptamers. For proof-of-principle devices, aptamers that have been previously demonstrated to function well on a variety of sensor platforms have been chosen as recognition elements. Consequently, the TBA and ATP aptamers have been incorporated into many sensor platforms [114, 115]. However, these targets are not relevant biomarkers for disease. With many sensor platforms now validated using these "prototype" aptamers the field has the opportunity to move forward to incorporate and optimize some of the many available clinically relevant aptamers for diagnostic applications.

Nucleic acids are more flexible polymers than polypeptides. Whereas there are two rotatable bonds between each amino acid side-chain in a polypeptide, there are six rotatable bonds between each base in a nucleic acid. This additional flexibility gives aptamers the property of ready structural rearrangement with target binding and enables their incorporation into many sensor platforms that rely on these rearrangements for creating signals. Thus, the fundamental principles underpinning antibody and aptamer-based sensors are different. Whereas antibody-based sensors rely on the uniformity of antibody structure and their bivalency, aptamer-based sensors rely on the flexibility of the monovalent aptamer structure and the structural changes that occur on target binding. Consequently, a single sensor format can be readily adapted to many antibodies, but each aptamer-sensor combination must be optimized to benefit from the unique structural change of the relevant aptamer. Thus, although there may be some applications for sensors that use the TBA and ATP model aptamers [116], their repeated use in developing and testing aptamer-based sensors has delayed the development of aptasensors for relevant biomarkers.

With the importance of aptamer structure and conformational changes for sensor development, a second challenge for reliable aptasensor development lies in the dearth of known aptamer tertiary structures. This deficiency results from several circumstances. First, biophysical determination of nucleic acid structure by conventional methods such as NMR and X-ray crystallography is more challenging than for proteins. Second, when structures can be determined by biophysical means, these are often only the structures of the target-bound aptamers because the apo-aptamer structures are too flexible to be reduced to a single structure. Third, new aptamers are being reported at an increasingly rapid rate that is much faster than their structures can be determined. Finally, small modifications of existing aptamers that can have large effects on aptamer structure are often made for their application in sensors. Thus, aptasensors must be developed with little information about the tertiary structure of the aptamer employed and how it changes with target binding. The most likely route to obtaining molecular structure for most aptamers will be in silico modeling. Although not yet demonstrated to be adequate for accurately predicting the structures of short nucleic acids and how they change with target binding, molecular modeling techniques are improving and their successful integration into models for sensor mechanisms could eventually result in dependable strategies for engineering new aptamers and integrating them into sensors.

Many aptamer selection protocols require the availability of purified target molecules [117]. Protein target molecules are usually expressed as recombinant proteins by prokaryotic or eukaryotic cell cultures and then purified, frequently by affinity chromatography based on a capture tag. Like for obtaining antibodies, the protein targets must be pure. Difficulties can come if the recombinant protein is not post-translationally modified similarly to the native protein. For example, many biomarkers found in the blood are glycosylated in the native form, but the recombinant proteins produced by bacterial cells are not glycosylated. Due to their steric hindrance or by their altering the protein structure in the region of the aptamer epitope these modifications can make regions of a native eukaryotic protein inaccessible to aptamers generated against the recombinant protein equivalent expressed in prokaryotic cells [118]. An early screen for aptamers selected to bind non-glycosylated recombinant proteins should be to determine if they bind the native glycosylated protein. Approaches to selecting for aptamers that recognize glycan structure in the context of the protein will also be useful [119].

Cell-SELEX avoids the complication of the target lacking the native posttranslational modifications by selecting against the cell surface protein target in situ [120, 121]. However, the identities of targets obtained by Cell-SELEX are often not known. As well, Cell-SELEX is performed with cell lines that are different from normal or in situ cancer cells and that are cultured under conditions that differ from those in the body. In particular, cultured cells have adjusted to exist in serum, which is not present in vivo and to an environment of much higher oxygen content than cells in situ, both of which conditions might result in altered protein expression on the cell surface.

Another consideration for aptasensors is that they frequently must function with the target (analyte) in a complex sample matrix. Many sensors have been demonstrated to function well in simple buffers, while the most common biomedical samples (blood, serum, urine and saliva) are complex with many potentially interfering substances. Aptamer target binding affinity can be sensitive to the matrix in the range of dilutions commonly used to detect the analyte, such as 10 or 50% serum or urine [122, 123]. For biofluids, sampling methods must also be considered with a view to minimal invasiveness and small sample volumes.

As aptasensors are developed that avoid the pitfalls discussed here, these sensors will take their place beside the antibody-based assays and provide new capabilities such as continuous analyte monitoring and inexpensive devices that can be distributed to small clinics throughout the world and yet be connected by Bluetooth and other options to send their results to central clinics and distant physicians for improved monitoring of patients in rural and other isolated locations.

5. Prospects of aptasensors in next-generation diagnosis

Medical diagnostics is moving towards a future of individualized patient care. Individuals vary greatly in their response to specific drugs and in the rate at which these drugs are removed from the system [124–128]. Pharmacokinetic parameters have been reported to vary daily in the same patient [129] or with time of drug exposure [130] and between individuals depending on disease state [131–133], age [134], genetics [130, 135–139], concurrent medication [137], body fat composition [140], and even circadian rhythm [141]. Adherence to a drug intake regime is also a factor, particularly for care of the young and the aged [142–144]. For these many reasons, individualized diagnostics are considered a clinical necessity for improved patient treatment and for establishing effective therapeutic windows [145].

With the push towards individualized medicine as a desirable future approach for optimal patient care comes the need to move some diagnostics out of the clinic into the home. For this purpose, reliable inexpensive sensors that might be linked by wireless connections to clinical centers would be optimal. Some personalized diagnostics has long been available in the home. These include pregnancy tests and glucose monitors, which are based on antibody and enzymatic reactions. However, these are not linked to the larger medical care network. The course is now set for a huge expansion of personalized diagnostics that do not require trained operators on site, but that can transmit information to clinical specialists who can monitor a patient's condition off-site and continuously. This is a niche for which aptasensors can provide a large diversity of options with their potential for long shelf-live under ambient conditions, simplicity of operation, ability to be designed for continuous use or repeated use, and compatibility with wearable sensor formats.

Aptasensors can also be applied to address the acute need for diagnostics during infectious disease epidemics by their placement in clinics located in isolated regions of the world and in individual physician's offices that are distant from major well-equipped hospitals and clinical centers. With the additional capability of Bluetooth communication, centers of disease control can be quickly updated regarding the spread of infections, which will enable central authorities to rapidly initiate effective means of controlling a potential epidemic.

A future of diagnostics for individualized medicine and for the control of infectious disease outbreaks over vast regions will result from many cross-disciplinary collaborations that are already underway, which include experts in molecular biology, virology, medicine, engineering, diagnostics and other disciplines. With this effort, many of the aptasensors that are still now at the proof-of concept stage, are expected to become major contributors to a future of improved personalized health care for all people, including those living in remote regions, and will help to stem future outbreaks of infectious diseases.

Author details

Soma Banerjee[1*] and Marit Nilsen-Hamilton[1,2]

1 Aptalogic Inc., Ames, IA, USA

2 Iowa State University, Ames, IA, USA

*Address all correspondence to: soma@aptalogic.com

References

[1] Ellington AD, Szostak JW. In vitro selection of RNA molecules that bind specific ligands. Nature. 1990;**346**: 818-822. DOI: 10.1038/346818a0

[2] Robertson DL, Joyce GF. Selection in vitro of an RNA enzyme that specifically cleaves single-stranded DNA. Nature. 1990;**344**:467-468. DOI: 10.1038/344467a0

[3] Tuerk C, Gold L. Systematic evolution of ligands by exponential enrichment: RNA ligands to bacteriophage T4 DNA polymerase. Science. 1990;**249**:505-510

[4] Dhiman A, Kalra P, Bansal V, Bruno JG, Sharma TK. Aptamer-based point-of-care diagnostic platforms. Sensors and Actuators B: Chemical. 2017;**246**: 535-553. DOI: 10.1016/j.snb.2017.02.060

[5] Banerjee J, Nilsen-Hamilton M. Aptamers: Multifunctional molecules for biomedical research. Journal of molecular medicine (Berlin, Germany). 2013;**91**:1333-1342. DOI: 10.1007/s00109-013-1085-2

[6] Wandtke T, Woźniak J, Kopiński P. Aptamers in diagnostics and treatment of viral infections. Viruses. 2015;**7**:751. DOI: 10.3390/v7020751

[7] Bala J, Chinnapaiyan S, Dutta RK, Unwalla H. Aptamers in HIV research diagnosis and therapy. RNA Biology. 2018;**15**:327-337. DOI: 10.1080/15476286.2017.1414131

[8] Zhou J, Rossi J. Aptamers as targeted therapeutics: Current potential and challenges. Nature Reviews Drug Discovery. 2016;**16**:181. DOI: 10.1038/nrd.2016.199

[9] Tombelli S, Minunni M, Mascini M. Aptamers-based assays for diagnostics, environmental and food analysis. Biomolecular Engineering. 2007;**24**: 191-200. DOI: 10.1016/j.bioeng.2007.03.003

[10] Tombelli S, Mascini M. Aptamers as molecular tools for bioanalytical methods. Current Opinion in Molecular Therapeutics. 2009;**11**:179-188

[11] Geiger A, Burgstaller P, von der Eltz H, Roeder A, Famulok M. RNA aptamers that bind L-arginine with sub-micromolar dissociation constants and high enantioselectivity. Nucleic Acids Research. 1996;**24**:1029-1036. DOI: 10.1093/nar/24.6.1029

[12] Jenison RD, Jennings SD, Walker DW, Bargatze RF, Parma D. Oligonucleotide inhibitors of P-selectin-dependent neutrophil-platelet adhesion. Antisense and Nucleic Acid Drug Development. 1998;**8**:265-279. DOI: 10.1089/oli.1.1998.8.265

[13] Tombelli S, Minunni M, Luzi E, Mascini M. Aptamer-based biosensors for the detection of HIV-1 Tat protein. Bioelectrochemistry. 2005;**67**:135-141. DOI: 10.1016/j.bioelechem.2004.04.011

[14] Ruslinda AR, Tanabe K, Ibori S, Wang X, Kawarada H. Effects of diamond-FET-based RNA aptamer sensing for detection of real sample of HIV-1 Tat protein. Biosensors and Bioelectronics. 2013;**40**:277-282. DOI: 10.1016/j.bios.2012.07.048

[15] Liss M, Petersen B, Wolf H, Prohaska E. An aptamer-based quartz crystal protein biosensor. Analytical Chemistry. 2002;**74**:4488-4495

[16] Rohrbach F, Karadeniz H, Erdem A, Famulok M, Mayer G. Label-free impedimetric aptasensor for lysozyme detection based on carbon nanotube-modified screen-printed electrodes. Analytical Biochemistry. 2012;**421**: 454-459. DOI: 10.1016/j.ab.2011.11.034

[17] Liang G, Man Y, Jin X, Pan L, Liu X. Aptamer-based biosensor for label-free detection of ethanolamine by electrochemical impedance spectroscopy. Analytica Chimica Acta. 2016;**936**:222-228. DOI: 10.1016/j.aca.2016.06.056

[18] Karimipour M, Heydari-Bafrooei E, Sanjari M, Johansson MB, Molaei M. A glassy carbon electrode modified with TiO_2(200)-rGO hybrid nanosheets for aptamer based impedimetric determination of the prostate specific antigen. Microchimica Acta. 2018;**186**:33. DOI: 10.1007/s00604-018-3141-7

[19] Ehsani A, Mahjani MG, Hosseini M, Safari R, Moshrefi R, Mohammad SH. Evaluation of *Thymus vulgaris* plant extract as an eco-friendly corrosion inhibitor for stainless steel 304 in acidic solution by means of electrochemical impedance spectroscopy, electrochemical noise analysis and density functional theory. Journal of Colloid and Interface Science. 2017;**490**:444-451. DOI: 10.1016/j.jcis.2016.11.048

[20] Kongsuphol P, Ng HH, Pursey JP, Arya SK, Wong CC, Stulz E, et al. EIS-based biosensor for ultra-sensitive detection of TNF-α from non-diluted human serum. Biosensors and Bioelectronics. 2014;**61**:274-279. DOI: 10.1016/j.bios.2014.05.017

[21] Grieshaber D, MacKenzie R, Vörös J, Reimhult E. Electrochemical biosensors—Sensor principles and architectures. Sensors (Basel, Switzerland). 2008;**8**:1400-1458. DOI: 10.3390/s80314000

[22] da Costa TH, Song E, Tortorich RP, Choi JW. A paper-based electrochemical sensor using inkjet-printed carbon nanotube electrodes. ECS Journal of Solid State Science and Technology. 2015;**4**:S3044-S3047. DOI: 10.1149/2.0121510jss

[23] Maddaus A, Curley P, Griswold MA, Costa BD, Hou S, Jeong KJ, et al. Design and fabrication of bio-hybrid materials using inkjet printing. Biointerphases. 2016;**11**:041002. DOI: 10.1116/1.4966164

[24] Khan NI, Maddaus AG, Song E. A low-cost inkjet-printed aptamer-based electrochemical biosensor for the selective detection of lysozyme. Biosensors. 2018;**8**:7. DOI: 10.3390/bios8010007

[25] Konopka K, Lee NS, Rossi J, Duzgunes N. Rev-binding aptamer and CMV promoter act as decoys to inhibit HIV replication. Gene. 2000;**255**:235-244

[26] Bai J, Banda N, Lee NS, Rossi J, Akkina R. RNA-based anti-HIV-1 gene therapeutic constructs in SCID-hu mouse model. Molecular Therapy. 2002;**6**:770-782

[27] Chaloin L, Lehmann MJ, Sczakiel G, Restle T. Endogenous expression of a high-affinity pseudoknot RNA aptamer suppresses replication of HIV-1. Nucleic Acids Research. 2002;**30**:4001-4008. DOI: 10.1093/nar/gkf522

[28] Khati M, Schuman M, Ibrahim J, Sattentau Q, Gordon S, James W. Neutralization of infectivity of diverse R5 clinical isolates of human immunodeficiency virus type 1 by gp120-binding 2'F-RNA aptamers. Journal of Virology. 2003;**77**:12692-12698. DOI: 10.1128/jvi.77.23.12692-12698.2003

[29] Dey AK, Khati M, Tang M, Wyatt R, Lea SM, James W. An aptamer that neutralizes R5 strains of human immunodeficiency virus type 1 blocks gp120-CCR5 interaction. Journal of Virology. 2005;**79**:13806-13810. DOI: 10.1128/JVI.79.21.13806-13810.2005

[30] Davydova A, Vorobjeva M, Pyshnyi D, Altman S, Vlassov V, Venyaminova A. Aptamers against

pathogenic microorganisms. Critical Reviews in Microbiology. 2016;**42**:847-865. DOI: 10.3109/1040841x.2015.1070115

[31] Zhang Y, Yu Z, Jiang F, Fu P, Shen J, Wu W, et al. Two DNA aptamers against Avian Influenza H9N2 virus prevent viral infection in cells. PLOS One. 2015;**10**:e0123060. DOI: 10.1371/journal.pone.0123060

[32] Binaymotlagh R, Hajareh Haghighi F, Aboutalebi F, Mirahmadi-Zare SZ, Hadadzadeh H, Nasr-Esfahani MH. Selective chemotherapy and imaging of colorectal and breast cancer cells by a modified MUC-1 aptamer conjugated to a poly(ethylene glycol)-dimethacrylate coated Fe_3O_4–AuNCs nanocomposite. New Journal of Chemistry. 2019;**43**: 238-248. DOI: 10.1039/c8nj04236e

[33] Gold L, Ayers D, Bertino J, Bock C, Bock A, Brody EN, et al. Aptamer-based multiplexed proteomic technology for biomarker discovery. PLOS One. 2010;**5**:e15004. DOI: 10.1371/journal.pone.0015004

[34] Zhou G, Latchoumanin O, Bagdesar M, Hebbard L, Duan W, Liddle C, et al. Aptamer-based therapeutic approaches to target cancer stem cells. Theranostics. 2017;**7**: 3948-3961. DOI: 10.7150/thno.20725

[35] Shinde SB, Fernandes CB, Patravale VB. Recent trends in in-vitro nanodiagnostics for detection of pathogens. Journal of Controlled Release. 2012;**159**:164-180. DOI: 10.1016/j.jconrel.2011.11.033

[36] Kaittanis C, Santra S, Perez JM. Emerging nanotechnology-based strategies for the identification of microbial pathogenesis. Advanced Drug Delivery Reviews. 2010;**62**:408-423. DOI: 10.1016/j.addr.2009.11.013

[37] Mendoza-Gallegos RA, Rios A, Garcia-Cordero JL. An affordable and portable thermocycler for real-time PCR made of 3D-printed parts and off-the-shelf electronics. Analytical Chemistry. 2018;**90**:5563-5568. DOI: 10.1021/acs.analchem.7b04843

[38] van Grootveld R, van Dam GJ, de Dood C, de Vries JJC, Visser LG, Corstjens PLAM, et al. Improved diagnosis of active Schistosoma infection in travellers and migrants using the ultra-sensitive in-house lateral flow test for detection of circulating anodic antigen (CAA) in serum. European Journal of Clinical Microbiology and Infectious Diseases. 2018;**37**:1709-1716. DOI: 10.1007/s10096-018-3303-x

[39] Pan Q, Zhang XL, Wu HY, He PW, Wang F, Zhang MS, et al. Aptamers that preferentially bind type IVB Pili and inhibit human monocytic-cell invasion by *Salmonella enterica* Serovar Typhi. Antimicrobial Agents and Chemotherapy. 2005;**49**:4052-4060. DOI: 10.1128/aac.49.10.4052-4060.2005

[40] Zelada-Guillén GA, Riu J, Düzgün A, Rius FX. Immediate detection of living bacteria at ultralow concentrations using a carbon nanotube based potentiometric aptasensor. Angewandte Chemie International Edition. 2009;**48**:7334-7337. DOI: 10.1002/anie.200902090

[41] Chang YC, Yang CY, Sun RL, Cheng YF, Kao WC, Yang PC. Rapid single cell detection of *Staphylococcus aureus* by aptamer-conjugated gold nanoparticles. Scientific Reports. 2013;**3**:1863-1863. DOI: 10.1038/srep01863

[42] Bruno JG, Phillips T, Carrillo MP, Crowell R. Plastic-adherent DNA aptamer-magnetic bead and quantum dot Sandwich assay for *Campylobacter* detection. Journal of Fluorescence. 2008;**19**:427. DOI: 10.1007/s10895-008-0429-8

[43] Dwivedi HP, Smiley RD, Jaykus LA. Selection and characterization of DNA aptamers with binding selectivity to *Campylobacter jejuni* using whole-cell SELEX. Applied Microbiology and Biotechnology. 2010;**87**:2323-2334. DOI: 10.1007/s00253-010-2728-7

[44] Duan N, Ding X, He L, Wu S, Wei Y, Wang Z. Selection, identification and application of a DNA aptamer against Listeria monocytogenes. Food Control. 2013;**33**:239-243. DOI: 10.1016/j.foodcont.2013.03.011

[45] Ohk SH, Koo OK, Sen T, Yamamoto CM, Bhunia AK. Antibody–aptamer functionalized fibre-optic biosensor for specific detection of Listeria monocytogenes from food. Journal of Applied Microbiology. 2010;**109**:808-817. DOI: 10.1111/j.1365-2672.2010.04709.x

[46] Suh SH, Choi SJ, Dwivedi HP, Moore MD, Escudero-Abarca BI, Jaykus L-A. Use of DNA aptamer for sandwich type detection of Listeria monocytogenes. Analytical Biochemistry. 2018;**557**:27-33. DOI: 10.1016/j.ab.2018.04.009

[47] Suh SH, Dwivedi HP, Choi SJ, Jaykus LA. Selection and characterization of DNA aptamers specific for Listeria species. Analytical Biochemistry. 2014;**459**:39-45. DOI: 10.1016/j.ab.2014.05.006

[48] Suh SH, Jaykus LA. Nucleic acid aptamers for capture and detection of Listeria spp. Journal of Biotechnology. 2013;**167**:454-461. DOI: 10.1016/j.jbiotec.2013.07.027

[49] Duan N, Wu S, Chen X, Huang Y, Wang Z. Selection and identification of a DNA aptamer targeted to *Vibrio parahemolyticus*. Journal of Agricultural and Food Chemistry. 2012;**60**:4034-4038. DOI: 10.1021/jf300395z

[50] Duan N, Ding X, Wu S, Xia Y, Ma X, Wang Z, et al. In vitro selection of a DNA aptamer targeted against *Shigella dysenteriae*. Journal of Microbiological Methods. 2013;**94**:170-174. DOI: 10.1016/j.mimet.2013.06.016

[51] Hamula CLA, Le XC, Li XF. DNA aptamers binding to multiple prevalent M-types of *Streptococcus pyogenes*. Analytical Chemistry. 2011;**83**: 3640-3647. DOI: 10.1021/ac200575e

[52] Vivekananda J, Kiel JL. Anti-Francisella tularensis DNA aptamers detect tularemia antigen from different subspecies by aptamer-linked immobilized sorbent assay. Laboratory Investigation. 2006;**86**:610. DOI: 10.1038/labinvest.3700417

[53] Wang KY, Zeng YL, Yang XY, Li WB, Lan XP. Utility of aptamer-fluorescence in situ hybridization for rapid detection of *Pseudomonas aeruginosa*. European Journal of Clinical Microbiology and Infectious Diseases. 2011;**30**:273-278. DOI: 10.1007/s10096-010-1074-0

[54] Bruno JG, Carrillo MP. Development of aptamer beacons for rapid presumptive detection of bacillus spores. Journal of Fluorescence. 2012;**22**:915-924. DOI: 10.1007/s10895-011-1030-0

[55] Bruno JG, Kiel JL. In vitro selection of DNA aptamers to anthrax spores with electrochemiluminescence detection. Biosensors and Bioelectronics. 1999;**14**:457-464

[56] Ospina-Villa JD, Lopez-Camarillo C, Castanon-Sanchez CA, Soto-Sanchez J, Ramirez-Moreno E, Marchat LA. Advances on aptamers against protozoan parasites. Genes (Basel). 2018;**9**:584. DOI: 10.3390/genes9120584

[57] Tang XL, Hua Y, Guan Q, Yuan CH. Improved detection of deeply invasive candidiasis with DNA aptamers specific

binding to (1 → 3)--D-glucans from *Candida albicans*. European Journal of Clinical Microbiology and Infectious Diseases. 2016;**35**:587-595. DOI: 10.1007/s10096-015-2574-8

[58] Kensch O, Connolly BA, Steinhoff HJ, McGregor A, Goody RS, Restle T. HIV-1 reverse transcriptase-pseudoknot RNA aptamer interaction has a binding affinity in the low picomolar range coupled with high specificity. The Journal of Biological Chemistry. 2000;**275**:18271-18278. DOI: 10.1074/jbc.M001309200

[59] de Soultrait VR, Lozach PY, Altmeyer R, Tarrago-Litvak L, Litvak S, Andreola ML. DNA aptamers derived from HIV-1 RNase H inhibitors are strong anti-integrase agents. Journal of Molecular Biology. 2002;**324**:195-203

[60] Kim SJ, Kim MY, Lee JH, You JC, Jeong S. Selection and stabilization of the RNA aptamers against the human immunodeficiency virus type-1 nucleocapsid protein. Biochemical and Biophysical Research Communications. 2002;**291**:925-931. DOI: 10.1006/bbrc.2002.6521

[61] Dey AK, Griffiths C, Lea SM, James W. Structural characterization of an anti-gp120 RNA aptamer that neutralizes R5 strains of HIV-1. RNA. 2005;**11**:873-884. DOI: 10.1261/rna.7205405

[62] Urvil PT, Kakiuchi N, Zhou DM, Shimotohno K, Kumar PK, Nishikawa S. Selection of RNA aptamers that bind specifically to the NS3 protease of hepatitis C virus. European Journal of Biochemistry. 1997;**248**:130-138

[63] Fukuda K, Vishnuvardhan D, Sekiya S, Hwang J, Kakiuchi N, Taira K, et al. Isolation and characterization of RNA aptamers specific for the hepatitis C virus nonstructural protein 3 protease. European Journal of Biochemistry. 2000;**267**:3685-3694

[64] Chen F, Hu Y, Li D, Chen H, Zhang XL. CS-SELEX generates high-affinity ssDNA aptamers as molecular probes for hepatitis C virus envelope glycoprotein E2. PLOS One. 2009;**4**:e8142. DOI: 10.1371/journal.pone.0008142

[65] Gopinath SC, Misono TS, Kawasaki K, Mizuno T, Imai M, Odagiri T, et al. An RNA aptamer that distinguishes between closely related human influenza viruses and inhibits haemagglutinin-mediated membrane fusion. The Journal of General Virology. 2006;**87**:479-487. DOI: 10.1099/vir.0.81508-0

[66] Park SY, Kim S, Yoon H, Kim KB, Kalme SS, Oh S, et al. Selection of an antiviral RNA aptamer against hemagglutinin of the subtype H5 avian influenza virus. Nucleic Acid Therapeutics. 2011;**21**:395-402. DOI: 10.1089/nat.2011.0321

[67] Bai C, Lu Z, Jiang H, Yang Z, Liu X, Ding H, et al. Aptamer selection and application in multivalent binding-based electrical impedance detection of inactivated H1N1 virus. Biosensors and Bioelectronics. 2018;**110**:162-167. DOI: 10.1016/j.bios.2018.03.047

[68] Gopinath SCB, Hayashi K, Kumar PKR. Aptamer that binds to the gD protein of herpes simplex virus 1 and efficiently inhibits viral entry. Journal of Virology. 2012;**86**:6732-6744. DOI: 10.1128/JVI.00377-12

[69] Chen HL, Hsiao WH, Lee HC, Wu SC, Cheng JW. Selection and characterization of DNA aptamers targeting all four serotypes of dengue viruses. PLOS One. 2015;**10**:e0131240. DOI: 10.1371/journal.pone.0131240

[70] Kim DTH, Bao DT, Park H, Ngoc NM, Yeo SJ. Development

of a novel peptide aptamer-based immunoassay to detect Zika virus in serum and urine. Theranostics. 2018;**8**:3629-3642. DOI: 10.7150/thno.25955

[71] Shubham S, Hoinka J, Banerjee S, Swanson E, Dillard JA, Lennemann NJ, et al. A 2′FY-RNA Motif defines an aptamer for Ebolavirus secreted protein. Scientific Reports. 2018;**8**:12373. DOI: 10.1038/s41598-018-30590-8

[72] Saraf N, Villegas M, Willenberg BJ, Seal S. Multiplex viral detection platform based on a aptamers-integrated microfluidic channel. ACS Omega. 2019;**4**:2234-2240. DOI: 10.1021/acsomega.8b03277

[73] Zelada-Guillén GA, Tweed-Kent A, Niemann M, Göringer HU, Riu J, Rius FX. Ultrasensitive and real-time detection of proteins in blood using a potentiometric carbon-nanotube aptasensor. Biosensors and Bioelectronics. 2013;**41**:366-371. DOI: 10.1016/j.bios.2012.08.055

[74] Li L, Yuan Y, Chen Y, Zhang P, Bai Y, Bai L. Aptamer based voltammetric biosensor for *Mycobacterium tuberculosis* antigen ESAT-6 using a nanohybrid material composed of reduced graphene oxide and a metal-organic framework. Microchimica Acta. 2018;**185**:379. DOI: 10.1007/s00604-018-2884-5

[75] Daniel C, Roupioz Y, Gasparutto D, Livache T, Buhot A. Solution-phase vs surface-phase aptamer-protein affinity from a label-free kinetic biosensor. PLOS One. 2013;**8**:e75419. DOI: 10.1371/journal.pone.0075419

[76] Gosai A, Hau Yeah BS, Nilsen-Hamilton M, Shrotriya P. Label free thrombin detection in presence of high concentration of albumin using an aptamer-functionalized nanoporous membrane. Biosensors and Bioelectronics. 2019;**126**:88-95. DOI: 10.1016/j.bios.2018.10.010

[77] Zhai L, Wang T, Kang K, Zhao Y, Shrotriya P, Nilsen-Hamilton M. An RNA aptamer-based microcantilever sensor to detect the inflammatory marker, mouse lipocalin-2. Analytical Chemistry. 2012;**84**:8763-8770. DOI: 10.1021/ac3020643

[78] Kang KH, Sachan A, Nilsen-Hamilton M, Shrotriya P. Aptamer functionalized microcantilever sensors for cocaine detection. Langmuir. 2011;**27**:14696-14702. DOI: 10.1021/la202067y

[79] Dupuis NF, Holmstrom ED, Nesbitt DJ. Molecular-crowding effects on single-molecule RNA folding/unfolding thermodynamics and kinetics. Proceedings of the National Academy of Sciences of the United States of America. 2014;**111**:8464-8469. DOI: 10.1073/pnas.1316039111 1316039111 [pii]

[80] Paudel BP, Rueda D. Molecular crowding accelerates ribozyme docking and catalysis. Journal of the American Chemical Society. 2014;**136**:16700-16703. DOI: 10.1021/ja5073146

[81] Nakano SI, Sugimoto N. The structural stability and catalytic activity of DNA and RNA oligonucleotides in the presence of organic solvents. Biophysical Reviews. 2016;**8**:11-23. DOI: 10.1007/s12551-015-0188-0

[82] Sun K, Xia N, Zhao L, Liu K, Hou W, Liu L. Aptasensors for the selective detection of alpha-synuclein oligomer by colorimetry, surface plasmon resonance and electrochemical impedance spectroscopy. Sensors and Actuators B: Chemical. 2017;**245**:87-94. DOI: 10.1016/j.snb.2017.01.171

[83] Hansen JA, Wang J, Kawde AN, Xiang Y, Gothelf KV, Collins G. Quantum-dot/aptamer-based ultrasensitive multi-analyte electrochemical biosensor. Journal of the American Chemical Society.

2006;**128**:2228-2229. DOI: 10.1021/ja060005h

[84] Nagarkatti R, Bist V, Sun S, Fortes de Araujo F, Nakhasi HL, Debrabant A. Development of an aptamer-based concentration method for the detection of *Trypanosoma cruzi* in blood. PLOS One. 2012;**7**:e43533. DOI: 10.1371/journal.pone.0043533

[85] Kaur H, Bruno JG, Kumar A, Sharma TK. Aptamers in the therapeutics and diagnostics pipelines. Theranostics. 2018;**8**:4016-4032. DOI: 10.7150/thno.25958

[86] Song MS, Sekhon S, Shin WR, Kim H, Min J, Ahn JY, et al. Detecting and discriminating Shigella sonnei using an aptamer-based fluorescent biosensor platform. Molecules. 2017;**22**:825

[87] Song S, Wang L, Li J, Fan C, Zhao J. Aptamer-based biosensors. TrAC Trends in Analytical Chemistry. 2008;**27**:108-117. DOI: 10.1016/j.trac.2007.12.004

[88] Wang RE, Zhang Y, Cai J, Cai W, Gao T. Aptamer-based fluorescent biosensors. Current Medicinal Chemistry. 2011;**18**:4175-4184

[89] Swensen JS, Xiao Y, Ferguson BS, Lubin AA, Lai RY, Heeger AJ, et al. Continuous, real-time monitoring of cocaine in undiluted blood serum via a microfluidic, electrochemical aptamer-based sensor. Journal of the American Chemical Society. 2009;**131**:4262-4266. DOI: 10.1021/ja806531z

[90] Yoon H, Kim JH, Lee N, Kim BG, Jang J. A novel sensor platform based on aptamer-conjugated polypyrrole nanotubes for label-free electrochemical protein detection. Chembiochem. 2008;**9**:634-641. DOI: 10.1002/cbic.200700660

[91] Polonschii C, David S, Tombelli S, Mascini M, Gheorghiu M. A novel low-cost and easy to develop functionalization platform. Case study: Aptamer-based detection of thrombin by surface plasmon resonance. Talanta. 2010;**80**:2157-2164. DOI: 10.1016/j.talanta.2009.11.023

[92] Zhang X, Yadavalli VK. Surface immobilization of DNA aptamers for biosensing and protein interaction analysis. Biosensors and Bioelectronics. 2011;**26**:3142-3147. DOI: 10.1016/j.bios.2010.12.012

[93] Treitz G, Gronewold TMA, Quandt E, Zabe-Kühn M. Combination of a SAW-biosensor with MALDI mass spectrometric analysis. Biosensors and Bioelectronics. 2008;**23**:1496-1502. DOI: 10.1016/j.bios.2008.01.013

[94] Chen X, Pan Y, Liu H, Bai X, Wang N, Zhang B. Label-free detection of liver cancer cells by aptamer-based microcantilever biosensor. Biosensors and Bioelectronics. 2016;**79**:353-358. DOI: 10.1016/j.bios.2015.12.060

[95] Kumar YVVA, Renuka RM, Achuth J, Mudili V, Poda S. Development of a FRET-based fluorescence aptasensor for the detection of aflatoxin B1 in contaminated food grain samples. RSC Advances. 2018;**8**:10465-10473. DOI: 10.1039/c8ra00317c

[96] Emrani AS, Danesh NM, Lavaee P, Ramezani M, Abnous K, Taghdisi SM. Colorimetric and fluorescence quenching aptasensors for detection of streptomycin in blood serum and milk based on double-stranded DNA and gold nanoparticles. Food Chemistry. 2016;**190**:115-121. DOI: 10.1016/j.foodchem.2015.05.079

[97] Jin B, Wang S, Lin M, Jin Y, Zhang S, Cui X, et al. Upconversion nanoparticles based FRET aptasensor

for rapid and ultrasenstive bacteria detection. Biosensors and Bioelectronics. 2017;**90**:525-533. DOI: 10.1016/j.bios.2016.10.029

[98] Nasirian V, Chabok A, Barati A, Rafienia M, Arabi MS, Shamsipur M. Ultrasensitive aflatoxin B1 assay based on FRET from aptamer labelled fluorescent polymer dots to silver nanoparticles labeled with complementary DNA. Microchimica Acta. 2017;**184**:4655-4662. DOI: 10.1007/s00604-017-2508-5

[99] Lu X, Wang C, Qian J, Ren C, An K, Wang K. Target-driven switch-on fluorescence aptasensor for trace aflatoxin B1 determination based on highly fluorescent ternary CdZnTe quantum dots. Analytica Chimica Acta. 2019;**1047**:163-171. DOI: 10.1016/j.aca.2018.10.002

[100] Khan IM, Zhao S, Niazi S, Mohsin A, Shoaib M, Duan N, et al. Silver nanoclusters based FRET aptasensor for sensitive and selective fluorescent detection of T-2 toxin. Sensors and Actuators B: Chemical. 2018;**277**:328-335. DOI: 10.1016/j.snb.2018.09.021

[101] Bruno JG, Carrillo MP, Phillips T, Hanson D, Bohmann JA. DNA aptamer beacon assay for C-telopeptide and handheld fluorometer to monitor bone resorption. Journal of Fluorescence. 2011;**44**:2021. DOI: 10.1007/s10895-011-0903-6

[102] Xiao Y, Lubin AA, Heeger AJ, Plaxco KW. Label-free electronic detection of thrombin in blood serum by using an aptamer-based sensor. Angewandte Chemie (International Ed. in English). 2005;**44**:5456-5459. DOI: 10.1002/anie.200500989

[103] Radi AE, Acero Sánchez JL, Baldrich E, O'Sullivan CK. Reagentless, reusable, ultrasensitive electrochemical molecular beacon aptasensor. Journal of the American Chemical Society. 2006;**128**:117-124. DOI: 10.1021/ja053121d

[104] Bai L, Chen Y, Bai Y, Chen Y, Zhou J, Huang A. Fullerene-doped polyaniline as new redox nanoprobe and catalyst in electrochemical aptasensor for ultrasensitive detection of *Mycobacterium tuberculosis* MPT64 antigen in human serum. Biomaterials. 2017;**133**:11-19. DOI: 10.1016/j.biomaterials.2017.04.010

[105] Kumari P, Lavania S, Tyagi S, Dhiman A, Rath D, Anthwal D, et al. A novel aptamer-based test for the rapid and accurate diagnosis of pleural tuberculosis. Analytical Biochemistry. 2019;**564-565**:80-87. DOI: 10.1016/j.ab.2018.10.019

[106] Chang BY, Park SM. Electrochemical impedance spectroscopy. Annual Review of Analytical Chemistry. 2010;**3**:207-229. DOI: 10.1146/annurev.anchem.012809.102211

[107] Figueroa-Miranda G, Feng L, Shiu SCC, Dirkzwager RM, Cheung YW, Tanner JA, et al. Aptamer-based electrochemical biosensor for highly sensitive and selective malaria detection with adjustable dynamic response range and reusability. Sensors and Actuators B: Chemical. 2018;**255**:235-243. DOI: 10.1016/j.snb.2017.07.117

[108] Cheung YW, Kwok J, Law AWL, Watt RM, Kotaka M, Tanner JA. Structural basis for discriminatory recognition of *Plasmodium* lactate dehydrogenase by a DNA aptamer. Proceedings of the National Academy of Sciences. 2013;**110**:15967. DOI: 10.1073/pnas.1309538110

[109] Arroyo-Currás N, Scida K, Ploense KL, Kippin TE, Plaxco KW. High surface area electrodes generated via electrochemical roughening improve the signaling of electrochemical aptamer-based biosensors. Analytical

Chemistry. 2017;**89**:12185-12191. DOI: 10.1021/acs.analchem.7b02830

[110] Barros WRP, Wei Q, Zhang G, Sun S, Lanza MRV, Tavares AC. Oxygen reduction to hydrogen peroxide on Fe_3O_4 nanoparticles supported on printex carbon and graphene. Electrochimica Acta. 2015;**162**:263-270. DOI: 10.1016/j.electacta.2015.02.175

[111] Cheon HJ, Lee SM, Kim SR, Shin HY, Seo YH, Cho YK, et al. Colorimetric detection of MPT64 antibody based on an aptamer adsorbed magnetic nanoparticles for diagnosis of tuberculosis. Journal of Nanoscience and Nanotechnology. 2019;**19**:622-626. DOI: 10.1166/jnn.2019.15905

[112] Wang CH, Wu JJ, Lee GB. Screening of highly-specific aptamers and their applications in paper-based microfluidic chips for rapid diagnosis of multiple bacteria. Sensors and Actuators B: Chemical. 2019;**284**: 395-402. DOI: 10.1016/j.snb.2018.12.112

[113] Lee BH, Kim SH, Ko Y, Park JC, Ji S, Gu MB. The sensitive detection of ODAM by using sandwich-type biosensors with a cognate pair of aptamers for the early diagnosis of periodontal disease. Biosensors and Bioelectronics. 2019;**126**:122-128. DOI: 10.1016/j.bios.2018.10.040

[114] Zhang Z, Oni O, Liu J. New insights into a classic aptamer: Binding sites, cooperativity and more sensitive adenosine detection. Nucleic Acids Research. 2017;**45**:7593-7601. DOI: 10.1093/nar/gkx517

[115] Baird GS. Where are all the aptamers? American Journal of Clinical Pathology. 2010;**134**:529-531. DOI: 10.1309/AJCPFU4CG2WGJJKS

[116] Jiang Y, Ma W, Ji W, Wei H, Mao L. Aptamer superstructure-based electrochemical biosensor for sensitive detection of ATP in rat brain with in vivo microdialysis. The Analyst. 2019. DOI: 10.1039/c8an02077a

[117] Lakhin AV, Tarantul VZ, Gening LV. Aptamers: Problems. Solutions and Prospects. Acta Naturae. 2013;**5**:34-43

[118] Liu J, Cao Z, Lu Y. Functional nucleic acid sensors. Chemical Reviews. 2009;**109**:1948-1998. DOI: 10.1021/cr030183i

[119] Díaz-Fernández A, Miranda-Castro R, de-los-Santos-Álvarez N, Rodríguez EF, Lobo-Castañón MJ. Focusing aptamer selection on the glycan structure of prostate-specific antigen: Toward more specific detection of prostate cancer. Biosensors and Bioelectronics. 2019;**128**:83-90. DOI: 10.1016/j.bios.2018.12.040

[120] Kaur H. Recent developments in cell-SELEX technology for aptamer selection. Biochimica et Biophysica Acta (BBA)—General Subjects. 2018;**1862**:2323-2329. DOI: 10.1016/j.bbagen.2018.07.029

[121] Chen M, Yu Y, Jiang F, Zhou J, Li Y, Liang C, et al. Development of cell-SELEX technology and its application in cancer diagnosis and therapy. International Journal of Molecular Sciences. 2016;**17**. DOI: 10.3390/ijms17122079

[122] Baaske P, Wienken CJ, Reineck P, Duhr S, Braun D. Optical thermophoresis for quantifying the buffer dependence of aptamer binding. Angewandte Chemie International Edition. 2010;**49**:2238-2241. DOI: 10.1002/anie.200903998

[123] Sachan A, Ilgu M, Kempema A, Kraus GA, Nilsen-Hamilton M. Specificity and ligand affinities of the cocaine aptamer: Impact of structural features and physiological NaCl. Analytical Chemistry.

2016;**88**:7715-7723. DOI: 10.1021/acs.analchem.6b01633

[124] Babalik A, Ulus IH, Bakirci N, Kuyucu T, Arpag H, Dagyildiz L, et al. Pharmacokinetics and serum concentrations of antimycobacterial drugs in adult Turkish patients. The International Journal of Tuberculosis and Lung Disease. 2013;**17**:1442-1447. DOI: 10.5588/ijtld.12.0771

[125] Casu GS, Hites M, Jacobs F, Cotton F, Wolff F, Beumier M, et al. Can changes in renal function predict variations in beta-lactam concentrations in septic patients? International Journal of Antimicrobial Agents. 2013;**42**:422-428. DOI: 10.1016/j.ijantimicag.2013.06.021

[126] Varatharajan S, Panetta JC, Abraham A, Karathedath S, Mohanan E, Lakshmi KM, et al. Population pharmacokinetics of Daunorubicin in adult patients with acute myeloid leukemia. Cancer Chemotherapy and Pharmacology. 2016;**78**:1051-1058. DOI: 10.1007/s00280-016-3166-8

[127] Andriguetti NB, Raymundo S, Antunes MV, Perassolo MS, Verza SG, Suyenaga ES, et al. Pharmacogenetic and pharmacokinetic dose individualization of the taxane chemotherapeutic drugs paclitaxel and docetaxel. Current Medicinal Chemistry. 2017;**24**:3559-3582. DOI: 10.2174/0929867324666170623093445

[128] Motohashi S, Mino Y, Hori K, Naito T, Hosokawa S, Furuse H, et al. Interindividual variations in aprepitant plasma pharmacokinetics in cancer patients receiving cisplatin-based chemotherapy for the first time. Biological and Pharmaceutical Bulletin. 2013;**36**:676-681

[129] Contin M, Alberghini L, Candela C, Benini G, Riva R. Intrapatient variation in antiepileptic drug plasma concentration after generic substitution vs stable brand-name drug regimens. Epilepsy Research. 2016;**122**:79-83. DOI: 10.1016/j.eplepsyres.2016.02.012

[130] Pisupati J, Jain A, Burckart G, Hamad I, Zuckerman S, Fung J, et al. Intraindividual and interindividual variations in the pharmacokinetics of mycophenolic acid in liver transplant patients. Journal of Clinical Pharmacology. 2005;**45**:34-41. DOI: 10.1177/0091270004270145

[131] Albrecht D, Turakhia MP, Ries D, Marbury T, Smith W, Dillon D, et al. Pharmacokinetics of tecarfarin and warfarin in patients with severe chronic kidney disease. Thrombosis and Haemostasis. 2017;**117**:2026-2033. DOI: 10.1160/TH16-10-0815

[132] Deitchman AN, Derendorf H. Measuring drug distribution in the critically ill patient. Advanced Drug Delivery Reviews. 2014;**77**:22-26. DOI: 10.1016/j.addr.2014.08.014

[133] Pena MA, Horga JF, Zapater P. Variations of pharmacokinetics of drugs in patients with cirrhosis. Expert Review of Clinical Pharmacology. 2016;**9**:441-458. DOI: 10.1586/17512433.2016.1135733

[134] Salem F, Abduljalil K, Kamiyama Y, Rostami-Hodjegan A. Considering age variation when coining drugs as high versus low hepatic extraction ratio. Drug Metabolism and Disposition. 2016;**44**:1099-1102. DOI: 10.1124/dmd.115.067595

[135] Nieuweboer AJ, Smid M, de Graan AM, Elbouazzaoui S, de Bruijn P, Eskens FA, et al. Role of genetic variation in docetaxel-induced neutropenia and pharmacokinetics. The Pharmacogenomics Journal. 2016;**16**:519-524. DOI: 10.1038/tpj.2015.66

[136] Xie XC, Li J, Wang HY, Li HL, Liu J, Fu Q, et al. Associations of UDP-glucuronosyltransferases

polymorphisms with mycophenolate mofetil pharmacokinetics in Chinese renal transplant patients. Acta Pharmacologica Sinica. 2015;**36**: 644-650. DOI: 10.1038/aps.2015.7

[137] Jiang F, Kim HD, Na HS, Lee SY, Seo DW, Choi JY, et al. The influences of CYP2D6 genotypes and drug interactions on the pharmacokinetics of venlafaxine: Exploring predictive biomarkers for treatment outcomes. Psychopharmacology (Berlin). 2015;**232**:1899-1909. DOI: 10.1007/s00213-014-3825-6

[138] Bertholee D, Maring JG, van Kuilenburg ABP. Genotypes affecting the pharmacokinetics of anticancer drugs. Clinical Pharmacokinetics. 2017;**56**:317-337. DOI: 10.1007/s40262-016-0450-z

[139] Elewa H, Wilby KJ. A review of pharmacogenetics of antimalarials and associated clinical implications. European Journal of Drug Metabolism and Pharmacokinetics. 2017;**42**:745-756. DOI: 10.1007/s13318-016-0399-1

[140] Wong AL, Seng KY, Ong EM, Wang LZ, Oscar H, Cordero MT, et al. Body fat composition impacts the hematologic toxicities and pharmacokinetics of doxorubicin in Asian breast cancer patients. Breast Cancer Research and Treatment. 2014;**144**:143-152. DOI: 10.1007/s10549-014-2843-8

[141] Janukonyte J, Parkner T, Lauritzen T, Christiansen JS, Laursen T. Circadian variation in the pharmacokinetics of steady state continuous subcutaneous infusion of growth hormone in adult growth hormone deficient patients. Growth Hormone and IGF Research. 2013;**23**:256-260. DOI: 10.1016/j.ghir.2013.09.002

[142] Muzzarelli S, Brunner-La Rocca H, Pfister O, Foglia P, Moschovitis G, Mombelli G, et al. Adherence to the medical regime in patients with heart failure. European Journal of Heart Failure. 2010;**12**:389-396. DOI: 10.1093/eurjhf/hfq015

[143] Suttorp M, Bornhauser M, Metzler M, Millot F, Schleyer E. Pharmacology and pharmacokinetics of imatinib in pediatric patients. Expert Review of Clinical Pharmacology. 2018;**11**:219-231. DOI: 10.1080/17512433.2018.1398644

[144] Bar-Dayan Y, Shotashvily T, Boaz M, Wainstein J. Error in drugs consumption among older patients. American Journal of Therapeutics. 2017;**24**:e701-e705. DOI: 10.1097/MJT.0000000000000400

[145] Dearling JLJ, Packard AB. Molecular imaging in nanomedicine—A developmental tool and a clinical necessity. Journal of Controlled Release. 2017;**261**:23-30. DOI: 10.1016/j.jconrel.2017.06.011

Chapter 5

Systematic Deletion of Type III Secretion System Effectors in Enteropathogenic *E. coli* Unveils the Role of Non-LEE Effectors in A/E Lesion Formation

Massiel Cepeda-Molero, Stephanie Schüller, Gad Frankel and Luis Ángel Fernández

Abstract

Enteropathogenic *E. coli* (EPEC) is a diarrheagenic human pathogen. The hallmark of EPEC infection is the formation of the attaching and effacing (A/E) lesion in the intestinal epithelial cells, characterized by the effacement of brush border microvilli and the intimate bacterial attachment to the enterocyte in actin-rich pedestal-like structures. The locus of enterocyte effacement (LEE) in the EPEC genome encodes a type III protein secretion system (T3SS) that translocates multiple effector proteins into the host cell to subvert cellular functions for the benefit of the pathogen. These effectors are encoded both within and outside the LEE. In vitro cell culture infections have shown that LEE effectors are required for intimate bacterial attachment to the epithelial cells, whereas non-LEE effectors mostly play a role in modulating inflammation and cell apoptosis in the gut epithelium. We constructed a set of EPEC mutant strains harboring deletions in the complete repertoire of genes encoding T3SS effectors. Infection of human intestinal in vitro organ cultures (IVOC) with these mutant strains surprisingly revealed that non-LEE effectors are also needed to induce efficient A/E lesion formation in the intestinal mucosal tissue.

Keywords: A/E lesion, EPEC, effectors, infection, IVOC, T3SS

1. Introduction

1.1 Enteropathogenic *E. coli*

Enteropathogenic *E. coli* (EPEC) was the first pathotype of *E. coli* to be associated with human disease and is a major cause of acute and chronic diarrhea in infants [1, 2]. The low microbial density of the small bowel caused by the forceful peristalsis in this part of the intestine is overcome by EPEC, which can successfully colonize the small intestine of humans [3, 4]. EPEC primarily affects children younger than 2 years old; however some outbreaks of EPEC infection in healthy adults have been associated with large inoculum ingestion [5]. The mechanism of transmission of EPEC is the fecal-oral route. In the 1940s and 1950s, EPEC was an important cause of diarrhea in developed countries with a mortality of 50% during outbreaks, but nowadays the infection by EPEC in industrial countries has a limited importance. In contrast, in low-income countries, EPEC is still an major cause of infant diarrhea [5, 6].

1.2 Hallmark of EPEC gastrointestinal infection

The phenotype that defines EPEC infection is the attaching and effacing (A/E) lesion [2, 7]. By adhering to intestinal epithelial cells, EPEC subverts cytoskeletal processes of the host cell and produces the histopathological feature of the A/E lesion. This lesion, which was first described in 1980 [8], is characterized by the intimate attachment of the bacteria to the intestinal epithelial cells and elongation and effacement of the brush border microvilli. Later on it was shown that infection is also associated with cytoskeletal rearrangements, including the accumulation of polymerized F-actin in pedestal-like structures underneath the attached bacteria [9] (**Figure 1**). EPEC together with enterohemorrhagic *E. coli* (EHEC) and *Citrobacter rodentium* (CR) is a member of the A/E family of bacterial pathogens that colonize the gastrointestinal tract via the A/E lesion. EPEC and EHEC are important human pathogens, while CR is a mouse-restricted pathogen [10–13].

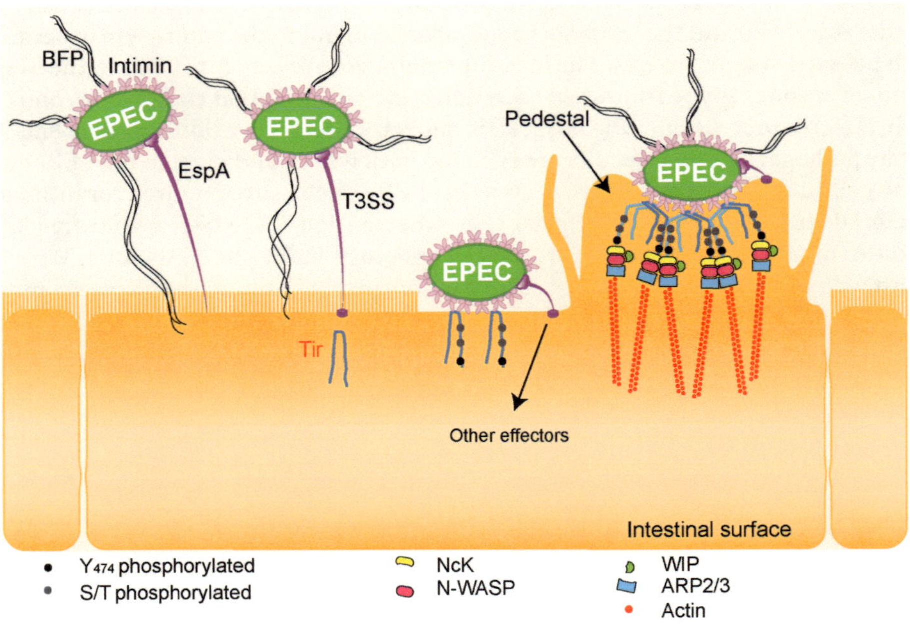

Figure 1.
Localized adherence (LA), intimate attachment, and EPEC A/E lesion formation in the intestinal epithelial surface. At an early stage, EPEC interacts in a non-intimate manner with the intestinal surface mainly through the BFP and EspA filament. After assembly of the translocation pore, EPEC injects translocated intimin receptor (Tir). Ser/Thr phosphorylation of Tir induces its anchoring in the enterocyte plasma membrane, leaving the TirM region exposed for the interaction with intimin. Subsequent Tir-intimin interaction triggers actin polymerization and pedestal formation underneath the attached bacterium. Tir phosphorylation of residue Y474 engages the host adaptor NcK, which later recruits N-WASP and WIP. N-WASP recruits the ARP2/3 complex, which induces actin nucleation and polymerization.

2. EPEC virulence factors

2.1 A pathogenicity island called LEE

The ability of EPEC to induce A/E lesions is related to a pathogenicity island (PAI) of 35 kb called the locus of enterocyte effacement (LEE) [14]. The LEE comprises 41 genes organized in 5 principal operons (LEE1-LEE5) and several smaller transcriptional units (**Figure 2**) [15, 16]. Orthologues of LEE are also found in other members of A/E pathogens [11]. The LEE encodes all the structural proteins necessary for the assembly of a filamentous type III secretion system (T3SS) injectisome on the bacterial cell envelope [17, 18]. The LEE also encodes transcriptional regulators (Ler, GrlR, and GrlA), translocator proteins (EspA, EspB, and EspD), six secreted effector proteins (including the translocated intimin receptor), the outer membrane protein intimin, molecular chaperones, and a lytic transglycosylase (EtgA) [19]. The mechanism of LEE regulation is complex and depends on environmental conditions, quorum sensing (QS), and several transcriptional regulators encoded within and outside the LEE [20, 21].

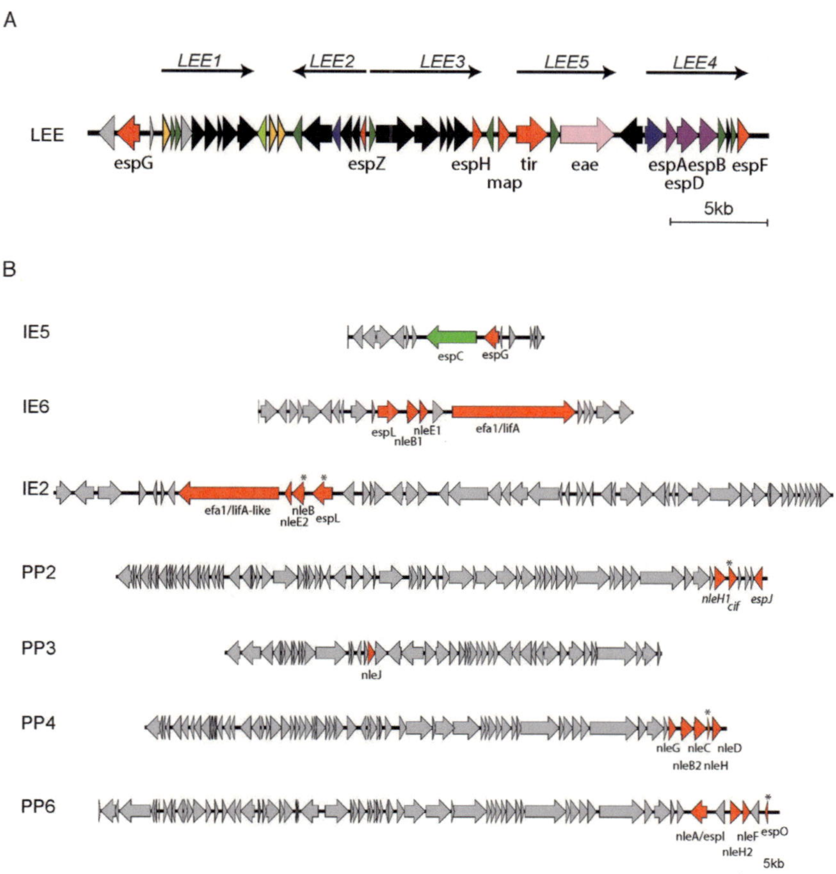

Figure 2.
Effectors of EPEC E2348/69. (A) Representation of the LEE island and effector genes espG, espZ, espH, map, tir, and espF. (B) Non-LEE effectors located outside the LEE are localized in integrative elements (IEs) and prophages (PPs). Effector genes are labeled in red. Pseudogenes are specified with asterisk. Scale of 5 kb is indicated at the bottom. Figure from [32].

2.2 The type III secretion system

The type III secretion system is a macromolecular transport apparatus that is used by many gram-negative bacterial pathogens (e.g., *Shigella*, *Yersinia*, *Salmonella*) to translocate virulence proteins, called effectors, into the cytosol of infected cells, thereby subverting host cellular functions for the benefit of the pathogen [22]. Since pathogens use this transport apparatus to inject proteins into the host cells, this structure is also known as the injectisome. The EPEC T3SS mediates the translocation of multiple effector proteins during infection. Some of them are encoded in the LEE, whereas others are encoded outside of the LEE being generally referred to as non-LEE effectors (Nle) [23, 24]. EspA filaments link the tip of the injectisome in the bacterial cell wall to a 3–5 nm translocation pore, formed in the plasma membrane of infected cells by the translocator proteins EspB and EspD (**Figure 1**) [25, 26].

2.3 Bundle-forming pilus (BFP)

Typical EPEC is endowed with a plasmid called pMAR2 which contains a 14-gene operon encoding the type IV pilus BFP [27, 28]. The BFP is a rope-like bundle, which allows EPEC to form microcolonies in a pattern called localized adherence and also mediates the initial interaction of bacteria with host cell surfaces (**Figure 1**) [29–31].

3. EPEC pathogenesis

EPEC tightly regulates its virulence genes in response to environmental conditions such as temperature [16], the increase of the pH of the small intestine [33, 34], and some hormones which are released during stress conditions [20]. Upon EPEC interaction with enterocytes, EspB and EspD proteins are inserted into the host cell membrane and assemble to form a translocation pore [25, 26]. EPEC then injects its own receptor called Tir, which is integrated into the plasma membrane in a hairpin loop topology, with the loop facing the outside of the cell where it serves as a receptor for the bacterial adhesin intimin [35–37]. Tir-intimin interaction induces clustering and dimerization of Tir, and this activates a signal cascade that starts with the phosphorylation of serine/threonine residues and leads to actin polymerization and pedestal formation underneath the attached bacterium [10, 38]. The most critical event for actin polymerization is the phosphorylation of the cytoplasmic Tir residue Y474 [39]. This induces a signal cascade which recruits the host cell adaptor Nck and N-WASP required to engage and activate the actin-nucleating ARP2/3 complex, which produces the actin nucleation and polymerization. Actin polymerization drives membrane protrusion and pedestal formation [10, 40] (**Figure 1**). Through the T3SS injectisome, EPEC translocates LEE-encoded effector proteins and additional effectors localized in mobile genetic element outside the LEE (Nle).

4. LEE effectors

Six effector proteins (EspG, EspZ, EspH, Map, Tir, and EspF) are encoded in the LEE island (**Figure 2**). Most of these, except EspZ, have important functions destabilizing the physiology of the intestinal epithelium, triggering cytoskeleton reorganization, inducing cytotoxicity and electrolyte imbalance which lead to

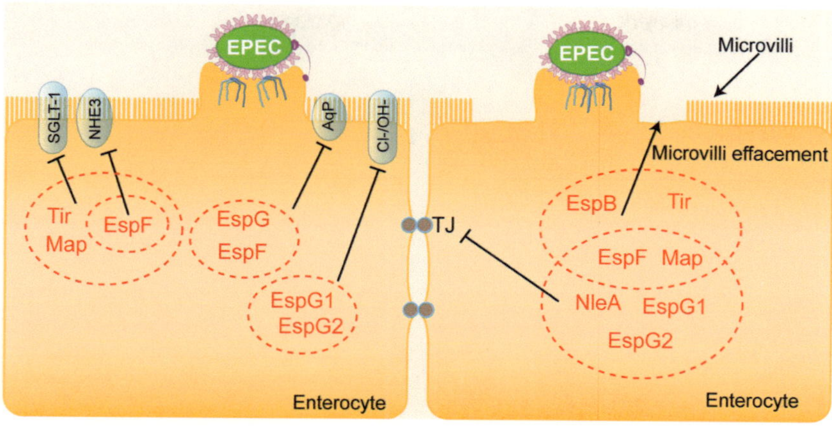

Figure 3.
EPEC effector proteins altering epithelial cell function and inducing water loss and diarrhea. Tir, map, and EspF inhibit the sodium-D-glucose transporter. EspF reduces expression of the sodium-hydrogen exchanger NHE3. EspG and EspF induce mislocalization of aquaporins (AqP). EspG1/EspG2 alters membrane targeting of the Cl-/OH-exchanger. EspF, map, NleA, EspG1, and EspG2 disrupt tight junction complexes (TJ). EspB, Tir, EspF, and map induce microvilli effacement.

diarrhea [11, 41]. The rapid onset of EPEC-induced diarrhea is likely induced by the cooperative action of Tir, Map, and EspF, which inhibits the sodium-D-glucose transporter (SGLT-1), the major water pump of the small intestine responsible for about 70% of the total fluid uptake [42]. In addition, Map and EspF reduce Na + absorption by the sodium-hydrogen exchanger (NHE3) [43], and EspG1/2 proteins alter the membrane targeting of the Cl-/OH-exchanger (DRA), resulting in reduced Cl-uptake. These processes result in the accumulation of salts in the gut lumen, which drives water loss from the mucosa [44].

Inhibition of endosomal trafficking by EspG1/2 reduces the level of cell surface receptors [45]. In addition, EspF and EspG induce mislocalization of aquaporins (AQP), thereby reducing epithelial water absorption [46]. Furthermore, EspB, Tir, EspF, and Map induce microvillus effacement, and this reduction of absorptive surface likely exacerbates EPEC diarrhea [47]. While EspF and Map synergistically disrupt TJs [48], EspG1/2 induces microtubule disruption contributing to TJ disruption [49]. The effector protein NleA also disrupts TJs by blocking the delivery of new TJ proteins [49–51]. The disruption of TJs increases intestinal permeability and thereby likely contributes to EPEC-induced diarrhea [52] (**Figure 3**).

5. Non-LEE effectors

In EPEC prototype strain E2348/69, 17 functional Nle effectors are encoded in different integrative elements and prophages, frequently associated in gene clusters, with some effectors having duplicated gene copies and/or paralogs in different clusters [53] (**Figure 2**). EPEC infection is characterized by a weak inflammatory response [54]. Previous studies have shown that most Nle effectors and some LEE effectors inhibit the host immune response, which favors bacterial survival (**Figure 4**). Although NleF and NleH2 activate the NF-κB inflammatory pathway during early infection (ref), EPEC translocates several effectors that dampen the pro-inflammatory pathways of the cell [11]. Thus, a large number of Nle effectors inhibit host inflammation by different mechanisms, such as inhibition of the NF-κB (NleB, C, E, and H) and MAPK proinflammatory pathways (NleC and D) [55–58], inhibition

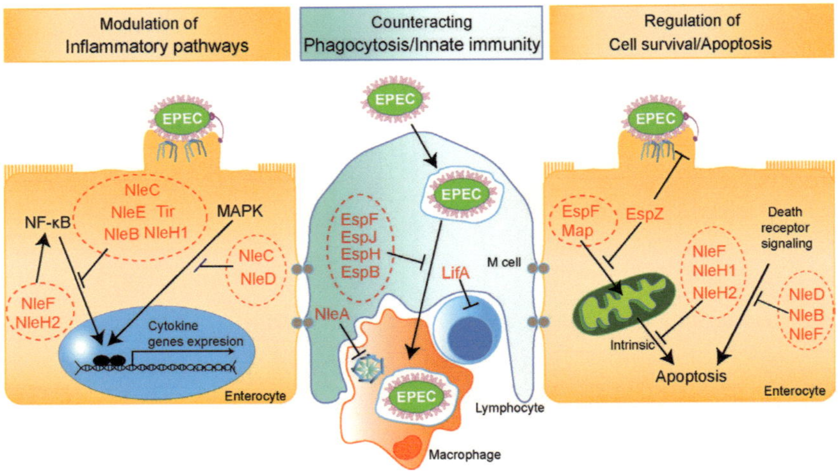

Figure 4.
Schematic representation of multifunctional and overlapping effectors to control host immune response. The NF-κB proinflammatory pathway is activated by NleF and NleH2 and is inhibited by NleE, NleB, NleH1, Tir, and NleC. NleC and NleD inhibit the MAPK proinflammatory pathway. EspF, EspJ, EspH, and EspB prevent macrophage phagocytosis. NleA disrupts inflammasome activation, and LifA inhibits IL-2 and IL-4 production and lymphocyte proliferation. While EspF and map induce intrinsic apoptosis, EspZ counteracts these effects by stabilizing mitochondrial membrane potential. NleH1/NleH2 and NleF inhibit intrinsic apoptosis, and NleF, NleD, and NleB counteract extrinsic apoptosis.

of the canonical (NleA) and noncanonical (NleF) inflammasomes [59], and inhibition of proliferation of lymphocytes and interleukin production (LifA) [60, 61].

The control of the epithelial cell death response to microbial infection is pivotal for pathogens and the host. Pathogens that are colonizing the epithelium need to prevent cell death to preserve their replicative foothold; by contrast, the host needs to eliminate infected cells in order to minimize tissue damage [62]. During infection of the intestinal epithelial cells, surface properties of EPEC are recognized by cell surface death receptors and induce extrinsic apoptotic pathways, while T3SS effectors (Map and EspF) trigger cytochrome c release, activation of caspases, and downstream intrinsic apoptotic pathways [11, 24]. Interestingly, early stages of apoptosis can be observed during EPEC infection, but late stages are not evident because EPEC translocates effector proteins that antagonize these pro-apoptotic effects. NleD and NleB interfere with the pro-apoptotic death receptor signaling and disrupt the downstream extrinsic apoptosis [63, 64]. NLeH1/2 and EspZ also inhibit intrinsic apoptosis and promote host cell survival [65–67] (**Figure 4**). NleF directly inhibits caspases involved in both intrinsic and extrinsic apoptosis pathways, including caspases 4, 8, and 9 [68]. In addition, EspZ localizes to the cytoplasmic side of the plasma membrane at the site of bacterial attachment and interacts with the translocator protein EspD. It has been proposed that EspZ indirectly prevents cell death by downregulating protein translocation and protecting cells from an overdose of effector proteins. Consistently, a Δ*espZ* mutant was found to be highly cytotoxic [69]. EPEC effectors are injected in a regulated manner to guarantee the success of infection. While the pro-survival effector EspZ is translocated at the early stages of infection, the pro-apoptotic effectors EspF and Map follow later [70].

6. Classical methodologies to study effector functions

Most research on EPEC effectors has been conducted by generating deletion mutants in a single or a few effector genes that are later complemented with multicopy plasmids overexpressing the effector(s). In addition, ectopic expression of individual effectors by plasmid transfection of the host cell has been applied. Both situations are prone to effector overexpression resulting in nonphysiological levels of effectors inside the host cell, which could alter effector activities. In addition, effectors often have synergistic and overlapping functions that cannot be fully appreciated by single mutations and individual transfection experiments [11, 54]. In order to overcome these limitations, we employed a marker-less gene deletion strategy to delete the whole repertoire of known effector genes found in the genome of the prototypical EPEC strain E2348/69 [32]. The genome engineering method for sequential deletion of EPEC effectors was based on the marker-less gene deletion technique described by Posfai et al. [71] and is illustrated in **Figure 5**.

Using this strategy, a set of EPEC mutants with sequential deletions of effectors was generated (**Table 1**), ultimately resulting in strains expressing only Tir and EspZ (EPEC2), Tir (EPEC1), and the effector-less strain EPEC0 (**Table 1**). This approach proved to be effective to specifically modify the genome of EPEC E2348/69, avoiding the introduction of unintended alterations in the genome and leaving no sequence "scars" or antibiotic resistance genes in the chromosome as demonstrated by whole-genome sequencing [32]. Besides, the deletion mutant strains showed normal growth and maintained functional T3SS injectisomes. In addition, they allowed the translocation of individual effectors from single-copy chromosomal genes under endogenous regulation, showing the expected phenotypes without the background of the other effectors [32]. Hence these mutant

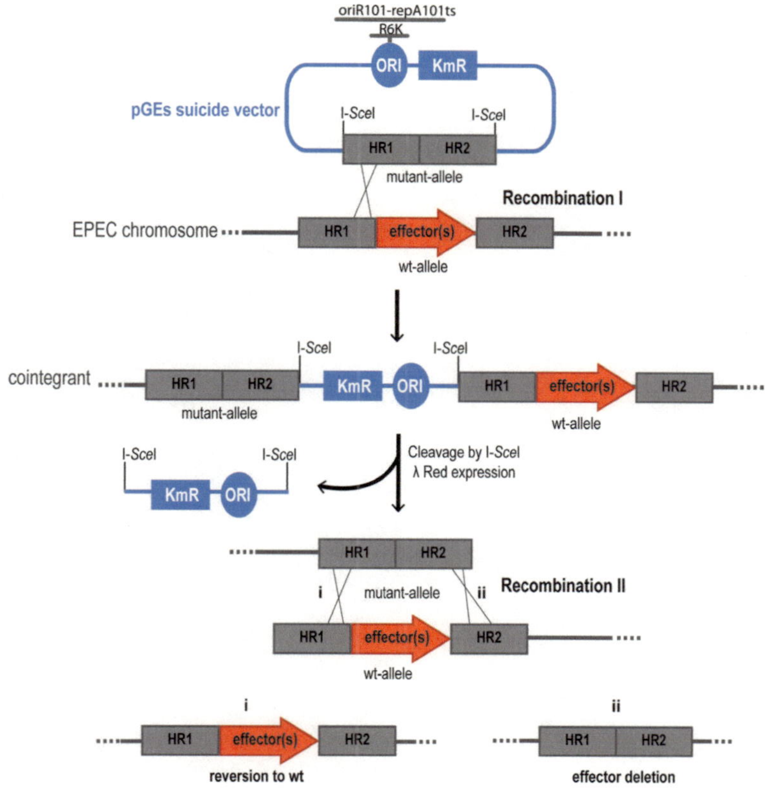

Figure 5.
Marker-less gene deletion strategy of EPEC effector genes. Deletions using pGE-suicide plasmids with I-SceI sites and mutant alleles assembled by fusing homology regions (HRs) flanking the targeted effector gene(s). Co-integrants are identified by the Kanamycin resistance phenotype. Expression of the I-SceI in vivo from helper plasmid induces double-strand brakes that are repaired by homologous recombination. Depending on the HRs involved in this second recombination, either the WT or the mutant allele can be obtained. Figure from [32].

Strain	Effector genes remaining*
WT	All
EPEC2	*espZ* and *tir*
EPEC1	*tir*
EPEC0	None
EPEC9	*espZ + tir + IE2 + IE5 + IE6 + PP2 + PP3 + PP4 + PP6*
EPEC2-LEE+	*espZ + tir + map + espH + espF + espG*

**Encoded effectors in the indicated IEs and PPs.*

Table 1.
EPEC mutant strains generated with the marker-less deletion strategy.

strains are an excellent tool to investigate the role of individual effectors and specific combinations maintaining physiological protein levels in the context of infection.

7. LEE effectors are sufficient for intimate adhesion of EPEC to the epithelial cells in vitro

When EPEC bacteria adhere in vitro to cultured cells, there is an accumulation of actin filaments in the cytoplasm beneath the adherent bacteria, due to a signal cascade triggered by intimin-Tir interaction [35, 38]. Using the effector deletion

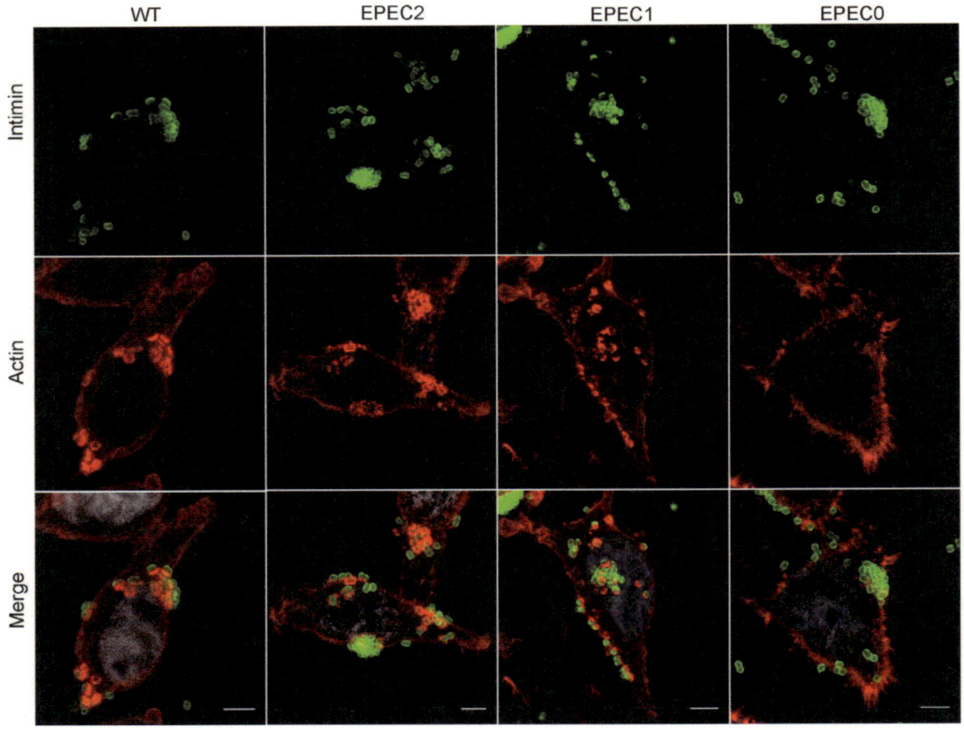

Figure 6.
Infection of HeLa cells with EPEC WT and effector mutant strains. Immunofluorescence confocal microscopy of HeLa cells infected with EPEC WT, EPEC2, EPEC1, and EPEC0 for 1.5 h using a MOI of 200. EPEC is labeled with anti-intimin-280 serum (green), actin is stained with TRITC phalloidin (red), and cell nuclei are labeled with DAPI (gray). Actin polymerization beneath adherent bacteria is observed in EPEC WT, EPEC2, and EPEC1. Scale bar 5 μm. Figure from [32].

mutants of EPEC, we demonstrated that the LEE effector Tir along with intimin is necessary and sufficient to induce these cytoskeletal rearrangements during in vitro infection of HeLa cells. Strains EPEC2 (bearing EspZ and Tir) and EPEC1 (bearing only Tir) were able to induce actin-pedestal formation underneath attached bacteria similar to the EPEC wild type (WT) (**Figure 6**). As expected because of the essential role of Tir in this process, infection of HeLa cells with the effector-less mutant EPEC0 did not induce any actin-pedestal formation (**Figure 6**). These data demonstrate that the individual translocation of Tir by EPEC1 is sufficient to trigger actin pedestals in HeLa cells and that non-LEE effectors are dispensable for this phenotype during in vitro infection of cultured cells.

8. Non-LEE effectors are required for efficient A/E lesion formation in intestinal tissue

EPEC pathogenic mechanisms have been widely investigated by in vitro infection of cultured epithelial cell lines, albeit in most cases these cells are non-polarized and are not from intestinal origin. In addition, EPEC infection studies in vivo are hindered because EPEC is a human-restricted pathogen [72]. A surrogate model established to investigate A/E pathogenesis in vivo is the mouse pathogen *Citrobacter rodentium* (CR) [12, 13]. Although *Citrobacter* infection in vitro requires Tir phosphorylation for actin-pedestal formation in cell lines, Tir phosphorylation-deficient mutants still colonize the mouse gut and induce A/E lesion formation and crypt hyperplasia typical of CR infection [73]. This result highlights the necessity of a model for EPEC infection closer to the in vivo conditions in the human gut. A good established model to study EPEC-host interactions is the infection of in vitro cultured human intestinal biopsies, which allows the formation of A/E lesions undistinguishable from those observed in vivo in biopsies of patients with EPEC-induced diarrhea [4, 36, 74]. Similar to results obtained in CR-infected mice, Tir phosphorylation was not necessary for EPEC A/E lesion formation in human intestinal biopsies [75]. Surprisingly, when EPEC2 and EPEC1 deletion mutants were used to infect human duodenal biopsies, none of the infected biopsies showed A/E lesions (**Table 2** and **Figure 7**), which contrasts with the pedestal formation observed in HeLa cells. Thus, intimin and Tir are not sufficient to induce A/E lesions in the intestinal tissue, and the IVOC model was used to identify additional LEE or non-LEE effector(s) required for A/E lesion formation. For this purpose, two additional effector mutant strains were tested: EPEC2-LEE+ (carrying all LEE effectors) and EPEC9 (carrying EspZ, Tir, and

Strain	Effector genes remaining*	Biopsies with A/E lesions positive/total (%)
WT	All	13/17 (76)
EPEC2	*espZ* and *tir*	0/6 (0)
EPEC1	*tir*	0/6 (0)
EPEC0	None	0/5 (0)
EPEC9	*espZ + tir + IE2 + IE5 + IE6 + PP2 + PP3 + PP4 + PP6*	5/6 (83)
EPEC2-LEE+	*espZ + tir + map + espH + espF + espG*	0/5 (0)

**Encoded effectors in the indicated IEs and PPs.*

Table 2.
Human duodenal biopsies infected by EPEC WT and EPEC effector mutants.

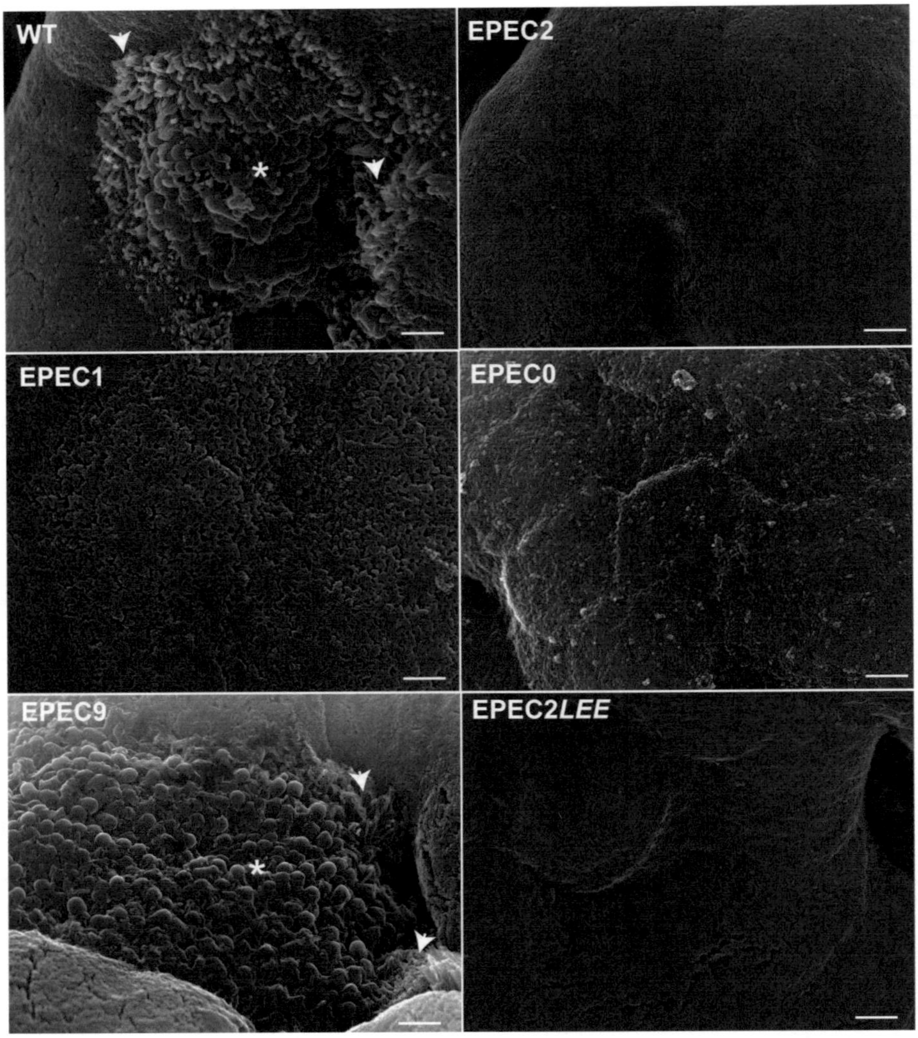

Figure 7.
Scanning electron microscopy of human duodenal biopsies infected with EPEC WT and mutant strains EPEC2, EPEC1, EPEC0, EPEC9, and EPEC2-LEE+. EPEC WT and EPEC9 induce characteristic A/E lesions with bacterial microcolony formation (asterisk) and microvilli elongation around bacterial colonies (arrowheads). In contrast, biopsies infected with EPEC2, EPEC1, EPEC0, and EPEC2-LEE+ lack adherent bacteria and A/E lesions and show a normal microvillous brush border. Scale bar 2 μm. Figure from [32].

all non-LEE effectors). Whereas infection with EPEC2-LEE+ did not reveal A/E lesions, infection with EPEC9 induced A/E lesions to a similar level as the wild-type strain (**Table 2** and **Figure 7**). It was previously reported that the LEE island is sufficient to confer the A/E phenotype to *E. coli* K-12 in the infection of colon carcinoma cell lines [76]. However, our results indicate that the LEE is not sufficient for A/E lesion formation in human mucosal tissue and that non-LEE effectors are required [32].

9. Conclusions and future perspectives

The marker-less gene deletion strategy enabled the generation of effector-less strains of EPEC O127:H6 using the prototypical strain E2348/69 [32]. Given the conservation of the recombination machinery among *E. coli* strains, it is likely that this strategy could be applied to other A/E pathogens, *E. coli* pathogens, and other bacteria. The effector mutant strains can be useful to study the role of individual effectors and of combinations of effectors in pathogenesis. An individual effector or a defined combination can be inserted in the effector-less strains in their endogenous genomic loci to obtain physiological expression levels and regulation. In cell culture infections, all EPEC effector mutant strains carrying intimin and Tir were able to trigger actin-rich pedestal-like structures underneath attached bacteria. On the other hand, when the infection was performed in human intestinal tissues, translocation of Tir alone was insufficient to induce A/E lesions. Furthermore, an EPEC deletion mutant maintaining all LEE effectors and devoid of all non-LEE effectors (EPEC2-LEE+) was still unable to induce A/E lesions in human intestinal biopsies. In contrast, an EPEC strain producing the complete repertoire of non-LEE effectors and devoid of LEE effectors, except Tir and EspZ, formed A/E lesions in intestinal tissue at wild-type levels [32]. Thus, these experiments revealed that non-LEE effectors are needed for A/E lesion formation in human intestinal tissue.

In addition to their potential for basic studies, the EPEC effector mutant strains may have different applications. For instance, EPEC (and other pathogenic) strains lacking multiple effectors are likely to be strongly attenuated, but they maintain the external antigenicity of the wild-type strain. Thus, an EPEC mutant strain with a functional T3SS and the minimum set of effectors necessary to colonize the intestinal surface could be a good vaccine candidate. Further, EPEC mutant strains with the ability to attach to the human intestine could also be engineered to translocate heterologous protein antigens to generate protection against other enteric pathogens causing diarrhea, including EHEC strains [77–80]. Lastly, the EPEC effector mutant strains may also have the potential to deliver therapeutic proteins to the intestinal epithelium, for instance, to combat inflammation and autoimmune disorders in the gastrointestinal tract [81].

Acknowledgements

We acknowledge support for the publication fee by the CSIC Open Access Publication Support Initiative through its Unit of Information Resources for Research (URICI).

Author details

Massiel Cepeda-Molero[1], Stephanie Schüller[2], Gad Frankel[3] and Luis Ángel Fernández[1*]

1 Department of Microbial Biotechnology, Centro Nacional de Biotecnología, Consejo Superior de Investigaciones Científicas (CSIC), Madrid, Spain

2 Norwich Medical School, University of East Anglia, Norwich, United Kingdom

3 Department of Life Sciences, MRC Centre for Molecular Bacteriology and Infection, Imperial College, London, United Kingdom

*Address all correspondence to: lafdez@cnb.csic.es

References

[1] Kaper JB, Nataro JP, Mobley HL. Pathogenic *Escherichia coli*. Nature Reviews Microbiology. 2004;**2**:123-139

[2] Chen HD, Frankel G. Enteropathogenic *Escherichia coli*: Unravelling pathogenesis. FEMS Microbiology Reviews. 2005;**29**(1):83-98

[3] Tlaskalova-Hogenova H et al. Commensal bacteria (normal microflora), mucosal immunity and chronic inflammatory and autoimmune diseases. Immunology Letters. 2004;**93**(2-3):97-108

[4] Hicks S et al. Role of intimin and bundle-forming pili in enteropathogenic *Escherichia coli* adhesion to pediatric intestinal tissue in vitro. Infection and Immunity. 1998;**66**(4):1570-1578

[5] Nataro JP, Kaper JB. Diarrheagenic *Escherichia coli*. Clinical Microbiology Reviews. 1998;**11**(1):142-201

[6] Ochoa TJ et al. New insights into the epidemiology of enteropathogenic *Escherichia coli* infection. Transactions of the Royal Society of Tropical Medicine and Hygiene. 2008;**102**(9):852-856

[7] Garmendia J, Frankel G, Crepin VF. Enteropathogenic and enterohemorrhagic *Escherichia coli* infections: Translocation, translocation, translocation. Infection and Immunity. 2005;**73**(5):2573-2585

[8] Ulshen MH, Rollo JL. Pathogenesis of *Escherichia coli* gastroenteritis in man--another mechanism. The New England Journal of Medicine. 1980;**302**(2):99-101

[9] Knutton S et al. Actin accumulation at sites of bacterial adhesion to tissue culture cells: Basis of a new diagnostic test for enteropathogenic and enterohemorrhagic *Escherichia coli*. Infection and Immunity. 1989;**57**(4):1290-1298

[10] Lai Y et al. Intimate host attachment: Enteropathogenic and enterohaemorrhagic *Escherichia coli*. Cellular Microbiology. 2013;**15**(11):1796-1808

[11] Wong AR et al. Enteropathogenic and enterohaemorrhagic *Escherichia coli*: Even more subversive elements. Molecular Microbiology. 2011;**80**(6):1420-1438

[12] Collins JW et al. Citrobacter rodentium: Infection, inflammation and the microbiota. Nature Reviews. Microbiology. 2014;**12**(9):612-623

[13] Mundy R et al. Citrobacter rodentium of mice and man. Cellular Microbiology. 2005;**7**(12):1697-1706

[14] McDaniel TK et al. A genetic locus of enterocyte effacement conserved among diverse enterobacterial pathogens. Proceedings of the National Academy of Sciences of the United States of America. 1995;**92**(5):1664-1668

[15] Elliott SJ et al. The locus of enterocyte effacement (LEE)-encoded regulator controls expression of both LEE- and non-LEE-encoded virulence factors in enteropathogenic and enterohemorrhagic *Escherichia coli*. Infection and Immunity. 2000;**68**(11):6115-6126

[16] Yerushalmi G et al. Dynamics of expression and maturation of the type III secretion system of enteropathogenic *Escherichia coli*. Journal of Bacteriology. 2014;**196**(15):2798-2806

[17] Daniell SJ et al. The filamentous type III secretion translocon of enteropathogenic *Escherichia coli*. Cellular Microbiology. 2001;**3**(12):865-871

[18] Gaytán MO et al. Type three secretion system in attaching and effacing pathogens. Frontiers in Cellular and Infection Microbiology. 2016;**6**:129

[19] García-Gómez E et al. The muramidase EtgA from enteropathogenic *Escherichia coli* is required for efficient type III secretion. Microbiology. 2011;**157**(Pt 4):1145-1160

[20] Franzin FM, Sircili MP. Locus of enterocyte effacement: A pathogenicity island involved in the virulence of enteropathogenic and enterohemorrhagic subjected to a complex network of gene regulation. BioMed Research International. 2015;**2015**:534738

[21] Furniss RCD, Clements A. Regulation of the locus of enterocyte effacement in attaching and effacing pathogens. Journal of Bacteriology. 2018;**200**(2):e00336-17

[22] Portaliou AG et al. Type III secretion: Building and operating a remarkable nanomachine. Trends in Biochemical Sciences. 2016;**41**(2):175-189

[23] Raymond B et al. Subversion of trafficking, apoptosis, and innate immunity by type III secretion system effectors. Trends in Microbiology. 2013;**21**(8):430-441

[24] Santos AS, Finlay BB. Bringing down the host: Enteropathogenic and enterohaemorrhagic *Escherichia coli* effector-mediated subversion of host innate immune pathways. Cellular Microbiology. 2015;**17**(3):318-332

[25] Ide T et al. Characterization of translocation pores inserted into plasma membranes by type III-secreted Esp proteins of enteropathogenic *Escherichia coli*. Cellular Microbiology. 2001;**3**(10):669-679

[26] Luo W, Donnenberg MS. Interactions and predicted host membrane topology of the enteropathogenic *Escherichia coli* translocator protein EspB. Journal of Bacteriology. 2011;**193**(12):2972-2980

[27] Stone KD et al. A cluster of fourteen genes from enteropathogenic *Escherichia coli* is sufficient for the biogenesis of a type IV pilus. Molecular Microbiology. 1996;**20**(2):325-337

[28] Brinkley C et al. Nucleotide sequence analysis of the enteropathogenic *Escherichia coli* adherence factor plasmid pMAR7. Infection and Immunity. 2006;**74**(9):5408-5413

[29] Ramboarina S et al. Structure of the bundle-forming pilus from enteropathogenic *Escherichia coli*. The Journal of Biological Chemistry. 2005;**280**(48):40252-40260

[30] Hyland RM et al. The bundlin pilin protein of enteropathogenic *Escherichia coli* is an N-acetyllactosamine-specific lectin. Cellular Microbiology. 2008;**10**(1):177-187

[31] Saldana Z et al. The *Escherichia coli* common Pilus and the bundle-forming Pilus act in concert during the formation of localized adherence by Enteropathogenic *E. coli*. Journal of Bacteriology. 2009;**191**(11):3451-3461

[32] Cepeda-Molero M et al. Attaching and effacing (A/E) lesion formation by enteropathogenic *E. coli* on human intestinal mucosa is dependent on non-LEE effectors. PLoS Pathogens. 2017;**13**(10):e1006706

[33] Fallingborg J. Intraluminal pH of the human gastrointestinal tract. Danish Medical Bulletin. 1999;**46**(3):183-196

[34] Shin S et al. An activator of glutamate decarboxylase genes regulates the expression of enteropathogenic *Escherichia coli* virulence genes through control of the plasmid-encoded

regulator, Per. Molecular Microbiology. 2001;**41**(5):1133-1150

[35] Kenny B et al. Enteropathogenic *E. coli* (EPEC) transfers its receptor for intimate adherence into mammalian cells. Cell. 1997;**91**(4):511-520

[36] Frankel G et al. Generation of *Escherichia coli* intimin derivatives with differing biological activities using site-directed mutagenesis of the intimin C-terminus domain. Molecular Microbiology. 1998;**29**(2):559-570

[37] Luo Y et al. Crystal structure of enteropathogenic *Escherichia coli* intimin-receptor complex. Nature. 2000;**405**(6790):1073-1077

[38] Frankel G, Phillips AD. Attaching effacing *Escherichia coli* and paradigms of Tir-triggered actin polymerization: Getting off the pedestal. Cellular Microbiology. 2008;**10**(3):549-556

[39] Devinney R et al. Tir tyrosine phosphorylation and pedestal formation are delayed in enteropathogenic *Escherichia coli* sepZ::TnphoA mutant 30-5-1(3). Infection and Immunity. 2001;**69**(1):559-563

[40] Goosney DL, de Grado M, Finlay BB. Putting *E. coli* on a pedestal: A unique system to study signal transduction and the actin cytoskeleton. Trends in Cell Biology. 1999;**9**(1):11-14

[41] Guttman JA, Finlay BB. Subcellular alterations that lead to diarrhea during bacterial pathogenesis. Trends in Microbiology. 2008;**16**(11):535-542

[42] Dean P et al. Potent diarrheagenic mechanism mediated by the cooperative action of three enteropathogenic *Escherichia coli*-injected effector proteins. Proceedings of the National Academy of Sciences of the United States of America. 2006;**103**(6):1876-1881

[43] Simpson N et al. The enteropathogenic *Escherichia coli* type III secretion system effector map binds EBP50/NHERF1: Implication for cell signalling and diarrhoea. Molecular Microbiology. 2006;**60**(2):349-363

[44] Gill RK et al. Mechanism underlying inhibition of intestinal apical Cl/OH exchange following infection with enteropathogenic *E. coli*. The Journal of Clinical Investigation. 2007;**117**(2):428-437

[45] Clements A et al. Enterohaemorrhagic *Escherichia coli* inhibits recycling endosome function and trafficking of surface receptors. Cellular Microbiology. 2014;**16**(11):1693-1705

[46] Guttman JA et al. Aquaporins contribute to diarrhoea caused by attaching and effacing bacterial pathogens. Cellular Microbiology. 2007;**9**(1):131-141

[47] Iizumi Y et al. The enteropathogenic *E. coli* effector EspB facilitates microvillus effacing and antiphagocytosis by inhibiting myosin function. Cell Host & Microbe. 2007;**2**(6):383-392

[48] Dean P, Kenny B. Intestinal barrier dysfunction by enteropathogenic *Escherichia coli* is mediated by two effector molecules and a bacterial surface protein. Molecular Microbiology. 2004;**54**(3):665-675

[49] Glotfelty LG et al. Enteropathogenic *E. coli* effectors EspG1/G2 disrupt microtubules, contribute to tight junction perturbation and inhibit restoration. Cellular Microbiology. 2014;**16**(12):1767-1783

[50] Thanabalasuriar A et al. The bacterial virulence factor NleA is required for the disruption of intestinal tight junctions by enteropathogenic *Escherichia coli*. Cellular Microbiology. 2010;**12**(1):31-41

[51] Kim J et al. The bacterial virulence factor NleA inhibits cellular protein secretion by disrupting mammalian COPII function. Cell Host & Microbe. 2007;**2**(3):160-171

[52] Croxen MA, Finlay BB. Molecular mechanisms of *Escherichia coli* pathogenicity. Nature Reviews. Microbiology. 2010;**8**(1):26-38

[53] Iguchi A et al. Complete genome sequence and comparative genome analysis of enteropathogenic *Escherichia coli* O127:H6 strain E2348/69. Journal of Bacteriology. 2009;**191**(1):347-354

[54] Dean P, Kenny B. The effector repertoire of enteropathogenic *E. coli*: Ganging up on the host cell. Current Opinion in Microbiology. 2009;**12**(1):101-109

[55] Baruch K et al. Metalloprotease type III effectors that specifically cleave JNK and NF-kappaB. The EMBO Journal. 2011;**30**(1):221-231

[56] Nadler C et al. The type III secretion effector NleE inhibits NF-kappaB activation. PLoS Pathogens. 2010;**6**(1):e1000743

[57] Sham HP et al. Attaching and effacing bacterial effector NleC suppresses epithelial inflammatory responses by inhibiting NF-{kappa} B and p38 mitogen-activated protein kinase activation. Infection and Immunity. 2011;**79**(9):3552-3562

[58] Gao X et al. NleB, a bacterial effector with glycosyltransferase activity, targets GADPH function to inhibit NF-κB activation. Cell Host & Microbe. 2013;**13**(1):87-99

[59] Yen H, Sugimoto N, Tobe T. Enteropathogenic *Escherichia coli* uses NleA to inhibit NLRP3 inflammasome activation. PLoS Pathogens. 2015;**11**(9):e1005121

[60] Abu-Median AB et al. Functional analysis of lymphostatin homologues in enterohaemorrhagic *Escherichia coli*. FEMS Microbiology Letters. 2006;**258**(1):43-49

[61] Klapproth JM et al. A large toxin from pathogenic *Escherichia coli* strains that inhibits lymphocyte activation. Infection and Immunity. 2000;**68**(4):2148-2155

[62] Kim M et al. Bacterial interactions with the host epithelium. Cell Host & Microbe. 2010;**8**(1):20-35

[63] Pearson JS et al. A type III effector antagonizes death receptor signalling during bacterial gut infection. Nature. 2013;**501**(7466):247-251

[64] Wong Fok Lung T et al. The cell death response to enteropathogenic *Escherichia coli* infection. Cellular Microbiology. 2014;**16**(12):1736-1745

[65] Roxas JL et al. The enteropathogenic *Escherichia coli*-secreted protein EspZ inhibits host cell apoptosis. Infection and Immunity. 2012;**80**(11):3850-3857

[66] Hemrajani C et al. NleH effectors interact with Bax inhibitor-1 to block apoptosis during enteropathogenic *Escherichia coli* infection. Proceedings of the National Academy of Sciences. 2010;**107**(7):3129-3134

[67] Royan SV et al. Enteropathogenic *E. coli* non-LEE encoded effectors NleH1 and NleH2 attenuate NF-kappaB activation. Molecular Microbiology. 2010;**78**(5):1232-1245

[68] Blasche S et al. The *E. coli* effector protein NleF is a Caspase inhibitor. PLoS One. 2013;**8**(3):e58937

[69] Berger CN et al. EspZ of enteropathogenic and enterohemorrhagic *Escherichia coli* regulates type III secretion system protein translocation. MBio. 2012;**3**(5):e00317-12

[70] Mills E et al. Real-time analysis of effector translocation by the type III secretion system of enteropathogenic *Escherichia coli*. Cell Host & Microbe. 2008;**3**(2):104-113

[71] Posfai G et al. Markerless gene replacement in *Escherichia coli* stimulated by a double-strand break in the chromosome. Nucleic Acids Research. 1999;**27**(22):4409-4415

[72] Hill SM, Phillips AD, Walker-Smith JA. Enteropathogenic *Escherichia coli* and life threatening chronic diarrhoea. Gut. 1991;**32**(2):154-158

[73] Deng W et al. Citrobacter rodentium translocated intimin receptor (Tir) is an essential virulence factor needed for actin condensation, intestinal colonization and colonic hyperplasia in mice. Molecular Microbiology. 2003;**48**(1):95-115

[74] Knutton S, Lloyd DR, McNeish AS. Adhesion of enteropathogenic *Escherichia coli* to human intestinal enterocytes and cultured human intestinal mucosa. Infection and Immunity. 1987;**55**(1):69-77

[75] Schüller S et al. Tir phosphorylation and Nck/N-WASP recruitment by enteropathogenic and enterohaemorrhagic *Escherichia coli* during ex vivo colonization of human intestinal mucosa is different to cell culture models. Cellular Microbiology. 2007;**9**(5):1352-1364

[76] McDaniel TK, Kaper JB. A cloned pathogenicity island from enteropathogenic *Escherichia coli* confers the attaching and effacing phenotype on *E. coli* K-12. Molecular Microbiology. 1997;**23**(2):399-407

[77] Riquelme-Neira R et al. Vaccination with DNA encoding truncated enterohemorrhagic *Escherichia coli* (EHEC) factor for adherence-1 gene (efa-1′) confers protective immunity to mice infected with *E. coli* O157:H7. Frontiers in Cellular and Infection Microbiology. 2016;**5**:104-104

[78] Szu SC, Ahmed A. Clinical studies of *Escherichia coli* O157:H7 conjugate vaccines in adults and young children. Microbiology Spectrum. 2014;**2**(6):1-7

[79] Rabinovitz BC et al. The intranasal vaccination of pregnant dams with Intimin and EspB confers protection in neonatal mice from *Escherichia coli* (EHEC) O157:H7 infection. Vaccine. 2016;**34**(25):2793-2797

[80] Marcato P et al. Recombinant Shiga toxin B-subunit-keyhole limpet hemocyanin conjugate vaccine protects mice from Shigatoxemia. Infection and Immunity. 2005;**73**(10):6523-6529

[81] Piñero-Lambea C, Ruano-Gallego D, Fernández LÁ. Engineered bacteria as therapeutic agents. Current Opinion in Biotechnology. 2015;**35**:94-102

Chapter 6

Management of *E. coli* Sepsis

Akshay Kumar, Ana Francesca Vommaro Leite, Lais Sanches Maekawa, Roopvir Kaur, Silas Jose Braz Filo, Purnadeo Persaud, Juber Dastagir Shaikh, Asim Kichloo and Nimisha Shiwalkar

Abstract

E. coli is the most common cause of urinary tract infections (UTIs) in humans and is a leading cause of enteric infections and systemic infections. The systemic infections include bacteremia, nosocomial pneumonia, cholecystitis, cholangitis, peritonitis, cellulitis, osteomyelitis, and infectious arthritis. *E. coli* is also the leading cause of neonatal meningitis.

Keywords: bacteremia, septicemia, septic shock, antimicrobial therapy

1. Introduction

Escherichia coli strains compose, physiologically part of the microflora of the gastrointestinal tract [1–4]. Belonging to the Enterobacteriaceae family, fermentative, non-sporulated and facultative anaerobic commensals, they are mainly from the large intestine [5, 6].

Despite being commensal microorganisms, they are the Gram-negatives which are most often a cause of human infections, having pathogenic strains that cause a wide variety of intestinal or extra-intestinal infections, such as urinary tract, intra-abdominal and soft tissue, sepsis, neonatal meningitis, gastrointestinal infection, and pneumonia, often leading to bacteremia [3, 7]. Although Gram-positive microorganisms have been increasing as a cause of sepsis due to the instrumentation of medical care—understood as the use of invasive devices or tools for the treatment or diagnosis of patients, and to infections associated with health care—*E. coli* continues to be an important and perhaps the most frequent cause of threatening infections in our environment [8, 9].

They are classified as Gram-negative bacteria and divided into 3 main groups: commensal lines, intestinal pathogenic lines (enteric or diarrhea) and extra-intestinal pathogenic lines [10].

Furthermore, Gram-negative bacteria produce large molecules consisting of a lipid and a polysaccharide, known as lipopolysaccharides (LPS), lipoglycans and endotoxin, which increases their pathogenicity in relation to Gram-positive bacteria [11].

2. Epidemiology

E. coli is one of the most commonly isolated bacteria in the bloodstream (responsible for approximately 20% of all clinically significant isolates) and is the

Gram-negative organism most frequently isolated in adult patients with bacteremia [12]. In the United States of America, *E. coli* sepsis was associated with approximately 40,000 deaths in 2001, a number that corresponds to 17% of all cases of sepsis [13].

Studies have shown an increasing incidence of *E. coli* early-onset sepsis in all age groups, overruling group B *Streptoccocus* for the last 10 years. Beyond that, *E. coli* resistant strains also increased equally in all age groups, with high resistance rates to first line antibiotics available (ampicillin and gentamicin).

Very low birth weight newborns remained the group with higher incidence (10.4 cases per 1000 live births) and mortality (35.3%). Systematic use of PCR increased *E. coli* early-onset sepsis diagnosis, mainly in the term newborn group. There was also an increase in resistant *E. coli* strains causing early-onset sepsis, with especially high resistance to ampicillin and gentamicin (92.8 and 28.6%, respectively) [14].

3. Risk factors

Several hospital-based studies have suggested that a number of comorbid illnesses, including diabetes, malignancy, chronic lung disease, cirrhosis and heart disease, may increase the risk of *E. coli* bacteremia. Previous researches have also identified age (very young and very elder), hospital acquisition, comorbid illnesses, presence of shock, non-urinary focus, and antimicrobial resistance in conjunction with inadequate treatment as being associated with higher rates of death [15–17].

Dialysis, solid organ transplantation and neoplastic disease were important risk factors for acquiring *E. coli* bacteraemia. Ciprofloxacin resistance and non-urinary focus were independently associated with an increased risk of death [18]. For males, urinary catheterization and incontinence were associated as risk factors to *Escherichia coli* bloodstream, and for females, cancer, renal failure, heart disease and urinary incontinence were risk factors reported [19]. Several risk factors which have significantly mortality due to *E. coli* bacteremia are age, severe sepsis or shock, non-urinary origin, Charlson index, inadequate empirical treatment (**Table 1**).

Mortality risk factor	*P*	**OR (95% CI)**
Age	0.03	1.04 (1–1.08)
Severe sepsis or shock	<0.0001	14.64 (6.14–30.86)
Non-urinary origin	0.013	2.78 (1.24–6.2)
Charlson index	0.006	1.31 (1.08–1.59)
Inadequate empirical treatment	0.006	2.98 (1.25–7.11)

Table 1.
Results of multivariate analyses examining risk factors for mortality associated with bacteraemia due to E. coli [15].

4. Pathogenesis

The human gastrointestinal tract is normally inhabited by *Escherichia coli*, which is why they are the bacterial species most commonly found in the isolation of fecal culture [20, 21]. By the time the strains acquire additional genetic material, they can become pathogenic and circulate widely throughout the body. Pathological clones are divided into two major groups: intestinal (among the most virulent enteric pathogens) and extraintestinal (less present, but not less dangerous) [22, 23].

4.1 Intestinal

4.1.1 EPEC

Typical enteropathogenic *Escherichia coli* (tEPEC) contains a virulence plasmid (pEAF) that encodes the bundle-forming pilus (BFP), the primary factor for colonization [24, 25]. In addition, EPEC carries the crossomic island of locus for enterocyte effacement, which features the eae gene, which is the encoder of a colonization factor in the outer membrane protein called intimin [26, 27]. Only the *E. coli* strain that has pEAF and the eae gene can be considered tEPEC, one that has only the eae gene and is called atypical EPEC (aETEC) [28].

The small intestine is the most likely place for EPEC infection to occur. For the onset of diseases, tEPEC obeys the following steps:

- Initial localized adhesion of organisms to enterocyte via BFP.

- Induction of signal transduction in the enterocyte by secretion of protein toxins.

- Development of intimin-mediated intimate adhesion to the enterocyte.

Around 20 protein toxins are injected directly into the target epithelial cell, made, together with the intimin, by the chromosomal island LEE and expressed by both tEPEC and aEPEC [29]. The complex nanomachine called type III secretion injector is the one that injects protein toxins. It is assumed that some modifications happen to the epithelial stem cells, which is physiologically absorbent, and through a pathological process, it becomes a secretory dynamo [30].

What is believed is that type III ejection toxins are responsible for binding to protein elements of the cell's signal transduction apparatus. This event is accompanied by the mobilization of calcium from the intracellular compartment, activation of protein kinase C, kinase light chain myosin and induction of protein phosphorylation by tyrosine. The rearrangement of cytoskeletal proteins is induced by effectors, which results in the classic lesion "attaching and erasing," changes in the secretion of water and electrolytes and increased permeability of the tight intestinal junctions [31].

4.1.2 ETEC

Enterotoxigenic *Escherichia coli* (ETEC) consists of ingestion of bacteria, intestinal colonization and production of virulence factors. Colonizing fimbriae (CFs) must be expressed by ETEC to allow the consolidation of the bacteria in the intestine [32].

After colonization, ETEC produces two classes of secretory toxins encoded by plasmids: heat-labile toxin (LT) and heat-stable toxin (ST). To be classified as ETEC, *E. coli* must contain one or both classes of toxins [33, 34].

LT toxin is related to Vibrio cholera toxins in terms of structure, function and mechanism. It works by stimulating adenylate cyclase and increasing adenosine intracellular cyclic monophosphate (AMP), a fact that stimulates chloride secretion from intestinal crypt cells and inhibits the absorption of sodium chloride at the ends of the villi. After that, the water secretion is free in the intestinal lumen, clinically developing watery diarrhea [35].

STa toxin, the only ST variant that causes disease in humans, activates cyclic GMP of enterocytes, leading to increased chloride secretion and decreased sodium chloride absorption. As a final result, the secretion of free water in the intestinal lumen clinically appears as watery diarrhea [36].

4.1.3 EHEC

Among the pathotypes that cause the most severe conditions, the strains classified as enterohemorrhagic (EHEC) stand out, which are the most common to cause disease in developed countries [29].

They are bacteria responsible for food infections and represent a risk to the health of the population, so they must be monitored frequently. Thus, good hygiene practices, as well as the use of quality tools, are extremely important to help reduce the risk of cross-contamination and human infection.

EHEC has the ability to attach itself to the host and to produce shiga-toxins, which gives the strain pathogenicity. The toxins produced by EHEC cause damage to the mucosa of the large intestine, where they are absorbed by reaching the bloodstream, which makes it possible to affect other organs, such as the kidneys [37]. An average of 5–10% of patients confirmed with EHEC infection develop potentially fatal complications, such as hemolytic uremic syndrome (HUS), which leads to sudden renal failure and hemolytic anemia [38].

Outbreaks are related to the ingestion of contaminated food and water, causing watery diarrhea and hemorrhagic colitis to those infected. The disease has a sudden onset with severe abdominal cramps and watery diarrhea that progresses to bloody, on average after 24 hours, lasting between 1 to 8 days.

The treatment consists of supportive therapy for fluid replacement, since the use of antibiotics is not indicated, as there is no proven efficacy. In fact, it could increase the risk of developing HUS, since the death of the bacteria would increase the release of toxins, predisposing to the syndrome [39].

4.1.4 EIEC

Enteroinvasive *E. coli* (EIEC) is very close to Shigella and develops a colitis similar to shigellosis. The intestinal cell is invaded by the EIEC which multiplies intracellularly and reaches the adjacent intestinal cells [40].

To differentiate Shigella from EIEC it is necessary to analyze the strains, those from EIEC ferment glucose and xylose, this differentiates them. Nucleic acid tests, including multiplexed panels, are used to detect organisms [41].

4.1.5 DAEC

Diffusely adherent *E. coli* is associated with diarrhea, which is characterized as watery and can become persistent in children between 1 and 5 years of age, occurring more frequently in developing and developed countries. In addition, this bacterium is also related to urinary tract infections and complications during the pregnancy period.

The pattern of diffuse adhesion in HEp-2 or HeLa cells is a characteristic that differentiates this pathotype from the others, although DAEC strains are quite heterogeneous. This adhesion is mediated by fimbrial and afimbrial adhesins, which can cause damage to microvilli due to the disorganization of the cytoskeleton. However, some strains produce an adhesin involved in diffuse adhesion (AIDA-I), instead of encoding the diffuse adhesion pattern, which is why they are called atypical DAEC [42].

In addition, DAEC can also provide a pro-inflammatory effect [43].

4.2 Extraintestinal

The type of *E. coli* responsible for the invasion, colonization and induction of diseases in body sites outside the gastrointestinal tract is the extraintestinal pathogenic *Escherichia coli* (ExPEC). It is noteworthy that diseases caused by ExPEC range from urinary tract infections, neonatal meningitis, sepsis, pneumonia, surgical site infections to infections in other extraintestinal sites, representing a burden in terms of medical costs and lost productivity [44].

Thereto, the ExPEC strains were isolated from food products, in particular raw meat and poultry, indicating that these organisms potentially represent a new class of foodborne pathogens [45].

4.2.1 Urosepsis

Almost 25% of sepsis cases originate from the urogenital tract. [46–48]. Considering this percentage, the most common pathogen that causes urinary tract infection (and, consequently, urosepsis) is *Escherichia coli* (50%) [49]. It is known that this condition is better managed with an interprofessional team of health professionals—a nephrologist, infectious disease expert, urologist, intensivist, a nurse and a pharmacist [50, 51]. The outcomes after urosepsis depend on the cause and severity of the infection, and if the patient has a complicating factor in the urinary tract that is identified and warrants treatment, it should be performed as soon as possible. As an example, the literature reveals Foley catheter placement to relieve urinary retention or stent placement to bypass an obstructing ureteral calculus causing urosepsis. Moreover, the prognosis also depends on the type of bacteria, antimicrobial resistance, and patient comorbidity.

In addition to early antibiotics, there are some important parts of the management of sepsis. Initial fluid resuscitation with crystalloid is still recommended at a minimum of 30 mL/kg. Consider early administration of vasopressor support to maintain a mean arterial pressure greater than 65 mm Hg. The first choice for vasopressor support in sepsis is norepinephrine (with epinephrine and vasopressin 2 and 3). Tight glucose control is also recommended, with corticosteroids and blood products being more controversial in the literature [52].

5. Antimicrobial resistance

Although *Escherichia coli* is one of the most-studied microorganisms worldwide, its characteristics are constantly changing. Elseways, one important global problem is the increase of antimicrobial resistance shown by bacteria, being considered as "threatens the achievements of modern medicine" [53, 54].

E. coli resistant strains increased equally in all age groups, with high resistance rates to our first line antibiotics (ampicillin and gentamicin), with relevant highlight in neonatal *E. coli* isolates from invasive infection [55]. **Table 2** shows the temporal trends for antibiotic resistance to *E. coli*.

Agent or phenotype [n (%)]	**1997** $n = 58$	**1998** $n = 49$	**1999** $n = 52$	**2000** $n = 83$	**2001** $n = 86$	**2002** $n = 70$	**2003** $n = 87$	**2004** $n = 122$	**2005 (January–June)** $n = 56$	**Total** $n = 663$	P
Ampicillin	27 (46.6)	24 (49)	24 (46.2)	50 (60.2)	54 (62.8)	46 (65.7)	55 (63.2)	70 (57.9)	35 (62.5)	385 (58.2)	0.02
Trimethoprim/ sulfamethoxazole	14 (24.1)	11 (22.4)	13 (25.0)	28 (33.7)	21 (24.4)	28 (40)	32 (36.8)	41 (33.6)	20 (35.7)	208 (31.4)	0.02
Ciprofloxacin	9 (15.5)	7 (14.3)	10 (19.2)	7 (8.4)	14 (16.3)	16 (22.9)	22 (25.3)	27 (22.1)	13 (23.2)	125 (18.9)	0.02
Amoxicillin/ clavulanate	9 (15.5)	4 (8.2)	9 (17.3)	16 (19.3)	8 (9.3)	7 (10)	11 (12.6)	15 (12.3)	20 (35.7)	99 (14.9)	0.1
Gentamicin	4 (6.9)	6 (12.2)	5 (9.6)	5 (6.0)	8 (9.3)	6 (8.6)	7 (8.0)	8 (6.6)	8 (14.3)	57 (8.6)	0.8
Piperacillin/ tazobactam	1 (1.7)	4 (8.2)	1 (1.9)	8 (9.6)	6 (7.0)	4 (5.7)	5 (5.7)	2 (1.6)	2 (3.6)	33 (5)	0.4
Cefotaxime	11	2 (4.1)	0	2 (2.4)	3 (3.5)	5 (7.1)	3 (3.4)	12 (9.8)	4 (7.1)	31 (4.7)	0.001
ESBL production	0	0	0	2 (2.4)	3 (3.5)	3 (4.3)	2 (2.3)	9 (7.4)	3 (5.4)	22 (3.3)	0.002
MDR	4 (6.9)	4 (8.2)	5 (9.6)	9 (10.8)	9 (10.5)	12 (17.1)	15 (17.2)	17 (13.9)	12 (21.4)	87 (13.1)	0.006

Table 2.
Number, yearly percentages, and P values for temporal trend of non-susceptible cases of E. coli bacteraemia.

6. Evaluation

The sepsis' diagnosis confirmation is done from the evaluation of the clinical status of the patient, analyzing some criteria. For adult patients, it is confirmed or a diagnosis of sepsis is made when two criteria are present: hyperthermia>38.3 °C or hypothermia <36°C, tachycardia>90 bpm, leukocytosis (>12,000 μL-1) or leukopenia (<4000 μL-1) or >10% bands, acutely altered mental status, tachypnea > 20 bpm, hyperglycemia (>120 mg/dl) in the absence of diabetes [56].

7. Clinical assessment and patient presentations

7.1 History and physical examination

Collect a careful history from patient, addressing information such as previous illnesses, surgeries, how long ago the symptoms started, if there are comorbidities, if it have traveled to a place recently and other details, added to a complete physical examination, which provides very relevant information and leads to a line of rationality, it is extremely important to start the development of a preliminary differential diagnosis of the patient's complaints.

All this information collected is recorded and saved in medical records, more recently, electronics, which are more organized, more readable and allows a better comparison, in relation to written records [57].

Some of the most frequent reasons that lead patients to go to a medical consultation are dyspnea, cough with or without hemoptysis and chest pain, as these symptoms can be indications of serious illnesses, it shows the importance of asking questions and exams in a way attentive and careful [58].

7.2 Presentations

7.2.1 Pneumonia

Ventilator-associated pneumonia (VAP) is the most common fatal hospital infection [59]. One of the bacteria most involved in the clinical picture in question is Enterobacteriaceae *Escherichia coli* [60, 61] and there is little awareness when it comes to the pathophysiology of *E. coli* pneumonia.

Studies show that these *E. coli* pathogenic islands (PAIs) are involved differently in the pathogenicity of the lung compared to those present in urinary tract and bloodstream infections [62]. In addition, research on mice has also shown that these isolated strains are highly virulent extra-intestinal pathogens that express virulence factors, representing potential targets for new therapy. A French national study also demonstrated that, despite the genomic and phylogenetic characteristics of *E. coli* pneumonia isolates from critically ill patients, they belong to the same extra-intestinal pathogen as *E. coli*, they have specific distinct characteristics when lungs [63].

7.2.2 Acute-bacterial meningitis

E. coli meningitis is rare in adult forms of the disease [64–66], but it is a frequent pathogen in the pediatric field [67]. Despite its rarity, it has a serious clinical course [64–66]. It is usually diagnosed based on clinical signs and cerebrospinal fluid (CSF) analysis.

Due to the severity of the disease, early diagnosis, adequate antibiotic treatment and hemodynamic control are essential [68].

E. coli meningitis follows a high degree of bacteraemia and invasion of the blood–brain barrier. With mortality rates ranging from 15 to 40%, Meningitis due to this bacterium leaves approximately 50% of survivors with some type of neurological sequelae [69–78].

Although the process is unknown, it is known that, for the onset of the disease, it is necessary to have an invasion of the blood–brain barrier by *E. coli*, which requires specific microbial and host factors such as specific signaling molecules for microbes and hosts. Thus, blocking these microbial and host factors that contribute to the invasion of the blood–brain barrier by *E. coli* is effective in preventing the penetration of *E. coli* into the brain.

With the complete discovery of this mechanism, it is likely that new targets for the prevention and therapy of *Escherichia coli* meningitis will be achieved [79].

Regarding treatment, it is currently known only that antimicrobial chemotherapy has limited efficacy [79–81].

7.2.3 Intra-abdominal infections

Intra-abdominal infections (IAI) are invasive and bacterial multiplications in the hollow organ walls and beyond. Usually, it is located in the abdominal cavity, in the retroperitoneum and in the abdominal organs, being a common complication in the post-surgical period [82]. In addition, they have a wide variety of pathological conditions, from appendicitis to fecal peritonitis, which makes IAI generally have a poor prognosis (especially in high-risk patients) and is an important cause of morbidity [83]. Mostly, the most common source of this infection is the appendix, followed by gastroduodenal perforations. The Gram-negative bacteria *E. coli* is the most common causative agent of IAI. Therefore, it is important to know that they have great sensitivity to imipenem, meropenem, mainly, and to amoxi-clavulanate, amikacin and piperacillin-tazobactam, next [84, 85]. However, amici-clavulanate is prescribed as a first-line drug in developing countries, due to cost factors [86].

7.2.4 Enteric infections

Although *E. coli* strains have been isolated as part of the normal beneficial flora of the intestine, some strains have developed pathogenic mechanisms to cause disease in humans and animals. One of these strains capable of causing diseases is enteric *Escherichia coli* (*E. coli*), comprising important pathogens, since they cause significant morbidity and mortality worldwide. Traditionally enteric *E. coli* was divided into 6 pathotypes, however two other divisions were proposed by several studies (as mentioned individually in topic 4) [87].

Although there are many etiological agents responsible for diarrhea, pathogenic *E. coli* is a major contributor. On the other hand, the onset and complications of enteric *E. coli* vary significantly, despite there are many common features in the pathogenic process of colonizing the intestinal mucosa and the onset of disease [88].

Outbreaks are common all over the world, with fatal consequences mainly in children under 5 years of age living in underdeveloped countries, where diarrheal diseases can lead to death more frequently [89].

The transmission of enteric *E. coli* is also a public health concern, related to the development of countries, since its transmission is through contaminated water and food. Thus, the seriousness in relation to the microorganism can be exemplified by national and international surveillance programs, created by developed countries that aim to constantly monitor outbreaks [90]. In developing countries ETEC,

EPEC and EAEC are considered to be the main causes of childhood diarrhea, and when left untreated, they have potentially fatal consequences. However. in developed countries, these infections are mild and self-limiting, with EHEC and, more recently, EAEC and STEAEC being the main *E. coli* pathotypes associated with food poisoning outbreaks [91, 92].

7.2.5 UTI

Among the most common types of bacterial infections that occur both in the community and in hospitals, urinary tract infections (UTI) stand out. Urinary tract infections can be associated with the hospital (HAUTIs) and the community (CAUTIs). In the case of CAUTIs, it is known whether women are the predominant group of patients.

Although the UTI is multifactorial, the main bacteria related to the diagnosis is *E. coli*, predominant in both community and nosocomial UTIs [93].

Co-trimoxazole (trimethoprim/sulfamethoxazole), nitrofurantoin, ciprofloxacin and ampicillin are the antibiotics commonly recommended for the treatment of UTIs. However, there is an overall increase in antibiotic resistance among pathogens in the urinary tract, which is a limitation on treatment options [94, 95].

Since the evidence suggests a significant relationship between the extensive use of antibiotics and antimicrobial resistance, it is necessary to prescribe and use antibiotics in order to reduce their complications and costs [96].

For this reason, in order to guide the selection of empirical therapy, surveillance of antibiotic resistance is crucial for determining the pattern of antimicrobial resistance [97].

8. Workup

8.1 Urine culture

It aims to check the presence of fungi and bacteria in the urine, being carried out from a urine sample, which was placed in Petri dishes. The urine culture is placed in an incubator (1–2 days) and if there is any microorganism in the tested material, colonies grow and are visible on the plate. When the result is positive for some bacteria, a test antibiogram is performed, which determines the type of antibiotic needed to act against the pathogen [98].

The culture of urine is important precisely because it allows the precise recognition of the bacteria and, consequently, the best antibiotic to be used [99].

As urine culture is most frequently requested when UTI is suspected, the most common bacteria found are *Escherichia coli* (between 47.5% and 56.4% of all urine culture) [100, 101].

8.2 Blood culture

Blood culture is part of the routine assessment of patients with suspected bloodstream infection, and is crucial to guide therapeutic intervention. The ideal method for collecting blood culture is venepuncture, since it increases diagnostic yield, and has lower rates of contamination, according to some studies [102].

Since the timing of blood culture collection does not influence the detection of clinically relevant microorganisms, most authorities recommend collecting several sets simultaneously or for a short period of time, with the exception of patients with endovascular infection who need documented continuous bacteremia [103, 104].

Two to four sets of blood samples should be collected, whenever possible, at independent locations [103–106]. For adults, the volume required for the examination varies between 40 and 160 mL of blood, and for babies and children, the volume is age-based and does not exceed 1% of the patient's total blood volume [103, 107].

The importance of blood culture, as well as urine, is related to the determination of the bacteria and the antibiogram, which directs the treatment to the best antibiotic to be used [108].

8.3 Localization of underlying abnormality

In some cases, it is possible to suspect a complicated urinary tract infection/urosepsis without being serious urological abnormalities. In such cases, there are some screening options that can be performed to assist in the management of the patient. Thus, simple abdominal radiography, intravenous urography, ultrasound, computed tomography and magnetic resonance imaging are cited [109].

8.4 Imaging exams

The anatomical identification of most areas of infection has become common with the development of high resolution cross-sectional images, which allow visualization of bacterial and viral metabolism, early diagnosis and treatment. Thus, the cross-sectional image was included as part of the routine investigation of unidentified infection sites and sources of sepsis. The trend is that the use of these images will become increasingly widespread and become part of standard clinical care in the near future [110].

8.4.1 Ultrasonography

When abdominal sepsis is suspected, ultrasound is a valuable tool. As it is a portable scanning technique, it is ideal for clinically unstable patients who cannot be transported to an examination room [110].

Ideal for the diagnosis of liver sepsis and gallbladder, ultrasound identifies and indicates the presence and location of intra-abdominal fluids (subphrenic space, in pericological calculations or pelvis) [110–113]. Intrahepatic fluids are also well visualized, and can even be drained percutaneously with ultrasound guidance [110].

The main obstacle for ultrasound responses is air interference, highlighted in loop regions of the intestine with intraluminal gas, since the USG image is darkened and makes it difficult to visualize interloop abscesses or peri-pancreatic collections. The intestine in patients with disease due to sepsis or recent intra-abdominal surgery is also capable of compromising the quality of the ultrasound [114].

8.4.2 CT scanning

The availability of CT scanners with multiple detectors allows rapid acquisition of images, making this method the most common in the diagnosis and detection of intra-abdominal abscesses [114, 115]. It is an interesting option especially for sick patients who have difficulty holding their breath, obese or with abdominal or chest bandages.

In addition, CT is essential in the diagnosis of interloop and retroperitoneal pathologies (including retroperitoneal abscesses or pancreatitis or intra-biliary stones), in addition to being highly sensitive in the detection of chest pathologies (pneumonia, pleural effusion and localized collections) [113, 115–117]. For intra-abdominal fluids and abscesses, CT showed a sensitivity of 90–100%, while ultrasound showed sensitivity between 80% and 85% [115, 118, 119].

Due to the contemporary contrast protocols available, it is possible to identify by CT even small infected collections [110].

8.4.3 Hybrid PET/MRI systems

With the development of hybrid cameras, the combination of PET and magnetic resonance imaging was introduced, which despite having interesting advantages and clinical applications, is still such an expensive tool.

The simultaneous acquisition of PET and magnetic resonance imaging can provide quantitative molecular functional information about the inflammatory lesion and precise location, in addition to anatomical changes with movement correction, improving the differential diagnosis and guiding anti-inflammatory treatment strategies.

Since MRI cannot visualize all parts of the body at once, the new hybrid technique may require collaboration between radiologists and nuclear medicine doctors to interpret the image and can be more expensive than PET/CT (capital and operational costs).

The functional image of inflammation and infection was mainly restricted to the flat image and SPECT, however, with the increasing development of PET radiopharmaceuticals, the detection and quantification of specific aspects of inflammatory processes became more sensitive. Precisely for this reason, there is an interesting potential in the application of hybrid whole body PET/MRI in the context of the investigation of infectious and inflammatory diseases [120].

8.5 Biologic scanning

Imaging technique that uses biological radionuclides to track hidden infections and improve the specificity of the infection diagnosis that allows the detection of early pathophysiological changes even when there are no apparent anatomical changes. When compared to ex vivo techniques (blood culture), in vivo biological screening is preferred since it is accurate, does not require a sterile environment and does not expose the health team to the risk of contamination by blood-borne pathogens.

This type of tool is used mainly in patients suspected of infection or abscess, but who have had negative results for the cross-sectional image. Thus, the use of marked leukocyte traffic allows a response to hidden sites, based on the recognition of white blood cells marked with radionuclides. The marked leukocytes travel to the infection sites and allow noninvasive images in areas of hidden infection, such as osteomyelitis, orthopedic prosthesis, endocarditis or inflammation and intestinal disease [110].

9. General management of sepsis and septic shock

9.1 Hemodynamic support

Adequate organ perfusion must be ensured. Hypotension should be managed initially with intravenous fluid administration and the goal should be maintenance of pulmonary capillary wedge pressure at 12–16 mm Hg or central venous pressure at 8–12 cm H2O. Urine output rate should be kept at greater than 0.5 mL/kg/hr. A mean arterial blood pressure of greater than 65 mmHg (systolic blood pressure greater than 90 mmHg) and a cardiac index of greater than or equal to 4 L/min/m^2 should be maintained. Vasopressor therapy should be initiated in the event of failure to achieve these goals with iv fluids alone. These include dopamine, dobutamine and norepinephrine [109].

9.2 Respiratory support

Ventilatory support should be provided for patients with progressive hypoxemia, hypercapnia, altered sensorium or respiratory muscle fatigue. A study of "early goal directed therapy" (EGDT) found that prompt resuscitation to maintain $SvO_2 > 70\%$ was associated with improved survival in patients of severe sepsis [121]. In this study, failure to maintain saturation after fluids and vasopressors was followed by erythrocyte infusion to raise hematocrit to 30%. Patients requiring mechanical ventilation should be adequately sedated and stress ulcer prophylaxis should be administered.

9.3 Metabolic support

Blood glucose levels should be maintained at less than 150 mg/dL during initial few days of severe sepsis and normoglycemic range could be targeted later. Frequent blood glucose monitoring should be done to avoid hypoglycemia in patients on intensive insulin therapy. Multi-organ dysfunction, if any should be managed. Disseminated intravascular coagulation, if accompanied by major bleeding, should be treated with fresh-frozen plasma and platelet transfusion. Hypercatabolic individuals with acute renal failure benefit substantially from hemodialysis or hemofiltration. Prophylaxis for deep vein thrombosis and nutritional supplementation should be undertaken [109].

10. Treatment of carbapenem-resistant Enterobacteriaceae

10.1 Monotherapy vs. combination therapy for treatment

Considering the limited knowledge about the combination of antibiotics, the susceptibility of these pathogens to drugs and the lack of evidence to support the routine use of combined antimicrobial therapy, the decision regarding the ideal therapy is the responsibility of medical professionals [122]. Regarding the most appropriate approach, it is prioritized in the literature that the optimization of antimicrobial therapy includes adaptation of the appropriate antibiotics in terms of class, dose, frequency, route and duration [123].

The combination of different antibiotics has been widely used by large centers when it comes to invasive infections by multi-resistant Gram-negative bacteria [122].

10.1.1 Positive and negative aspects of combination therapy for treatment

The various positive and negative aspects of combination therapy are depicted in **Table 3**.

Positive aspects of combination therapy for treatment	Negative aspects of combination therapy for treatment
1. Greater probability of choosing an effective agent and well-founded theoretical reasons to support its use 2. Considering the increase in mortality related to the delay in the establishment of treatment and delays in appropriate and effective antimicrobial treatment, it is prudent to initiate empirical broad-spectrum antimicrobial treatment in the first suspected infection in critically ill patients 3. Indicated for patients with compromised immune systems, previous ICU admissions or who have recently received broad-spectrum antibiotics [124]	1. Increased toxicity in treatment by combining antibiotics (nephrotoxicity and ototoxicity). In such cases, it is suggested to discontinue the old therapy and introduce a new one, based on the clinical evolution of the patient and the results of the culture and susceptibility profile 2. This type of therapy has not been shown to be effective by clinical data (meta-analyses performed with the evaluation of randomized clinical trials demonstrate that there was no difference in clinical results between the two strategies for definitive treatment of Gram-negative bacteria infections) [124]

Table 3.
Comparison of positive and negative aspects of combination therapy.

10.2 Colistin

Antibiotics such as colistin are the last resort to deal with infections by carbapenem-resistant Enterobacteriaceae (CREB), and when the pathogen does not respond to colistin, therapeutic options are severely restricted. Thus, it becomes necessary to restore the sensitivity of the pathogen to the drug [125].

The combination of colistin + salicylate + potent efflux pump inhibitor (BC1) has been documented with highly positive results, providing a connection between colistin and the efflux pump inhibitor (BC1), which prevents extrusion of colistin [126].

The reduction in affinity between the drug and Gram-negative bacteria is due to the modification of lipid A, linked to the appearance of the gene that confers resistance to bacteria, which is present in animals that receive colistin and are part of human food. Despite this, there is still no complete explanation of the mutation and resistance of Gram-negative bacteria (especially Enterobacteriaceae) in patients who received administered colistin [127].

10.3 Carbapenems

Due to the increased resistance of bacteria to cephalosporin (and aminopenicillins), the use of narrow-spectrum β-lactamases, especially carbapenems, has increased considerably, being the only β-lactamase antibiotics with proven effectiveness in serious infections due to ESBL-producing bacteria [128–130].

With the discovery of *E. coli* isolates capable of producing new b-lactamases, a new strain of *E. coli* was found capable of resisting the action of carbapenems, mediated by plasmids.

These enzymes are able to confer resistance to drugs of the class b-lactamases, and in relation to *E. coli* specifically, the main types of enzymes are CMY, CTX-M and NDM of b-lactamase [131].

10.4 Tigecycline and other tetracyclines

Tigecycline is a new expanded-spectrum antimicrobial agent in the glycylcycline class. Developed with the objective of overcoming the most common processes of bacterial resistance, the drug has emerged as a great therapeutic option in the treatment of serious infections, which endanger the patient's life, and which no longer respond to traditional antibiotics. The use of tigecycline is mainly interesting for the initial therapy of major infections, and is largely effective in the action against multi-resistant Gram-negative bacteria [132].

10.5 Aminoglycosides

Aminoglycosides are natural or semi-synthetic drugs obtained from actinomycetes, used as an antibiotic since the beginning of bacterial treatment. As it was replaced in the 1980s by cephalosporins, carbapenems and fluoroquinolones, aminoglycosides had little use.

With the increase in the number of cases of multidrug-resistant bacteria, aminoglycosides were again considered for their ability to synergize with a variety of other classes of antibacterials, improving the safety and effectiveness of the class through optimized dosing regimens, being broad-spectrum and quickly bactericidal.

Enzymatic modification by acetylation of an amino group, impaired uptake and phosphorylation of aminoglycosides are the most commonly reported processes that confer resistance to bacteria in relation to aminoglycosides [133].

10.6 Fosfomycin

Fosfomycin is an antibiotic from the 1969s, prescribed mainly in its oral form for the treatment of uncomplicated urinary tract infections (UTI), and considered as an option in the treatment of bacteria with advanced resistance, causing serious infections [134].

For *E. coli* NDM-producing strains, fosfomycin, colistin and tigecycline are more effective than other antibiotics [135].

The best pharmacological approach to *E. coli* infections resistant to carbapenems is still an obstacle to be overcome, since patients infected with this type of bacteria have more limited clinical results and when compared to patients infected with bacteria susceptible to drugs [136].

10.7 Duration of therapy

The duration of treatment for infection caused by *Escherichia coli* varies in the literature, but most patients require treatment for about 14–21 days [109]. For *E. coli* perinephric abscesses or prostatitis, it is recommended that the minimum antibiotic use time should be 6 weeks, intra-abdominal infections 14–21 days, and pneumonia 14 days (**Table 4**) [137].

Condition	General	Perinephric abscesses	Prostatitis	Intra-abdominal infections	Pneumonia
Duration	14–21 days	42 days	42 days	14–21 days	14 days

Table 4.
Recommended duration of antibiotic therapy depending upon the type of infection.

11. Special considerations

11.1 Urinary infections in diabetes mellitus

In general, infectious diseases occur more frequently and cause greater concern when dealing with diabetic patients. This occurs because the environment offered by the organism is rich in glucose, which favors immune dysfunction, including decreasing the antibacterial activity of the urine and its motility [138].

Moreover, when comparing *E. coli* isolated in the urine of diabetics and non-diabetics, the same virulence factors and the same resistance to antimicrobials are found, inferring that there is no difference in the causative bacteria. This way, what makes the prevalence of urinary infections to be higher in diabetic patients is the greater adhesion of *E. coli* bacteria to diabetic uroepithelial cells, the reduction of urinary cytokine secretion and the number of leukocytes [139].

Hence, to treat the disease, the most commonly prescribed antimicrobials are used—amoxicillin, nitrofurantoin, trimethoprim/sulfamethoxazole (TMP/SMX) and ciprofloxacin. It is understood that the same treatment choice used by nondiabetic patients can be made, depending only on the local resistance patterns of the commonly found uropathogens [140, 141].

Generally, most uropathogens have a high resistance to TMP/SMX, in addition, this antimicrobial can cause hypoglycemia, which makes it not a good first choice of treatment for this portion of patients [142].

As for the treatment, it is recommended to consider the urinary tract infection complicated, it is advisable to keep the treatment for a period of 7 to 14 days [143].

11.2 Acute pyelonephritis

Acute pyelonephritis is an infection located in the upper urinary tract, which accommodates either parenchyma and renal pelvis, with *Escherichia coli* being the most common etiological agent [144, 145].

Approximately 250,000 cases of this disease are reported each year, with more than 100,000 eventually requiring hospitalization [146].

In order to confirm the diagnosis of the disease, the patient's urine culture is performed before the start of antibiotic therapy [147]. In addition, it is recommended to perform a microbial susceptibility test in order to select the most appropriate antimicrobial regimen [148, 149].

If the diagnosis is uncertain or the patient is immunocompromised and suspected of having a hematogenic infection, blood culture analysis is requested [150, 151].

In the last few decades, there has been an increasing rate of resistance of *E. coli* bacteria to beta-lactam antibiotics of extended spectrum [152]. Thus, for patients with mild and uncomplicated acute pyelonephritis, fluoroquinolone is a good choice for initial outpatient antibiotic therapy, if the drug resistance rate is 10% or less in the community [153].

On the other hand, in cases of complicated infections, sepsis or failed outpatient treatment, hospital treatment is best indicated [154]. After antibiotic therapy, urine culture should be performed again after 1–2 weeks to conclude whether the treatment was successful or not [155].

11.3 Emphysematous pyelonephritis

Emphysematous pyelonephritis (EPN) is a severe necrotizing infection of the renal parenchyma and its surrounding tissues—resulting in the presence of gas in the renal parenchyma, collecting system or perinephric tissue—and is caused in 70% of cases by *Escherichia coli* (isolated in cultures of urine or pus from patients with the condition) [156].

The clinical evolution of EPN when not recognized and treated immediately can be serious and pose a risk to the patient's life. Another fact that should be mentioned is that up to 95% of the cases of EPN are underlyingly associated with uncontrolled diabetes mellitus [157, 158].

In addition to the risk of developing EPN primarily, the risk of developing secondary to an obstruction of the urinary tract is considerably relevant, about 25–40% can be considered as positive findings in EPN [159, 160].

The combination of percutaneous drainage (PCD) and medical management (MM) revealed a significant reduction in mortality rates [161, 162]. Thus, it is recommended that PCD be performed in patients with localized areas of gas and the presence of functional renal tissue. Another approach that can be used in association with treatment is emergency nephrectomy, classified as simple, radical or laparoscopic [163].

11.4 Renal abscess

Being caused by kidney stones, structural abnormality, history of urological surgery, trauma or any other cause of obstruction, renal abscess can also be related to pathogens [164]. The predominant organisms causing renal abscesses are Gram-negative organisms, and the most common is *Escherichia coli* [165–167].

Among the various intra-abdominal abscesses, renal abscess is a rare entity, especially in children and accounts for a number of cases of "missed diagnoses" [166, 168].

With regard to the symptoms of pediatric patients, the presentation of fever, flank pain, with or without a palpable mass, has been established in the literature; increased leukocyte count and increased erythrocyte sedimentation rate [169].

Early diagnosis is a key factor in the management of these patients, which can be aided by Ultrasound (USG). Drainage of pus and appropriate antibiotic therapy is the gold standard for treatment, being able to treat a great amount of cases. Thereby, the most successful combination of antibiotics was ceftriaxone, being associated with amikacin. Cases that cannot be resolved by the conventional approach can be treated with surgery, such as nephrectomy. Thus, complications such as extension of the peritoneal cavity, skin or chest can be avoided [166, 167].

11.5 Perinephric abscess

Perinephric abscess results from perirenal fatty necrosis, usually a complication of urological infection (more than 75%) [170]. Most of these abscesses have *Escherichia coli* as the main responsible, about 51.4% [171]. Perinephric abscess, when more diffuse, is capable of affecting the renal capsule and also Gerota's fascia [170]. Since the condition has an insidious onset of nonspecific protein symptoms, it is necessary for a clinical physician to maintain a high level of attention to avoid possible delay in diagnosis, since perinephric abscesses are associated with significant morbidity and mortality [172].

11.6 Renal papillary necrosis

Renal papillary necrosis (NPN) is a condition defined as ischemic necrobiosis of the papilla in the kidney medulla. Among several etiological factors important for the involvement of papillary necrosis, pyelonephritis due to bacterial uropathogens such as *E. coli* is one of those mentioned in the literature [173].

In order to improve the prognosis of the disease and reduce morbidity, the ideal is that the diagnosis of the disease is as early as possible. In this sense, it is clear that the radiological image is able to offer an early diagnosis and guidance in relation to the immediate treatment of papillary necrosis, thus minimizing the decline in renal function [174].

11.7 Prostatic abscess

Failure to respond to standard therapy for acute bacterial prostatitis can lead to complications, such as prostate abscess or fistula [175].

Acute bacterial prostatitis is a common and clinically important genitourinary disorder that has a higher incidence in patients with diabetes, cirrhosis and suppressed immune system. Usually caused by an ascending infection, it can also be triggered by organisms that cause other common genitourinary infections that may also be responsible for acute bacterial prostatitis. Being introduced during transrectal prostate biopsy, the clinical presentation ranges from mild symptoms of the lower urinary tract to total sepsis, and *Escherichia coli* is one of the main bacteria related to the clinical picture.

Regarding the therapeutic approach, oral or intravenous antibiotics are most effective in curing the infection. In this sense, the progression to chronic bacterial prostatitis is uncommon. It should be noted that special attention is needed in relation to immunosuppressed patients, whereas bacterial prostatitis in these patients may be caused by atypical infecting organisms and, therefore, may require additional therapies [176].

12. Prevention

It is already known that iron is an essential micronutrient for most bacteria and hosts, in this thought line, it is also known that there are relatively rare classical siderophilic pathogens that cause an increase in hepcidin in the body, responsible for the sequestration of iron for macrophages and enterocytes and, consequently hypoferremia [177–180]. So, current studies investigate if this mechanism used by the body against rare siderophilic bacteria, it also works for a wider set of bacteria. Results of these studies are shown to be positive, by demonstrating that excess iron allows rapid bacterial replication and spread, which means a susceptibility to infection caused by *E. coli* and that hepcidin is essential to protect against infections caused by *Escherichia coli*. [181, 182]. Thus, the use of hepcidin agonists promises to be an effective early intervention in patients with infections and dysregulated iron metabolism to avoid complications.

With regard to urinary tract infection, an effective preventive measure is the characterization and correction of the underlying genitourinary abnormalities that promote the infection. Another alternative mentioned in the literature is the future development of catheters whose material limits the growth of biofilm [109].

13. Conclusion

Early symptom recognition, followed by appropriate investigations, accurate diagnosis and early goal-directed therapy, is essential to improve results. Patient management includes an interprofessional team approach, with microbiologists, radiologists, surgeons and intensive care physicians [109].

Author details

Akshay Kumar[1*], Ana Francesca Vommaro Leite[2], Lais Sanches Maekawa[2], Roopvir Kaur[3], Silas Jose Braz Filo[2], Purnadeo Persaud[4], Juber Dastagir Shaikh[5], Asim Kichloo[6] and Nimisha Shiwalkar[7]

1 Medanta Hospital, Gurugram, India

2 University of Minas Gerais, Brazil

3 Government Medical College, Amritsar, India

4 Kansas City University, Kansas City, United States

5 MGM Hospital, Navi Mumbai, India

6 Department of Internal Medicine, Central Michigan University, United States

7 University of Texas Southwestern Medical Center, Dallas, Texas, United States

*Address all correspondence to: drakshay82@gmail.com

References

[1] Foxman B. The epidemiology of urinary tract infection. Nature Reviews. Urology. 2010;**7**:653-660

[2] Kaper JB, Nataro JP, Mobley HL. Pathogenic *Escherichia coli*. Nature Reviews. Microbiology. 2004;**2**:123-140

[3] Kim KS. Current concepts on the pathogenesis of *Escherichia coli* meningitis: Implications for therapy and prevention. Current Opinion in Infectious Diseases. 2012;**25**:273-278

[4] Leimbach A, Hacker J, Dobrindt U. *E. coli* as an all-rounder: The thin line between commensalism and pathogenicity. Current Topics in Microbiology and Immunology. 2013;**358**:3-32

[5] Berg RD. The indigenous gastrointestinal microflora. Trends in Microbiology. 1996;**4**:430-435

[6] Gordon DM, Cowling A. The distribution and genetic structure of *Escherichia coli* in Australian vertebrates: Host and geographic effects. Microbiology. 2003;**149**:3575-3586

[7] Wasiński B. Extra-intestinal pathogenic *Escherichia coli*-threat connected with food-borne infections. Annals of Agricultural and Environmental Medicine. 2019;**26**(4):532-537

[8] Donnenberg MS. Capítulo 220: Enterobacteriaceae. In: Mandell GL, Bennett JE, Dolin R, Blase MJ, editors. Mandell, Douglas y Bennett. Enfermedades Infecciosas. Principios y Práctica. 8va Edición ed. España: Elsevier; 2016. pp. 2640-2655

[9] Artico MJ, Rocchi M, Gasparotto A, Ocaña Carrizo V, Navarro M, Mollo V, et al. Bacteriemias de origen comunitario en pacientes adultos que acuden al servicio de urgencias de un hospital universitario. Revista Argentina de Microbiología. 2012;**44**(1):10-15

[10] Salyers AA, Whitt DD. Bacterial Pathogenesis: A Molecular Approach. Washington, DC: ASM Press; 2002

[11] Bertani B, Ruiz N. Function and biogenesis of lipopolysaccharides. EcoSal Plus. 2018;**8**(1)

[12] Mora-Rillo M, Fernández-Romero N, Navarro-San Francisco C, Díez-Sebastián J, Romero-Gómez MP, Fernández FA, et al. Impact of virulence genes on sepsis severity and survival in *Escherichia coli* bacteremia. Virulence. 2015;**6**(1):93-100

[13] Biran D, Ron EZ. Extraintestinal pathogenic *Escherichia coli*. Current Topics in Microbiology and Immunology. 2018;**416**:149-161

[14] Mendoza-Palomar N, Balasch-Carulla M, González-Di Lauro S, et al. *Escherichia coli* early-onset sepsis: Trends over two decades. European Journal of Pediatrics. 2017;**176**(9):1227-1234

[15] Peralta G, Sanchez MB, Garrido JC, et al. Impact of antibiotic resistance and of adequate empirical antibiotic treatment in the prognosis of patients with *Escherichia coli* bacteraemia. The Journal of Antimicrobial Chemotherapy. 2007;**60**:855-863

[16] Melzer M, Petersen I. Mortality following bacteraemic infection caused by extended spectrum beta-lactamase (ESBL) producing E. coli compared to non-ESBL producing *E. coli*. The Journal of Infection. 2007;**55**:254-259

[17] Gransden WR, Eykyn SJ, Phillips I, Rowe B. Bacteremia due to *Escherichia coli*: A study of 861 episodes. Reviews of Infectious Diseases. 1990;**12**:1008-1018

[18] Laupland KB, Gregson DB, Church DL, Ross T, Pitout JD.

Incidence, risk factors and outcomes of *Escherichia coli* bloodstream infections in a large Canadian region. Clinical Microbiology and Infection. 2008;**14**(11):1041-1047

[19] Jackson LA, Benson P, Neuzil KM, Grandjean M, Marino JL. Burden of community-onset *Escherichia coli* bacteremia in seniors. The Journal of Infectious Diseases. 2005;**191**:1523-1529

[20] Katouli M. Population structure of gut *Escherichia coli* and its role in development of extra-intestinal infections. Iranian Journal of Microbiology. 2010;**2**(2):59

[21] Thursby E, Juge N. Introduction to the human gut microbiota. Biochemical Journal. 2017;**474**(11):1823-1836

[22] Kaper JB, Nataro JP, Mobley HL. Pathogenic *Escherichia coli*. Nature Reviews. Microbiology. 2004;**2**:123-140

[23] Croxen MA, Finlay BB. Molecular mechanisms of *Escherichia coli* pathogenicity. Nature Reviews. Microbiology. 2010;**8**:26-38

[24] Trabulsi LR, Keller R, Tardelli Gomes TA. Typical and atypical enteropathogenic *Escherichia coli*. Emerging Infectious Diseases. 2002;**8**:508-513

[25] Bieber D, Ramer SW, Wu CY, Murray WJ, Tobe T, Fernandez R, et al. Type IV pili, transient bacterial aggregates, and virulence of enteropathogenic *Escherichia coli*. Science. 1998;**280**:2114-2118

[26] Frankel G, Phillips AD, Rosenshine I, Dougan G, Kaper JB, Knutton S. Enteropathogenic and enterohaemorrhagic *Escherichia coli*: More subversive elements. Molecular Microbiology. 1998;**30**:911-921

[27] Gomez-Duarte OG, Kaper JB. A plasmid-encoded regulatory region activates chromosomal eaeA expression in enteropathogenic *Escherichia coli*. Infection and Immunity. 1995;**63**:1767-1776

[28] Hu J, Torres AG. Enteropathogenic *Escherichia coli*: Foe or innocent bystander? Clinical Microbiology and Infection. 2015;**21**(8):729-734

[29] Gomes TAT, Elias WP, Scaletsky ICA, Guth BEC, Rodrigues JF, Piazza RMF, et al. Diarrheagenic *Escherichia coli*. Brazilian Journal of Microbiology. 2016;**47**(Suppl. 1):3-30

[30] Diepold A, Armitage JP. Type III secretion systems: The bacterial flagellum and the injectisome. Philosophical Transactions of the Royal Society, B: Biological Sciences. 2015;**370**(1679):20150020

[31] Berkes J, Viswanathan VK, Savkovic SD, Hecht G. Intestinal epithelial responses to enteric pathogens: Effects on the tight junction barrier, ion transport, and inflammation. Gut. 2003;**52**(3): 439-451

[32] Madhavan TP, Sakellaris H. Colonization factors of enterotoxigenic *Escherichia coli*. Advances in Applied Microbiology. 2015;**90**:155-197. DOI: 10.1016/bs.aambs.2014.09.003

[33] Loos M, Geens M, Schauvliege S, Gasthuys F, van der Meulen J, Dubreuil JD, et al. Role of heat-stable enterotoxins in the induction of early immune responses in piglets after infection with enterotoxigenic *Escherichia coli*. PLoS One. 2012;**7**:e41041

[34] Beltran AR, Carraro-Lacroix LR, Bezerra CN, Cornejo M, et al. *Escherichia coli* heat-stable enterotoxin mediates Na^{+}/H^{+} exchanger 4 inhibition involving cAMP in T84 human intestinal epithelial cells. PLoS One. 2015;**10**:e0146042

[35] Dorsey FC, Fischer JF, Fleckenstein JM. Directed delivery of heat-labile enterotoxin by enterotoxigenic *Escherichia coli*. Cellular Microbiology. 2006;**8**(9):1516-1527

[36] Sears CL, Kaper JB. Enteric bacterial toxins: Mechanisms of action and linkage to intestinal secretion. Microbiological Reviews. 1996;**60**(1):167

[37] Ceponis PJ, Riff JD, Sherman PM. Epithelial cell signaling responses to enterohemorrhagic *Escherichia coli* infection. Memórias do Instituto Oswaldo Cruz. 2005;**100**:199-203

[38] Boyer O, Niaudet P. Hemolytic uremic syndrome: New developments in pathogenesis and treatment. International Journal of Nephrology. 2011

[39] Polito MG, Kirsztajn GM. Microangiopatias trombóticas: púrpura trombocitopênica trombótica e síndrome hemolítico-urêmica. Brazilian Journal of Nephrology. 2010;**32**(3):303-315

[40] Ud-Din A, Wahid S. Relationship among *Shigella* spp. and enteroinvasive *Escherichia coli* (EIEC) and their differentiation. Brazilian Journal of Microbiology. 2014;**45**(4):1131-1138

[41] Tulloch EF, Ryan KJ, Formal SB, Franklin FA. Invasive enteropathic *Escherichia coli* dysentery: An outbreak in 28 adults. Annals of Internal Medicine. 1973;**79**(1):13-17

[42] Jafari A, Aslani MM, Bouzari S. *Escherichia coli*: A brief review of diarrheagenic pathotypes and their role in diarrheal diseases in Iran. Iranian Journal of Microbiology. 2012;**4**(3):102

[43] Mansan-Almeida R, Pereira AL, Giugliano LG. Diffusely adherent *Escherichia coli* strains isolated from children and adults constitute two different populations. BMC Microbiology. 2013;**13**(1):22

[44] Sarowska J, Futoma-Koloch B, Jama-Kmiecik A, Frej-Madrzak M, Ksiazczyk, et al. Virulence factors, prevalence and potential transmission of extraintestinal pathogenic Escherichia coli isolated from different sources: Recent reports. Gut Pathogens. 2019;**11**(1):10

[45] Smith JL, Fratamico PM, Gunther NW. Extraintestinal pathogenic *Escherichia coli*. Foodborne Pathogens and Disease. 2007;**4**(2):134-163

[46] Areda MA, Bailey CR, O'Mara D, Weiss CR. Transplant uretero-inguinal hernia resulting in urosepsis. Radiology Case Reports. 2019;**14**(1):14-17

[47] Thornton HV, Hammond A, Hay AD. Urosepsis: A growing and preventable problem? The British Journal of General Practice. 2018;**68**(675):493-494

[48] Fadel MG, Louis C, Tay A, Bolgeri M. Obstructive urosepsis secondary to ureteric herniation into the sciatic foramen. BML Case Reports. 2018 27;2018

[49] Scotland KB, Lange D. Prevention and management of urosepsis triggered by ureteroscopy. Research and Reports in Urology. 2018;**10**:43-49

[50] Heppner HJ, Yapan F, Wiedemann A. Urosepsis in geriatric patients. Aktuelle Urologie. 2016;**47**(1):54-59

[51] Goveas B. Urosepsis: A simple infection turns toxic. The Nurse Practitioner. 2017;**42**(7):53-54

[52] Porat A, Kesler S. Urosepsis. In: StatPearls [Internet]. Treasure Island, FL: StatPearls Publishing; 2020

[53] Vila J, Sáez-López E, Johnson JR, et al. *Escherichia coli*: An old friend with new tidings. FEMS Microbiology Reviews. 2016;**40**(4):437-463

[54] World Health Organization. Antimicrobial Resistance Global Report on Surveillance; 2014

[55] Friedman S, Shah V, Ohlsson A, Matlow AG. Neonatal *Escherichia coli* infections: Concerns regarding resistance to current therapy. Acta Paediatrica. 2000;**89**(6):686-689

[56] Stony Brook Medicine. Severe Sepsis/Septic Shock Recognition and Treatment Protocols. Surviving Sepsis Campaign. Society of Critical of Medicine. 2013

[57] Tsai J, Bond G. A comparison of electronic records to paper records in mental health centers. International Journal for Quality in Health Care. 2008;**20**:136-143

[58] Davis JL, Murray JF. History and physical examination. Murray and Nadel's Textbook of Respiratory Medicine. 2016;**263**

[59] Timsit JF, Esaied W, Neuville M, Bouadma L, Mourvllier B. Update on ventilator-associated pneumonia. F1000Research. 2017;**6**:2061

[60] Chastre J, Fagon JY. Ventilator-associated pneumonia. American Journal of Respiratory and Critical Care Medicine. 2002;**165**:867-903

[61] Messika J, Magdoud F, Clermont O, et al. Pathophysiology of *Escherichia coli* ventilator-associated pneumonia: Implication of highly virulent extraintestinal pathogenic strains. Intensive Care Medicine. 2012;**38**(12):2007-2016

[62] Phillips-Houlbracq M, Ricard JD, Foucrier A, Yoder-Himes D, Gaudry S, et al. Pathophysiology of *Escherichia coli* pneumonia: Respective contribution of pathogenicity islands to virulence. International Journal of Medical Microbiology. 2018;**308**(2):290-296

[63] La Combe B, Clermont O, Messika J, Eveillard M, Kouatchet A, Lasocki S, et al. Pneumonia-specific *Escherichia coli* with distinct phylogenetic and virulence profiles, France, 2012-2014. Emerging Infectious Diseases. 2019;**25**(4):710-718

[64] Sule AA, Tai D. Spontaneous *Escherichia coli* meningitis in an adult. Critical Care and Shock. 2007;**10**:148-150

[65] Mofredj A, Guerin JM, Leibinger F, Mamoudi R. Spontaneous *Escherichia coli* meningitis in an adult. Scandinavian Journal of Infectious Diseases. 2000;**32**:699-700

[66] Abhilash K, Gigin SV. A rare manifestation of *Escherichia coli* septicemia. International Journal of Research in Medical Sciences. 2013;**1**:299-300

[67] Shrestha RG, Tandukar S, Ansari S, Subedi A, Shrestha A, Poudel R, et al. Bacterial meningitis in children under 15 years of age in Nepal. BMC Pediatrics. 2015;**15**:94

[68] Ishida K, Noborio M, Nakamura M, Ieki Y, Sogabe T, Sadamitsu D. Spontaneous *Escherichia coli* bacterial meningitis mimicking heatstroke in an adult. Clinical Case Reports. 2016;**4**(4):323-326

[69] Chang CJ, Chang WN, Huang LT, Huang SC, Chang YC, et al. Bacterial meningitis in infants: The epidemiology, clinical features, and prognostic factors. Brain & Development. 2004;**26**:168-175

[70] Dawson KG, Emerson JC, Burns JL. Fifteen years of experience with bacterial meningitis. The Pediatric Infectious Disease Journal. 1999;**18**:816-822

[71] de Louvois J, Halket S, Harvey D. Neontal meningitis in England and Wales: Sequelae at 5 years of age. European Journal of Pediatrics. 2004;**7**:730-734

[72] Doctor BA, Newman N, Minich NM, Taylor HG, Fanaroff AA, Hack M. Clinical outcomes of neonatal meningitis in very-low birth-weight infants. Clinical Pediatrics (Phila). 2001;**40**:473-480

[73] Gladstone IM, Ehrenkranz RA, Edberg SC, Baltimore RS. A ten-year review of neonatal sepsis and comparison with the previous fifty-year experience. The Pediatric Infectious Disease Journal. 1990;**9**:819-825

[74] Holt DE, Halket S, de Louvois J, Harvey D. Neonatal meningitis in England and Wales: 10 years on. Archives of Disease in Childhood. Fetal and Neonatal Edition. 2001;**84**:F85-F89

[75] Kim KS. Acute bacterial meningitis in infants and children. The Lancet Infectious Diseases. 2010;**10**:32-42

[76] Klinger G, Chin C-N, Beyene J, Perlman M. Predicting the outcome of neonatal bacterial meningitis. Pediatrics. 2000;**106**:477-482

[77] Stevens JP, Eames M, Kent A, Halket S, Holt D, Harvey D. Long term outcome of neonatal meningitis. Archives of Disease in Childhood. Fetal and Neonatal Edition. 2003;**88**:F179-F184

[78] Unhanand M, Mustafa MM, McCracken GH Jr, Nelson JD. Gram-negative enteric bacillary meningitis: A twenty-one-year experience. The Journal of Pediatrics. 1993;**122**:15-21

[79] Kim KS. Comparison of cefotaxime, imipenem-cilastatin, ampicillin-gentamicin, and ampicillin chloramphenicol in the treatment of experimental *Escherichia coli* bacteremia and meningitis. Antimicrobial Agents and Chemotherapy. 1985;**28**:433-436

[80] McCracken GH Jr, Threlkeld N, Mize S, Baker CJ, Kapal SL, Fraingezicht I, et al. Moxalactam therapy for neonatal meningitis due to gram-negative sepsis enteric bacilli. Journal of the American Medical Association. 1984;**252**:1427-1432

[81] Kim KS. Human meningitis-associated *Escherichia coli*. EcoSal Plus. 2016;**7**(1). DOI: 10.1128/ecosalplus. ESP-0015-2015

[82] Bodmann KF. Complicated intra-abdominal infections: Pathogens, resistance. Recommendations of the Infectliga on antbiotic therapy. Der Chirurg; Zeitschrift fur alle Gebiete der operativen Medizen. 2010;**81**(1):38-49

[83] Sartelli M. A focus on intra-abdominal infections. World Journal of Emergency Surgery. 2010;**5**(1):9

[84] Chaithanya J, Ashwini RK, Rajagopalan S. Clinical and microbiological profile in intra-abdominal infection. International Surgery Journal. 2019;**6**(10):3608-3613

[85] Ruppé É, Woerther PL, Barbier F. Mechanisms of antimicrobial resistance in Gram-negative bacilli. Annals of Intensive Care. 2015;**5**:61

[86] Kot B. Antibiotic resistance among uropathogenic *Escherichia coli*. Polish Journal of Microbiology. 2019;**68**(4):403-415

[87] Clements A, Young JC, Constantinou N, Frankel G. Infection strategies of enteric pathogenic *Escherichia coli*. Gut Microbes. 2012;**3**(2):71-87

[88] Croxen MA, Law RJ, Scholz R, Keeney KM, Wlodarska M, Finlay BB. Recent advances in understanding enteric pathogenic *Escherichia coli*.

Clinical Microbiology Reviews. 2013;**26**(4):822-880

[89] World Health Organization. World Health Statistics 2012. Geneva, Switzerland: WHO Press; 2012

[90] Croxen MA, Law RJ, Scholz R, Keeney KM, Wlodarska M, Finlay BB. Recent advances in understanding enteric pathogenic *Escherichia coli*. Clinical Microbiology Reviews. 2013;**26**(4):822-880

[91] Clements A, Young JC, Constantinou N, Frankel G. Infection strategies of enteric pathogenic *Escherichia coli*. Gut Microbes. 2012;**3**(2):71-87

[92] Berger CN, Sodha SV, Shaw RK, Griffin PM, Pink D, Hand P, et al. Fresh fruit and vegetables as vehicles for the transmission of human pathogens. Environmental Microbiology. 2010;**12**:2385-2397

[93] Car J. Urinary tract infections in women: Diagnosis and management in primary care. BMJ. 2006;**332**(7533):94-97

[94] Alanazi MQ, Alqahtani FY, Aleanizy FS. An evaluation of *E. coli* in urinary tract infection in emergency department at KAMC in Riyadh, Saudi Arabia: Retrospective study. Annals of Clinical Microbiology and Antimicrobials. 2018;**17**(1)

[95] Lee JB, Neild GH. Urinary tract infection. Medicine. 2007;**35**(8):423-428

[96] Foxman B et al. Urinary tract infection: Self-reported incidence and associated costs. Annals of Epidemiology. 2000;**10**(8):509-515

[97] Leekha S, Terrell CL, Edson RS. General principles of antimicrobial therapy. Mayo Clinic Proceedings. 2011;**86**(2):156-167. DOI: 10.4065/mcp.2010.0639

[98] Understanding urine tests. InformedHealth.org [Internet]. Cologne, Germany: Institute for Quality and Efficiency in Health Care (IQWiG); 2006. 2010 Jul 27 [Updated 2019 Oct 24].

[99] Schmiemann G, Kniehl E, Gebhardt K, Matejczyk MM, Hummers-Pradier E. The diagnosis of urinary tract infection: A systematic review. Deutsches Ärzteblatt International. 2010;**107**(21):361-367

[100] Fonseca FLA, Santos PM, Belardo TMG, Fonseca ALA, Caputto LZ, Alves BCA. Análise de leucócitos em urina de pacientes com uroculturas positivas. Revista Brasileira de Análises Clínicas. 2016;**48**(3):258-261

[101] de Oliveira LCA. Prevalência de infecção do trato urinário em pacientes ambulatoriais e sua relação com os valores de nitrito e leucócitos. Revista Brasileira de Análises Clínicas. 2018;**50**(3):237-243

[102] Bryant JK, Strand CL. Reliability of blood cultures collected from intravascular catheter versus venipuncture. American Journal of Clinical Pathology. 1987;**88**:113-116

[103] Wilson M. Clinical and Laboratory Standards Institute. Principles and Procedures for Blood Cultures: Approved Guideline. Wayne, PA: Clinical and Laboratory Standards Institute; 2007

[104] Cockerill FR III, Wilson JW, Vetter EA, Goodman KM, Torgerson CA, Harmsen WS, et al. Optimal testing parameters for blood cultures. Clinical Infectious Diseases. 2004;**38**(12):1724-1730

[105] Lee A, Mirrett S, Reller LB, Weinstein MP. Detection of bloodstream infections in adults: How many blood cultures are needed? Journal of Clinical Microbiology. 2007;**45**(11):3546-3548

[106] Washington JA 2nd. Blood cultures: Principles and techniques. Mayo Clinic Proceedings. 1975;**50**(2):91

[107] Gaur AH, Giannini MA, Flynn PM, Boudreaux, et al. Optimizing blood culture practices in pediatric immunocompromised patients: Evaluation of media types and blood culture volume. The Pediatric Infectious Disease Journal. 2003;**22**(6):545-552

[108] Kirn TJ, Weinstein MP. Update on blood cultures: How to obtain, process, report, and interpret. Clinical Microbiology and Infection. 2013;**19**(6):513-520

[109] Kalra OP, Raizada A. Approach to a patient with urosepsis. Journal of Global Infectious Diseases. 2009;**1**(1):57-63

[110] Ady J, Fong Y. Imaging for infection: From visualization of inflammation to visualization of microbes. Surgical Infections. 2014;**15**(6):700-707

[111] Jones MW, Ferguson T. Gallbladder Imaging. In: StatPearls. StatPearls Publishing; 2019

[112] Doust BD, Quiroz F, Stewart JM. Ultrasonic distinction of abscesses from other intra-abdominal fluid collections. Radiology. 1977;**125**:213-218

[113] Korobkin M, Callen PW, Filly RA, et al. Comparison of computed tomography, ultrasonography, and gallium-67 scanning in the evaluation of suspected abdominal abscess. Radiology. 1978;**129**:89-93

[114] Filly RA. Detection of fetal malformations with ultrasonic B-scans. Birth Defects Original Article Series. 1979;**15**:45-49

[115] Wolverson MK, Jagannadharao B, Sundaram M, et al. CT as a primary diagnostic method in evaluating intraabdominal abscess. American Journal of Roentgenology. 1979;**133**:1089-1095

[116] Haaga JR, Alfidi RJ, Havrilla TR, et al. CT detection and aspiration of abdominal abscesses. American Journal of Roentgenology. 1977;**128**:465-474

[117] Silverstein W, Isikoff MB, Hill MC, Barkin J. Diagnostic imaging of acute pancreatitis: Prospective study using CT and sonography. American Journal of Roentgenology. 1981;**137**:497-502

[118] Knochel JQ, Koehler PR, Lee TG, Welch DM. Diagnosis of abdominal abscesses with computed tomography, ultrasound, and ^{111}In leukocyte scans. Radiology. 1980;**137**:425-432

[119] Roche J. Effectiveness of computed tomography in the diagnosis of intra-abdominal abscess: A review of 111 patients. The Medical Journal of Australia. 1981;**2**:85-88

[120] Glaudemans AW, Quintero AM, Signore A. PET/MRI in Infectious and Inflammatory Diseases: Will it be a Useful Improvement? 2012

[121] Otero RM, Nguyen HB, Huang DT, Gaieski DF, Goyal M, Gunnerson KJ, et al. Early goal directed therapy in severe sepsis and septic shock revisited: Concepts, controversies, and contemporary findings. Chest. 2006;**130**:1579-1595

[122] Tamma PD, Cosgrove SE, Maragakis LL. Combination therapy for treatment of infections with gram-negative bacteria. Clinical Microbiology Reviews. 2012;**25**(3):450-470

[123] Leekha S, Terrell CL, Edson RS. General principles of antimicrobial therapy. Mayo Clinic Proceedings. 2011;**86**(2):156-167. DOI: 10.4065/mcp.2010.0639

[124] Salomão R, Diament D, Rigatto O, Gomes B, Silva E, Carvalho NB, et al. Diretrizes para tratamento da sepse grave/choque séptico: Abordagem do agente infeccioso-controle do foco

infeccioso e tratamento antimicrobiano. Revista Brasileira de Terapia Intensiva. 2011;**23**(2):145-157

[125] Sheu CC, Chang YT, Lin SY, Chen YH, Hsueh PR. Infections caused by carbapenem-resistant Enterobacteriaceae: An update on therapeutic options. Frontiers in Microbiology. 2019;**10**:80

[126] Sundaramoorthy NS, Suresh P, Ganesan SS, GaneshPrasad A, Nagarajan S. Restoring colistin sensitivity in colistin-resistant *E. coli*: Combinatorial use of MarR inhibitor with efflux pump inhibitor. Scientific Reports. 2019;**9**(1):1-13

[127] Ezadi F, Ardebili A, Mirnejad R. Antimicrobial susceptibility testing for polymyxins: Challenges, issues, and recommendations. Journal of Clinical Microbiology. 2019;**57**(4)

[128] Hawkey Peter M, Livermore DM. Carbapenem antibiotics for serious infections. BMJ. 2012;**344**:e3236

[129] Shaikh S, Fatima J, Shakil S, Rizvi SM, Kamal MA. Antibiotic resistance and extended spectrum beta-lactamases: Types, epidemiology and treatment. Saudi Journal of Biological Sciences. 2015;**22**(1):90-101

[130] Pitout JD. Extraintestinal pathogenic *Escherichia coli*: An update on antimicrobial resistance, laboratory diagnosis and treatment. Expert Review of Anti-Infective Therapy. 2012;**10**:1165-1176

[131] Rossi F, Andreazzi D. Overview of tigecycline and its role in the era of antibiotic resistance. Brazilian Journal of Infectious Diseases. 2006;**10**(3):203-216

[132] Krause KM, Serio AW, Kane TR, Connolly LE. Aminoglycosides: An overview. Cold Spring Harbor Perspectives in Medicine. 2016;**6**(6):a027029

[133] Falagas ME, Vouloumanou EK, Samonis G, Vardakas KZ. Fosfomycin. Clinical Microbiology Reviews. 2016;**29**(2):321-347

[134] Ku YH, Chen CC, Lee MF, Chuang YC, Tang HJ, Yu WL. Comparison of synergism between colistin, fosfomycin and tigecycline against extended-spectrum β-lactamase-producing *Klebsiella pneumoniae* isolates or with carbapenem resistance. Journal of Microbiology, Immunology and Infection. 2017;**50**(6):931-939

[135] Zhou YF, Liu P, Zhang CJ, Liao XP, Sun J, Liu YH. Colistin combined with tigecycline: A promising alternative strategy to combat *Escherichia coli* harboring blaNDM–5 and mcr-1. Frontiers in Microbiology. 2020;**10**:2957

[136] Hamza NS, Khalil A. Resistant Gram-negative urinary tract bacterial infections. In: Urinary Tract Infection—The Result of the Strength of the Pathogen, or the Weakness of the Host. IntechOpen; 2017

[137] Giamarellou H. Multidrug-resistant Gram-negative bacteria: How to treat and for how long. International Journal of Antimicrobial Agents. 2010;**36**(Suppl 2):S50-S54

[138] Casqueiro J, Casqueiro J, Alves C. Infections in patients with diabetes mellitus: A review of pathogenesis. Indian Journal of Endocrinology and Metabolism. 2012;**16**(Suppl. 1):S27

[139] Geerlings SE. Urinary tract infections in patients with diabetes mellitus: Epidemiology, pathogenesis and treatment. International Journal of Antimicrobial Agents. 2008;**31**:54-57

[140] Fair WR, Crane DB, Peterson LJ, Dahmer C, Tague B, Amos W. Three-day treatment of urinary tract infections. The Journal of Urology. 1980;**123**:717-721

[141] Kavatha D, Giamarellou H, Alexiou Z, Vlachogiannis N, et al. Cefpodoxime-proxetil versus trimethoprim-sulfamethoxazole for short-term therapy of uncomplicated acute cystitis in women. Antimicrobial Agents and Chemotherapy. 2003;**47**:897-900

[142] Grigoryan L, Nash S, Zoorob R, et al. Qualitative analysis of primary care provider prescribing decisions for urinary tract infections. Antibiotics (Basel, Switzerland). 2019;**8**(2)

[143] Kang CI, Kim J, Park DW, Kim BN, et al. Clinical practice guidelines for the antibiotic treatment of community-acquired urinary tract infections. Infection & Chemotherapy. 2018;**50**(1):67-100

[144] Bass PF 3rd, Jarvis JA, Mitchell CK. Urinary tract infections. Primary Care. 2003;**30**:41-61

[145] Stamm WE, Hooton TM. Management of urinary tract infections in adults. The New England Journal of Medicine. 1993;**329**:1328-1334

[146] Hooton TM, Stamm WE. Diagnosis and treatment of uncomplicated urinary tract infection. Infectious Disease Clinics of North America. 1997;**11**:551-581

[147] Ramakrishnan K, Scheid DC. Diagnosis and management of acute pyelonephritis in adults. American Family Physician. 2005;**71**(5):933-942

[148] Johnson JR. Treatment and prevention of urinary tract infections. In: Mobley HLT, Warren JW, editors. Urinary Tract Infections: Molecular Pathogenesis and Clinical Management. Washington, DC: American Society for Microbiology Press; 1996. pp. 95-118

[149] Hooton TM, Stamm WE. Diagnosis and treatment of uncomplicated urinary tract infection. In: Moellering RC Jr, Andriole VT, eds. Infectious Disease Clinics of North America. 1997;**11**:551-581

[150] McMurray BR, Wrenn KD, Wright SW. Usefulness of blood cultures in pyelonephritis. The American Journal of Emergency Medicine. 1997;**15**:137-140

[151] Velasco M, Martinez JA, Moreno-Martinez A, Horcajada JP, Ruiz J, Barranco M, et al. Blood cultures for women with uncomplicated acute pyelonephritis: Are they necessary? Clinical Infectious Diseases. 2003;**37**:1127-1130

[152] Zahar JR, Lortholary O, Martin C, Potel G, Plesiat P, Nordmann P. Addressing the challenge of extended-spectrum beta-lactamases. Current Opinion in Investigational Drugs. 2009;**10**(2):172-180

[153] Colgan R, Williams M, Johnson JR. Diagnosis and treatment of acute pyelonephritis in women. American Family Physician. 2011;**84**(5):519-526

[154] Ramakrishnan K, Scheid DC. Diagnosis and management of acute pyelonephritis in adults. American Family Physician. 2005;**71**(5):933-942

[155] Delzell JEJR, Fitzsimmons A. Urinary tract infection. In: Reference Guide. 7th ed. The American Board of Family Practice; 2001

[156] Ubee SS, McGlynn L, Fordham M. Emphysematous pyelonephritis. BJU International. 2011;**107**(9):1474-1478

[157] Wan YL, Lee TY, Bullard MJ, Tsai CC. Acute gas-producing bacterial renal infection: Correlation between imaging findings and clinical outcome. Radiology. 1996;**198**:433-438

[158] Falagas ME, Alexiou VG, Giannopoulou KP, Siempos II. Risk factors for mortality in patients with

emphysematous pyelonephritis: A meta-analysis. The Journal of Urology. 2007;**178**:880-885

[159] Michaeli J, Mogle S, Perlberg S, Heiman S, Caine M. Emphysematous pyelonephritis. The Journal of Urology. 1984;**131**:203-207

[160] Godec CJ, Cass AS, Berkseth R. Emphysematous pyelonephritis in a solitary kidney. The Journal of Urology. 1980;**124**:119-121

[161] Khaira A, Gupta A, Rana DS, Gupta A, Bhalla A, Khullar D. Retrospective analysis of clinical profile, prognostic factors and outcomes of 19 patients of emphysematous pyelonephritis. International Urology and Nephrology. 2009;**41**:959-966

[162] Narlawar RS, Raut AA, Nagar A, Hira P, Hanchate V, Asrani A. Imaging features and guided drainage in emphysematous pyelonephritis: A study of 11 cases. Clinical Radiology. 2004;**59**:192-197

[163] Somani BK, Nabi G, Thorpe P, et al. Is percutaneous drainage the new gold standard in the management of emphysematous pyelonephritis? Evidence from a systematic review. The Journal of Urology. 2008;**179**:1844-1849

[164] Laufer J, Grisaru-Soen G, Portnoy O, Mor Y. Bilateral renal abscesses in a healthy child. The Israel Medical Association Journal. 2002;**4**(12):1150-1151

[165] Bhat YR. Renal subcapsular abscess. Indian Pediatrics. 2001;**44**:546-547

[166] Paul B, Agrawal A, Goyal RK. Renal abscess: A case of missed diagnosis. Journal of Indian Academy of Clinical Medicine. 2001;**2**(1):91-92

[167] Rai RS, Karan SC, Kayastha A. Renal & perinephric abscesses revisited. Medical Journal, Armed Forces India. 2007;**63**(3):223-225

[168] Baradkar VP, Mathur M, Kumar S. Renal abscess due to *Escherichia coli* in a child. Saudi Journal of Kidney Diseases and Transplantation. 2011;**22**(6):1215

[169] Chen CY, Kuo HT, Chang YJ, Wu KH, Yang WC, Wu HP. Clinical assessment of children with renal abscesses presenting to the pediatric emergency department. BMC Pediatrics. 2016;**16**(1):189

[170] Okafor CN, Onyeaso EE. Perinephric abscess. In: StatPearls. Treasure Island, FL: StatPearls Publishing; 2020

[171] Liu XQ, Wang CC, Liu YB, Liu K. Renal and perinephric abscesses in West China hospital: 10-year retrospective-descriptive study. World Journal of Nephrology. 2016;**5**(1):108-114

[172] Tsukagoshi D, Dinkovski B, Dasan S, Jethwa J. Perinephric abscess secondary to a staghorn calculus presenting as a subcutaneous abscess. Canadian Journal of Emergency Medicine. 2006;**8**(4):285-288

[173] Geller SA, de Campos F. Renal papillary necrosis. Autopsy & Case Reports. 2013;**3**(4):69-71

[174] Sutariya HC, Pandya VK. Renal papillary necrosis: Role of radiology. Journal of Clinical and Diagnostic Research. 2016;**10**(1):TD10-TD12

[175] Reddivari AKR, Mehta P. Prostate abscess. In: StatPearls. Treasure Island, FL: StatPearls Publishing; 2020

[176] Brede CM, Shoskes DA. The etiology and management of acute prostatitis. Nature Reviews. Urology. 2011;**8**(4):207-212

[177] Ganz T, Nemeth E. Regulation of iron acquisition and iron distribution

in mammals. Biochimica et Biophysica Acta. 2006;**1763**:690-699

[178] Ganz T, Nemeth E. Iron homeostasis in host defence and inflammation. Nature Reviews. Immunology. 2015;**15**:500-510

[179] Stefanova D, Raychev A, Deville J, Humphries R, Campeau S, Ruchala P, et al. Hepcidin protects against lethal *Escherichia coli* sepsis in mice inoculated with isolates from septic patients. Infection and Immunity. 2018;**86**(7):e00253-18

[180] Rodriguez R, Jung CL, Gabayan V, Deng JC, Ganz T, Nemeth E, et al. Hepcidin induction by pathogens and pathogen-derived molecules is strongly dependent on interleukin-6. Infection and Immunity. 2014;**82**:745-752

[181] Stefanova D, Raychev A, Deville J, Humphries R, Campeau S, Ruchala P, et al. Hepcidin protects against lethal *Escherichia coli* sepsis in mice inoculated with isolates from septic patients. Infection and Immunity. 2018;**86**(7):e00253-18

[182] Bulut Y. Abstract PCCLB-21: HEPCIDIN DEFICIENCY AND IRON OVERLOAD POTENTLY PROMOTE E. COLI SEPSIS IN MOUSE MODELS BUT THE HEPCIDIN AGONIST PR-73 IS PROTECTIVE. Pediatric Critical Care Medicine. 2018;**19**(6S):249

Chapter 7

Computational Studies of Drug Repurposing Targeting P-Glycoprotein-Mediated Multidrug-Resistance Phenotypes in Agents of Neglected Tropical Diseases

Nivedita Jaishankar, Sangeetha Muthamilselvan and Ashok Palaniappan

Abstract

Mammalian ABCB1 P-glycoprotein is an ATP- dependent efflux pump with broad substrate specificity associated with cellular drug resistance. Homologous to this role in mammalian biology, the P-glycoprotein of agents of neglected tropical diseases (NTDs) mediates the emergence of multidrug-resistance phenotypes. The clinical and socioeconomic implications of NTDs are exacerbated by the lack of research interest among Big Pharma for treating such conditions. This work aims to characterise P-gp homologues in certain agents of key NTDs, namely (1) Protozoa: *Leishmania major*, *Trypanosoma cruzi*; (2) Helminths: *Onchocerca volvulus*, *Schistosoma mansoni*. Based on structural modelling of the organismal P-gp homologues, potential antibiotics targeting these structures were identified based on similarity and repurposing of existing drugs. Docking studies of the Pgp receptor—antibiotic ligand complexes were carried out and the most tenable target-ligand conformations assessed. The interacting residues were identified, and binding pockets studied. The in silico studies yielded measurements of the relative efficacy of the new drugs, which need experimental validation. Our studies could lay the foundation for the development of effective synergistic or new therapies against key neglected tropical diseases. The potential mechanisms of multidrug resistance emergence in *E. coli* were examined.

Keywords: P-glycoprotein, neglected tropical diseases, multidrug resistance, homology modeling, receptor-ligand docking, differential ligand affinity, synergistic effects, leishmaniasis, trypanosomiasis, onchocerciasis, schistosomiasis

1. Introduction

1.1 Multidrug resistance (MDR)

Bacterial evolution has been constrained to respond to the selection pressure of antibiotics and combined with their reckless use and has led to the emergence of varied defenses against antimicrobial agents. The main mechanisms whereby the bacteria develop resistance to antimicrobial agents include enzymatic inactivation, modification of the drug target(s), and reduction of intracellular drug concentration by changes in membrane permeability or by the over expression of efflux pumps [1]. Multidrug resistance efflux pumps are recognized as an important component of resistance in both Gram-positive and Gram-negative bacteria [2]. Some bacterial efflux pumps may be selective for one substrate or transport antibiotics of different classes, conferring a multiple drug resistance (MDR) phenotype. With respect to efflux pumps, they provide a self-defense mechanism whereby antibiotics are extruded from the cell interior to the external environment. This results in sublethal drug concentrations at the active site that in turn may predispose the organism to the development of high-level target-based resistance [3]. Therefore, efflux pumps are viable antibacterial targets, and identification and development of potent efflux pump inhibitors is a promising and valid strategy potential therapeutic agents that can rejuvenate the activity of antibiotics that are no longer effective against bacterial pathogens. The world is searching for new tools to combat multidrug resistance.

1.2 P-glycoprotein

P-glycoprotein is a mammalian multidrug-resistance protein belonging to the ATP-binding cassette (ABC) superfamily [4]. It is an ATP-dependent efflux pump encoded by the MDR1 gene and is primarily found in epithelial cells lining the colon, small intestine, pancreatic ductules, bile ductules, kidney proximal tubes, the adrenal gland, and the blood-testis and the blood-brain barrier [5]. This efflux activity of P-glycoprotein, coupled with its wide substrate specificity, is responsible for the reduction in bioavailability of drugs as it extrudes all foreign substances such as drugs and xenobiotics out of the cells. ATP hydrolysis provides energy for the efflux of drugs from the inner leaflet of the cell membrane [6, 7]. This protein is believed to have evolved as a defense mechanism against toxic compounds and prevent their entry into the cytosol [8].

P-glycoprotein confers resistance to a wide range of structurally and functionally diverse compounds, which has resulted in the emergence of multidrug resistance in medically relevant microorganisms. The pharmacodynamic role of P-glycoprotein in parasitic helminths has widespread clinical and socioeconomic implications, exacerbating the problem of neglected tropical diseases (NTDs) whose causative agents are helminths and protozoa.

Sheps et al. [9] reported that 15 P-glycoproteins are present in *Caenorhabditis elegans*, and Laing et al. [10] reported that 10 homologous P-glycoproteins were present in *Haemonchus contortus*. A bioinformatic and phylogenetic study conducted by Bourguinat et al. [11] on the *Dirofilaria immitis* genome identified three orthologous ABC-B transporter genes. These genes are suspected to be responsible for the P-glycoprotein-mediated drug extrusion of melarsomine in *D. immitus* and other parasites.

1.3 Neglected tropical diseases

Neglected tropical diseases (NTDs) encompass 17 bacterial, parasitic, and viral diseases that prevail in tropical and subtropical conditions in 149 countries and affect more than 1 billion people worldwide, according to WHO.

1.3.1 Leishmaniasis

Leishmaniasis is a disease caused by parasites of the Leishmania type. It is spread by the bite of certain types of sandflies [12]. The disease can present in three main ways: cutaneous, mucocutaneous, or visceral leishmaniasis [13]. The cutaneous form presents with skin ulcers, whereas the mucocutaneous form presents with ulcers of the skin, mouth, and nose [12]. Leishmaniasis is transmitted by the bite of infected female phlebotomine sand flies [14] which can transmit the protozoa *Leishmania*.

Gammaro et al. [12] first reported that the overexpression of P-glycoprotein in *Leishmania* species was responsible for the drug resistance of the organisms against drugs such as methotrexate. The multidrug resistance has been associated with several ATP-binding cassette transporters including MRP1 (ABCC1) and P-glycoprotein (ABCB1). Wyllie et al. [15] demonstrated the presence of metal efflux pumps in the cell membrane of all *Leishmania* species. Soares et al. [16] reported that natural or synthetic modulators of human P-glycoprotein such as flavonoids restore sensitivity to pentamidine, sodium stibogluconate, and miltefosine by modulating intracellular drug concentrations.

1.3.2 Onchocerciasis

Onchocerciasis, also known as river blindness, is a disease caused by infection with the parasitic worm *Onchocerca volvulus* and is transmitted by the bite of an infected black fly of the *Simulium* type. Symptoms include severe itching, bumps under the skin, and blindness. It is the second most common cause of blindness due to infection, after trachoma, according to WHO. Usually, many bites are required before infection occurs. A vaccine against the disease does not exist. Prevention is by avoiding being bitten by flies.

Ivermectin (IVM) is a semisynthesized macrocyclic lactone that belongs to the avermectin class of compounds. It is administered en masse and but is effective only against microfilariae [17]. Bourguinat et al. [11] have found evidence of IVM resistance in *Onchocerca volvulus*. The clinical trial sampled patients before and after IVM treatment over a period of 3 years. The nodules collected from the patients contained IVM-resistant *O. volvulus* worms.

1.3.3 Schistosomiasis

Schistosomiasis is a disease caused by infection with one of the species of *Schistosoma* helminthic flatworms known as flukes belonging to the class Trematoda of the phylum Platyhelminthes. There are three main species of *Schistosoma* associated with human disease: *Schistosoma mansoni* and *Schistosoma japonicum* cause intestinal schistosomiasis, and *Schistosoma haematobium* causes genitourinary schistosomiasis. Other *Schistosoma* species have been recognized less commonly as agents of intestinal schistosomiasis in humans [18]. Pinto-Almeida et al. [19] demonstrated that drug resistance by *Schistosoma mansoni* to praziquantel (commonly employed drug) is mediated by efflux pump proteins, including P-glycoprotein and multidrug resistance-associated proteins.

1.3.4 Trypanosomiasis

The trypanosomiasis consists of a group of diseases caused by parasitic protozoa of the genus *Trypanosoma*. There are two main parasites such as *Trypanosoma brucei*, which causes the sleeping sickness or human African trypanosomiasis and *Trypanosoma cruzi*, which causes the Chagas' disease or American trypanosomiasis

[20]. These diseases are transmitted by several arthropod vectors such as *Glossina* and *Triatominae*. Chaga's disease causes 21,000 deaths per year mainly in Latin America [21]. Benznidazole and Nifurtimox, only available drugs, however, have limited efficacy in the advanced stages of the disease [22]. Liu et al. [23] and Rappa et al. [24] concluded that *Trypanosoma cruzi* develops resistance to the drugs after prolonged treatment. It was shown that this happens due to the overexpression of the MDR1 gene, at high levels of the drug, which accumulates in the cells over time. Campos et al. [25] demonstrated that the drug resistance is continued throughout the life cycle of the worm.

2. Methods

The methodology is essentially similar to that in our earlier study on P-glycoproteins in priority infectious agents [26].

2.1 Determining the full helminthic complement of efflux pump proteins homologous to mammalian P-glycoprotein

The protein sequence of the human P-glycoprotein (P08183) was obtained from the SWISS-PROT database. The position-specific iterated BLAST (PSI-BLAST) was performed against a search set of nonredundant protein sequences in the organism of interest, using hPGP as the query. Through a PSI-BLAST search, a large set of related proteins are compiled. It is used to identify distant evolutionary relationships between protein sequences [27]. In the algorithm, parameters were set with an E-value of 0.001, and the scoring matrix BLOSUM62 was used. This step was performed on all four organisms of interest (*Leishmania major*, *Onchocerca volvulus*, *Schistosoma mansoni*, and *Trypanosoma cruzi*). Hundreds of hits were obtained for P-glycoprotein, and these results were prioritized according to predetermined parameters such as medical relevance, annotation status, and the presence of conserved regions. Sequences having a high percentage of sequence identity and query coverage were prioritized. Specific UniProt searches of these protein sequences were performed using the accession number. The results were analyzed, and the P-glycoprotein sequence of each organism was finalized.

2.2 Multiple sequence alignment

The templates chosen for multiple sequence alignment (MSA) were 4M1M (*Mus musculus*), 4F4C (*Caenorhabditis elegans*), 3WME (*Cyanidioschyzon merolae*), 2HYD (*Staphylococcus aureus*), 3B5Z (*Salmonella enterica*). These five metazoan, algal, and bacterial templates were used due to their high sequence identity with the hPGP sequence. The target sequences and the five templates were aligned using ClustalX 2.1 [28]. MSA was performed in order to infer the homology and evolutionary relationship between the sequences of the biological data set. The clustering algorithm used was Neighbor Joining (NJ). The phylogenetic distance between the target sequence and the templates was calculated.

2.3 Homology modeling

The chosen P-glycoprotein sequences were used as target sequences for modeling using software such as SWISS-MODEL. SWISS-MODEL is an open-source, structural bioinformatics tool used for the automated comparative modeling of three-dimensional protein structures [29, 30]. Several P-glycoprotein structures were modeled for each organism, using multiple templates. The templates having high sequence similarity with the target sequences were given preference. The models were built, and the PDB files of the structures were obtained.

2.4 Structure validation

The validity of the structures was checked using Procheck, an open source tool used to assess the reliability of the protein structure. It is a part of the SWISS-MODEL server. The structures were refined using energy-minimization protocols, and the least energetic structure corresponding to each protein was chosen for docking studies. The criteria used to assess the quality of the structure include model geometry and the Ramachandran plot. The Ramachandran plot describes the rotation of the polypeptide backbone around the N-C_α (ϕ) and C-C_α (ψ) bonds. It provides an overview of the distribution of the torsion angles over the core, allowed, generous, and disallowed regions. The three main parameters used to select the structures were:

1. Overall Ramachandran value

2. Phylogenetic tree distance

3. Taxonomy

2.5 Creation of the ligand dataset

The ligand data set was created by surveying the literature to determine the drugs which the pathogenic helminths are both sensitive and resistant to. Drug resistance which was conferred via efflux pump activity was given importance. This set of ligands was created for each efflux pump, comprising known and potential antibiotics. The canonical *simplified molecular-input line-entry system* (SMILES) of each drug was retrieved from the PubChem database. The PDB model of each antibiotic was then generated using MarvinView by converting the canonical SMILES [31].

2.6 Protein and ligand preparation

The efflux pump proteins and ligands were individually docked using the AutoDock Version 4.2.6 suite of programs [32]. The software consists of two main programs: AutoGrid, which precalculates a set of grid points on the receptor, and AutoDock, which docks the ligand to the receptor through the grids. The PDB files of the P-glycoprotein structures and the ligands were modified through the addition of Gasteiger charges, followed by the addition and merging of hydrogen atoms to each structure. These modified structures were then saved as PDBQT files using the AutoDock tools. A uniform grid box was then defined and centered in the internal binding cavity of each P-glycoprotein structure, and the affinity maps were generated using AutoGrid. This procedure was repeated for each protein-drug complex.

2.7 Molecular docking of the helminthic efflux pumps with known and potential antibiotics

Each drug was individually docked with each target protein using AutoDock 4.2.6. The local search algorithm used was the Lamarckian genetic algorithm, set to its default parameters. The docking parameters were set to 250,000 cycles per run and 10 runs per protein-drug complex, to obtain the 10 best poses for each complex. The best pose was defined as the conformation having the least binding energy. The 10 poses obtained for each receptor-ligand pair were clustered at 2.0 Å r.m.s. to validate the convergence to the best pose. The AutoDock was run, and the PDBQT file of the best pose of each docked complex was generated.

The results were analyzed to verify whether the pathogenic strain could develop resistance to known antibiotics using efflux pump activity and if the novel antibiotics could be effective against the development of such resistance.

2.8 Calculation of differential ligand binding affinity

The differential binding affinities of the repurposed ligands were determined using the conventionally used drugs as a baseline. A lower value is indicative of a more stable complex. The differential affinity of the potential drug for a given efflux pump protein relative to the known drug is estimated as the difference in the binding energies of the known and potential drugs, as given by Eq. (1):

$$\Delta\Delta G_{invest.known} = \Delta G_{bind,potential} - \Delta G_{bind,known} \quad (1)$$

where $\Delta\Delta G_{invest.known}$ = differential ligand affinity, kcal/mol; ΔG_{bind} = free energy of binding, kcal/mol.

2.9 Identification of interacting residues in each docked complex

The best pose of each docked complex was viewed using RasMol [33], and all interacting residues within a radius of 4.5 Å of the ligand were selected. The PDBQT file of each restricted complex was saved as a PDB file. The interacting residues of each docked complex were then analyzed.

3. Results and discussion

Extensive literature searches on Neglected Tropical Diseases (NTDs) showed that leishmaniasis, onchocerciasis, schistosomiasis, and trypanosomiasis have started exhibiting multidrug resistance, mediated by P-glycoprotein efflux pumps [11, 12, 25, 34]. New drugs targeting NTD's are undergoing clinical trials [35–37], and efforts are being taken to uncover the mechanisms of drug resistance employed by the causative helminths.

The sequence identity of each helminthic P-glycoprotein with the human P-glycoprotein (hPGP) which was retrieved from the UniProt database (UniProt ID: P08183) was determined by running a PSI-BLAST.

3.1 Psi-blast analysis

The PSI-BLAST was performed on each target organism using hPGP as the query. The results were refined according to predetermined parameters such as medical relevance, annotation status, and the presence of conserved regions. The chosen efflux pump protein sequences were shown in **Table 1**.

The top hits of each PSI-BLAST were analyzed, and the hit having the highest Max Score was chosen only in the case of *Leishmania major* and *Onchocerca volvulus.* These protein sequences were fully annotated and had high sequence identities over a large portion of the protein sequence. The top hits of the PSI-BLAST of *Schistosoma mansoni* and *Trypanosoma cruzi* with hPGP yielded results having high

Organism	Name of protein	Sequence length	% of identity	Query coverage	Max score
Leishmania major	P-glycoprotein	1341	36%	98%	767
Onchocerca volvulus	P-glycoprotein	1278	37%	97%	776
Schistosoma mansoni	SMDR2	1254	40%	98%	889
Trypanosoma cruzi	P-glycoprotein	1034	29%	30%	79.7

Table 1.
PSI-BLAST results of the target organisms using hPGP as the query.

Max Scores, but low query coverage. These protein sequences were also found to be unannotated. For these reasons, the proteins which had a lower Max Score in comparison to other results, but satisfied other parameters, were chosen.

3.2 Template selection and multiple sequence alignment

Certain metazoan, algal and bacterial crystal structures shown in **Table 2** were selected as potential templates for homology modeling [38].

Each target protein sequence was aligned with the set of chosen templates using ClustalX 2.1. The MSA between Leishmania major and the 4M1M and 4F4C templates showed the highest sequence identity, as shown in **Figure 1**. Additionally, the phylogenetic distances between the sequences were calculated using the NJ algorithm (**Table 3**).

3.3 Homology modeling

The chosen P-glycoprotein sequences of the organisms were used as target sequences for homology modeling using the SWISS-MODELER. Each protein was modeled using several templates, and the predetermined templates were used if they were found to have a fairly high GMQE score. Each modeled structure was saved as a PDB file. The results are summarized in **Table 4**.

Global Model Quality Estimation (GMQE) is a score that provides an estimation of the quality of the alignment. It is expressed as a value between 0 and 1, where the reliability of the model is directly proportional to the score. The GMQE of the homology models are found to be (mostly) between 0.60 and 0.70 for all organisms, with the exception of *Trypanosoma cruzi*, which gave scores in the range 0.29–0.52.

The templates 4M1M, 4F4C, and 3WME were found to be comparatively more reliable. Hence, only the protein structures modeled using these templates were used for further validation studies.

Template	Organism
4M1M	*Mus musculus*
4F4C	*Canorhabditis elegans*
3WME	*Cyanidioschyzon merolae*
2HYD	*Staphylococcus aureus*
3B5Z	*Salmonella enteric*

Table 2.
Templates chosen for multiple sequence alignment.

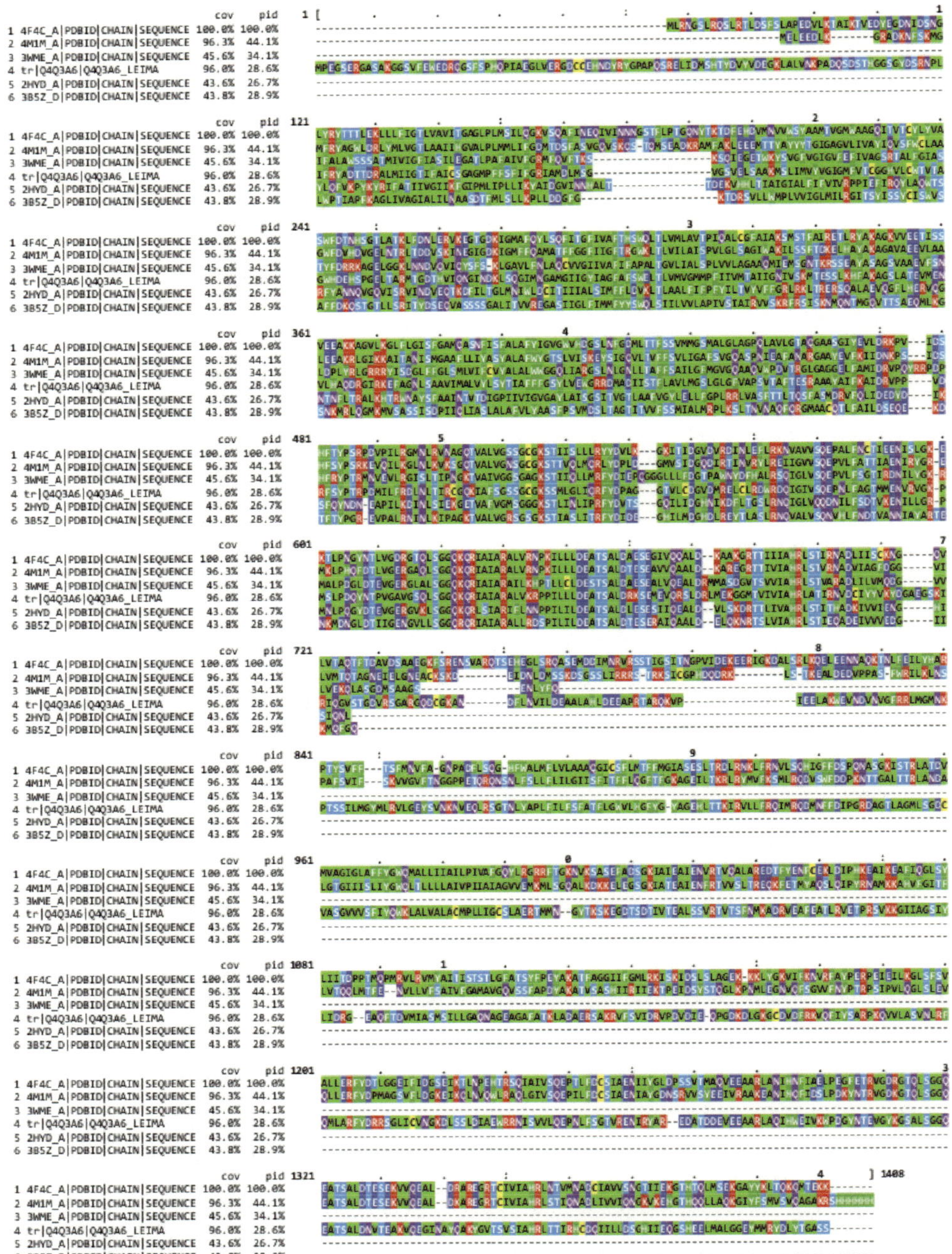

Figure 1.
Multiple sequence alignment of the target sequence P-glycoprotein of Leishmania major *(tr_Q4Q3A6) with the templates of interest.*

3.4 Structure validation

The quality of each structure was assessed using Procheck. Criteria such as model geometry and the Ramachandran plot were used to validate the structures. The PDB file of each structure was used to run the Procheck, and the Ramachandran plot values were obtained. The Ramachandran values are summarized in **Table 5**.

Template	Phylogenetic distance			
	Leishmania major	***Onchocerca volvulus***	***Schistosoma mansoni***	***Trypanosoma cruzi***
4M1M	0.685	0.648	0.646	**0.847***
4F4C	**0.642**	**0.638**	**0.605**	0.861
3WME	0.653	0.679	0.649	0.867
2HYD	0.72	0.709	0.694	0.826
3B5Z	0.731	0.707	0.698	0.841

The distance between T. cruzi and 4M1M is prioritized as the 4M1M and 4F4C templates were found to have higher sequence identity with the helminthic P-glycoproteins.
Bold values signify the final template in the case of each agent.

Table 3.
Phylogenetic distance matrix between the target sequence of each organism and the templates.

Organism	Template	Sequence identity	Query coverage	GMQE
Leishmania major	4F4C	34.43	0.91	0.64
	4M1M	36.09	0.90	0.65
	3WME	37.30	0.43	0.29
	4Q9I	36.20	0.90	0.65
	4KSC	38.25	0.90	0.64
	4KSB	38.25	0.90	0.64
	5KPJ	38.25	0.90	0.66
Onchocerca volvulus	4F4C	38.83	0.97	0.69
	4M1M	37.10	0.96	0.69
	3WME	31.75	0.44	0.33
	3G5U	38.77	0.95	0.66
	4KSB	38.77	0.95	0.67
	4Q9I	36.97	0.95	0.69
	4LSG	38.77	0.95	0.67
Schistosoma mansoni	4F4C	38.60	0.97	0.69
	4M1M	36.09	0.90	0.65
	3G5U	39.41	0.97	0.68
	3G60	42.11	0.95	0.68
	5KPJ	39.52	0.97	0.70
	4KSC	42.11	0.95	0.69
	4KSB	42.11	0.95	0.69
	4LSG	39.41	0.97	0.69
Trypanosoma cruzi	4F4C	15.11	0.81	0.45
	4M1M	14.36	0.78	0.29
	3WME	18.37	0.51	0.33
	4KSC	14.23	0.79	0.44
	3G5U	14.46	0.78	0.43
	5TSI	23.43	0.90	0.52
	4LSG	14.46	0.78	0.43

Table 4.
Homology modeling results.

The structures were finalized by analyzing overall Ramachandran value, Phylogenetic tree distance, and taxonomy parameters. The 4F4C template was found to suitable for all the organisms excluding *Leishmania major*, for which the 4M1M template was selected (**Table 5**).

3.4.1 Validation of the P-glycoprotein structure modeled using the 4M1M template for Leishmania major

The Ramachandran plots having a core region of at least 90% are prioritized for further studies. The core, allowed, generous and disallowed regions are colored and distinguished (**Figure 2**). The red, brown, and yellow regions represent the favored, allowed, and generously allowed regions.

A more comprehensive analysis of the structure is provided by other programs that generate other data such as Phi-Psi graphs and Chi1-Chi2 plots for each residue type. Each Phi-Psi plot provides an analysis of the torsion angle of each residue type. The red, brown, and yellow regions represent the favored, allowed, and generously allowed regions (shown in **Figure 3**).

The Chi1-Chi2 plot describes the side-chain torsion angles combinations for each amino acid [28]. The darker regions indicate a more favorable angle combination (shown **Figure 4**).

3.4.2 Validation of the P-glycoprotein structure modeled using the 4F4C template for Onchocerca volvulus, Schistosoma mansoni *and* Trypansoma cruzi

For all the three P-glycoproteins, the structures were modeled using the 4F4C template and as such and showed remarkable structural similarity with respect to the Ramachandran plot (90.8% in the core region), and residue torsion angles. **Figures 5–7** summarize this exercise.

3.5 Creation of the ligand dataset

Upon extensive survey of the literature, a comprehensive data set of the known and potential drugs was compiled (**Table 6**). The list of potential drugs comprises of both unapproved, investigational drugs that are undergoing phase trials, and FDA approved antibiotics. In this study, these known drugs have been repurposed for other helminthic diseases.

3.6 Molecular docking of the helminthic efflux pumps with known and potential antibiotics

The molecular docking was carried out using the AutoDock suite of tools. The search algorithm used was the Lamarckian Genetic Algorithm, and the docking parameters were set to 10 runs per protein-drug complex. Each docked complex yielded 10 poses, and the best pose was defined as the conformation possessing the least free binding energy.

*3.6.1 Molecular docking results of benznidazole with P-glycoprotein (*Leishmania major*)*

The drug benznidazole is docked with P-glycoprotein (*Leishmania major*), and their interaction is studied (**Table 7**). The best pose has a free binding energy of −5.00 kcal/mol. The clustering was performed at 2.0 *Å* r.m.s. to validate the convergence to the best pose. The clustering figure (**Figure 8**) shows closer peaks near −2.5 kcal/mol, whereas the least binding energy of the complex, that is, most

Template	Core region	Additionally allowed region	Generously allowed region	Disallowed region	No. of residues	Query coverage	Sequence identity	Phylogenetic distance
				Leishmania major				
4F4C	90.8	8.1	1.2	0	1250	0.43	34.43	0.685
4M1M	92	6.4	1.4	0.2	1188	0.9	36.09	0.642
3WME	94.6	4.6	0.4	0.4	573	0.43	36.46	0.653
				Onchocerca volvulus				
Template	**Core region**	**Additionally allowed region**	**Generously allowed region**	**Disallowed region**	**No. of residues**	**Query coverage**	**Sequence identity**	**Phylogenetic distance**
4F4C	90.8	8.1	1.2	0	1250	0.97	38.83	0.648
4M1M	91.1	7.6	1.2	0.1	571	0.95	37.1	0.638
3WME	93.1	5.8	0.6	0.6	1180	0.45	33.51	0.679
				Schistosoma mansoni				
Template	**Core region**	**Additionally allowed region**	**Generously allowed region**	**Disallowed region**	**No. of residues**	**Query coverage**	**Sequence identity**	**Phylogenetic distance**
4F4C	90.8	8.1	1.2	0	1250	0.97	38.6	0.646
4M1M	91.1	6.9	1.8	0.2	572	0.46	36.6	0.605
3WME	93.7	5.8	0.2	0.4	567	0.45	38.31	0.649
				Trypanosoma cruzi				
Template	**Core region**	**Additionally allowed region**	**Generously allowed region**	**Disallowed region**	**No. of residues**	**Query coverage**	**Sequence identity**	**Phylogenetic distance**
4F4C	90.8	8.1	1.2	0	1250	0.81	15.11	0.847
4M1M	86.0	11.9	1.8	0.2	1034	0.51	17.80	0.867
3WME	89.8	8	1.6	0.6	573	0.51	18.37	0.861

Table 5.
Justification of the template chosen for each organism using the Ramachandran plot values and the phylogenetic distance between the target protein and the template.

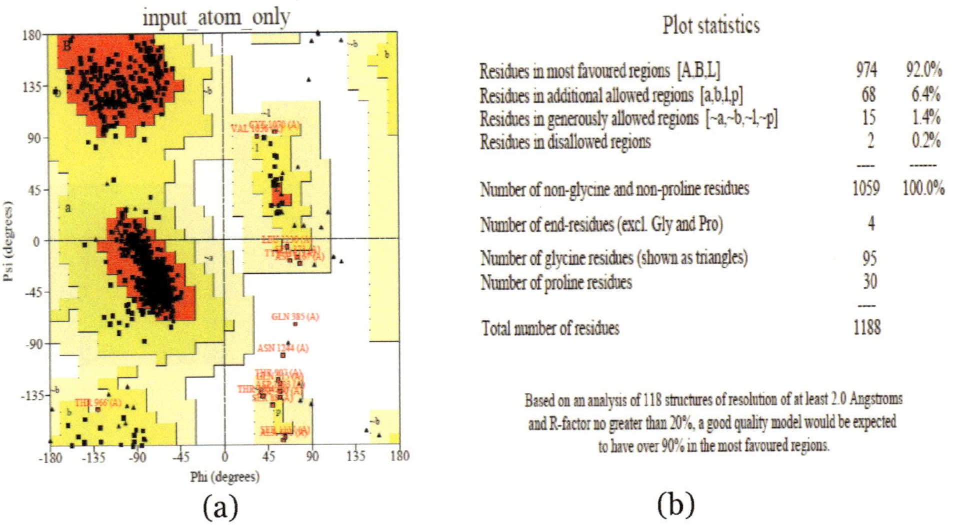

Figure 2.
*(a) The Ramachandran plot generated for P-glycoprotein (*Leishmania major*), modeled using the 4M1M template and (b) plot statistics of the P-glycoprotein (*Leishmania major*), modeled using the 4M1M template.*

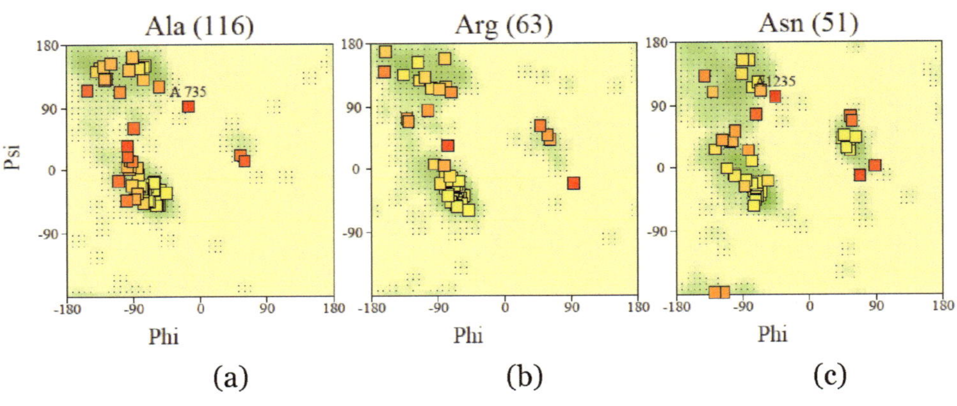

Figure 3.
Phi-psi plot of residues of the P-glycoprotein structure of Leishmania major, *modeled using the 4M1M template (a) Ala, (b) Arg, and (c) Asn.*

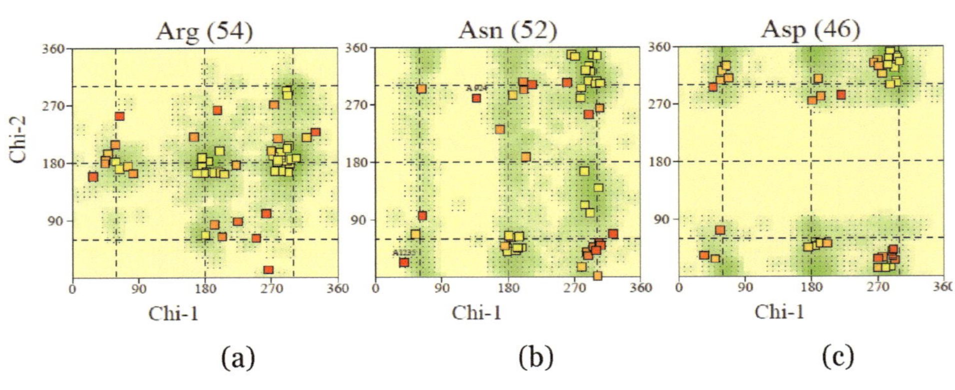

Figure 4.
Chi1-Chi2 plot of residues of the P-glycoprotein structure of Leishmania major, *modeled using the 4M1M template (a) Arg, (b) Asn and (c) Asp.*

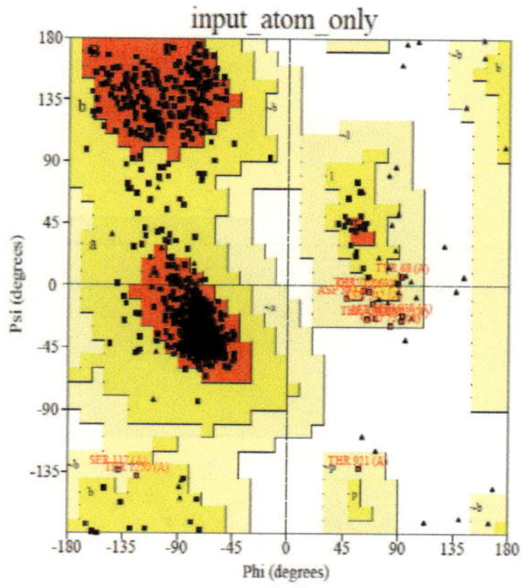

(a)

Plot statistics

Residues in most favoured regions [A,B,L]	1011	90.8%
Residues in additional allowed regions [a,b,l,p]	90	8.1%
Residues in generously allowed regions [~a,~b,~l,~p]	13	1.2%
Residues in disallowed regions	0	0.0%
	----	------
Number of non-glycine and non-proline residues	1114	100.0%
Number of end-residues (excl. Gly and Pro)	5	
Number of glycine residues (shown as triangles)	98	
Number of proline residues	33	

Total number of residues	1250	

Based on an analysis of 118 structures of resolution of at least 2.0 Angstroms and R-factor no greater than 20%, a good quality model would be expected to have over 90% in the most favoured regions.

(b)

Figure 5.
*(a) The Ramachandran plot generated for P-glycoprotein (*Onchocerca volvulus*), modeled using the 4F4C template and (b) plot statistics of the P-glycoprotein (*Onchocerca volvulus*), modeled using the 4F4C template.*

S. no.	Drug	PubChem CID	3-D Structure
1.	Albendazole	2082	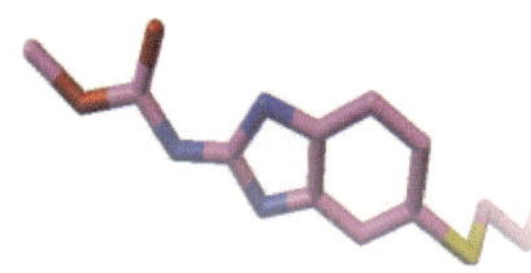
2.	Amphotericin B	5280965	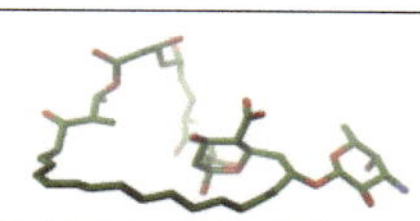
3.	Artesunate	65664	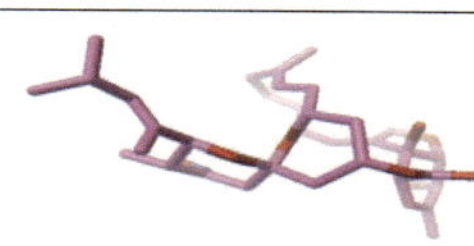
4.	Benznidazole	5798	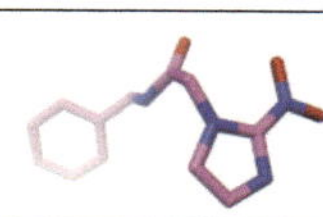
5.	Cladosporin	13990016	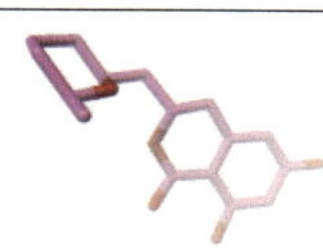
6.	Dapsone	2955	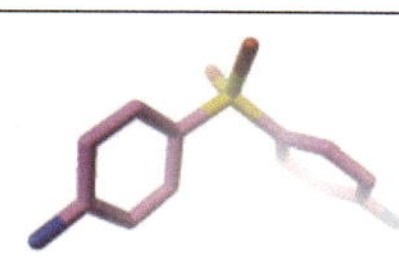
7.	Diethylcarbamazine	15432	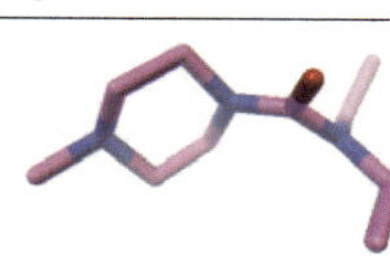
8.	Emodepside	6918632	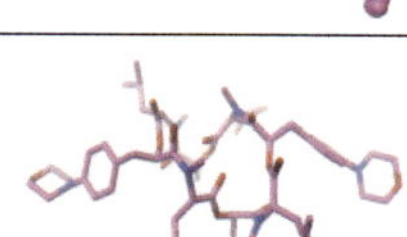

S. no.	Drug	PubChem CID	3-D Structure
9.	Fexinidazole	68792	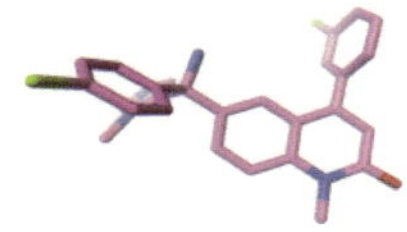
10.	Flubendazole	35802	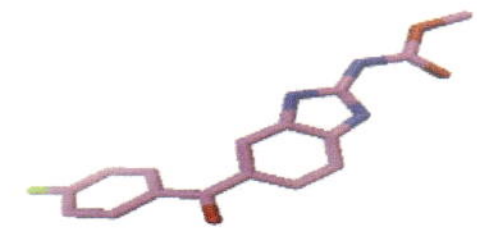
11.	Fluconazole	3365	
12.	Furozan	67517	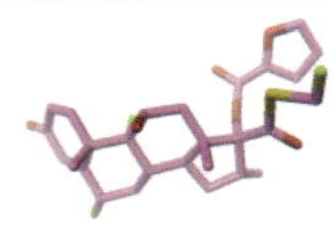
13.	Imatinib	5291	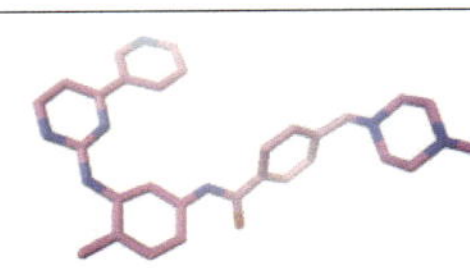
14.	Ivermectin	6321424	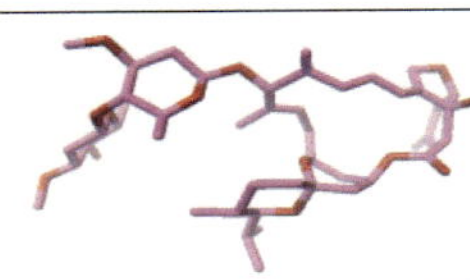
15.	Jaspamide	9831636	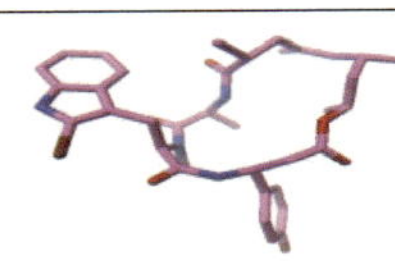
16.	Mebendazole	4030	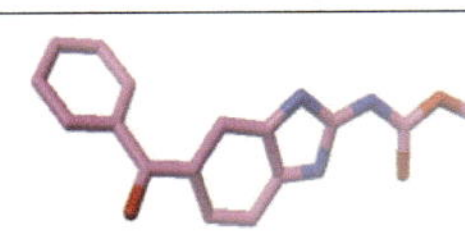
17.	Metrifonate	5853	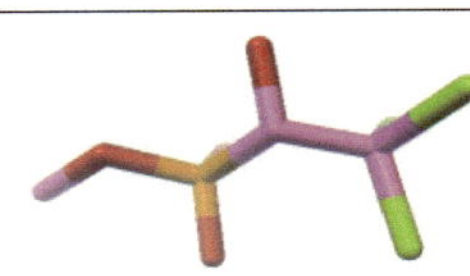
18.	Miltefosine	3599	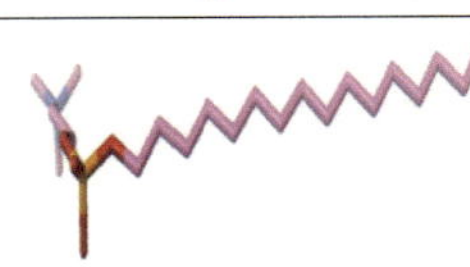
19.	Moxidectin	9832912	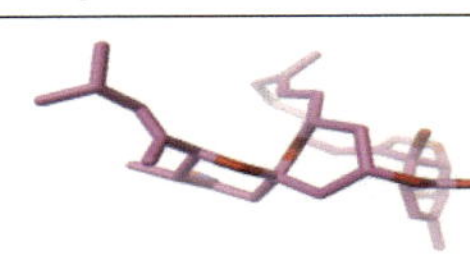
20.	Niclosamide	4477	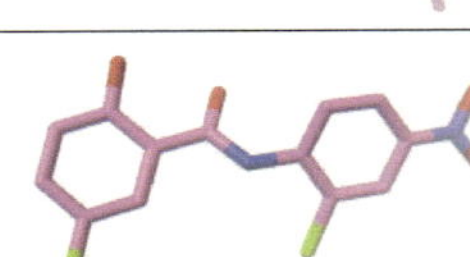

S. no.	Drug	PubChem CID	3-D Structure
21.	Nifurtimox	6842999	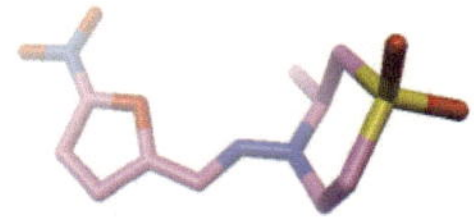
22.	Oxamniquine	4612	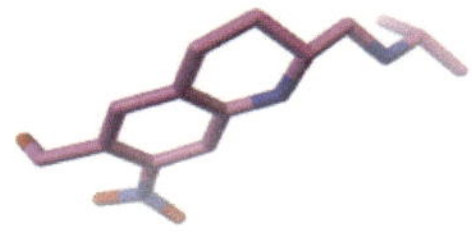
23.	Paromomycin	165580	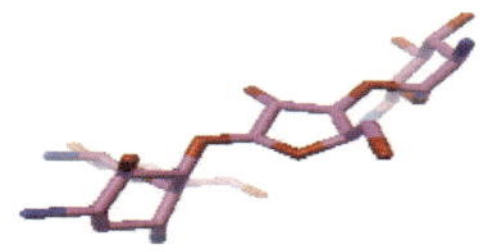
24.	Pentamidine	4735	
25.	Posaconazole	468595	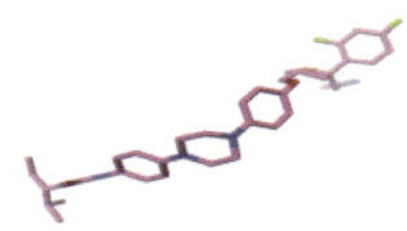
26.	Praziquantel	4891	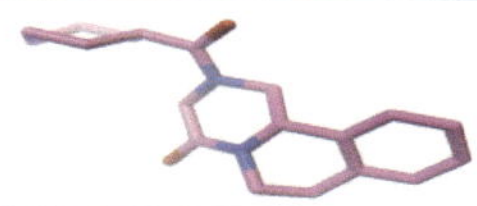
27.	Ravuconazole	467825	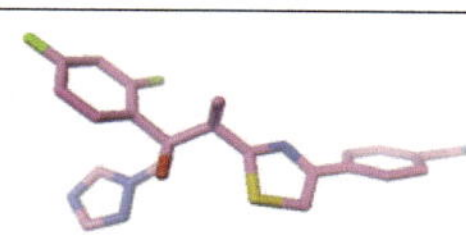
28.	Sodium stibogluconate	76968133	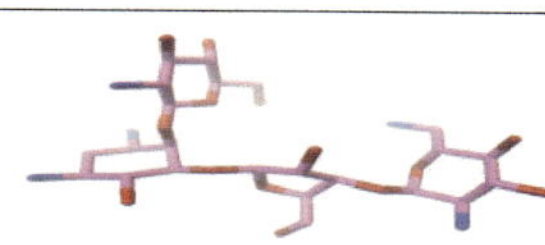
29.	Suramin	5361	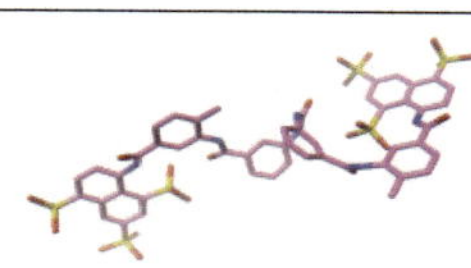
30.	Terbinafine	1549008	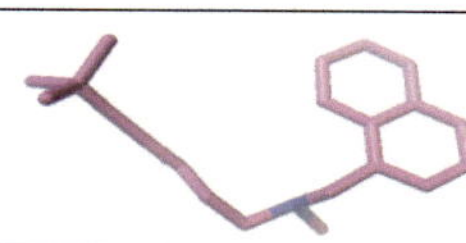
31.	Thiabendazole	5430	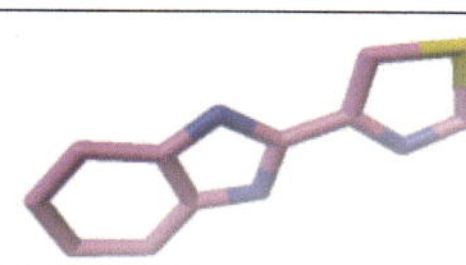
32.	Tipifarnib	159324	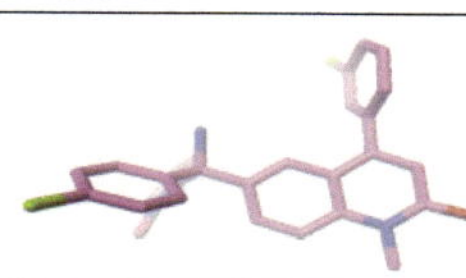

3D structures of the drugs are visualized using python molecular viewer (PMV-1.5.6).

Table 6.
PubChem compound ID and 3D structure of the ligands used for docking studies.

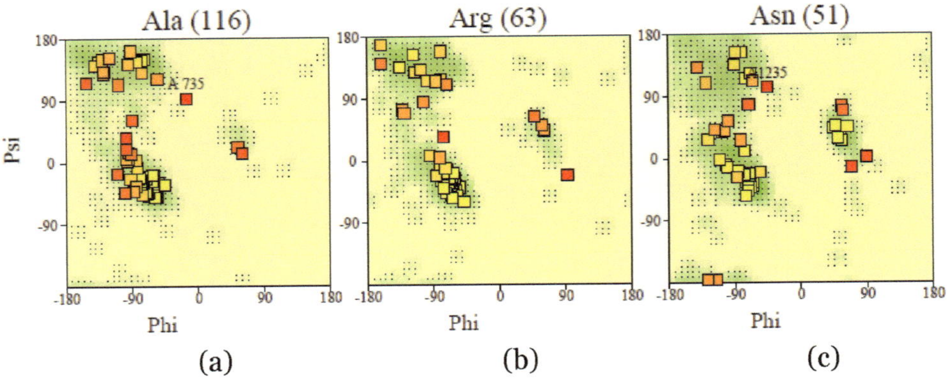

Figure 6.
Phi-psi plot of residues of the P-glycoprotein structure of Onchocerca volvulus, *modeled using the 4F4C template (a) Ala, (b) Arg, and (c) Asn.*

Rank of complex	Free binding energy (kcal/mol)
1	−5.00
2	−4.84
3	−4.2
4	−4.41
5	−3.77
6	−3.48
7	−2.96
8	−2.64
9	−2.54
10	−2.48

Table 7.
*Interaction of the drug benznidazole with P-glycoprotein (*Leishmania major*).*

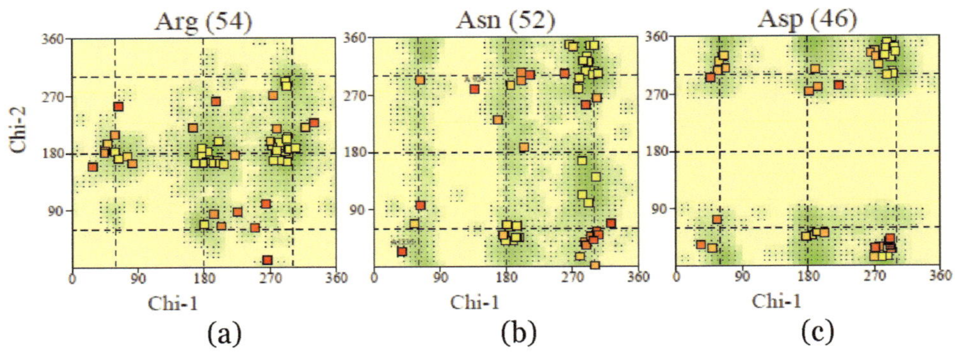

Figure 7.
Chi1-Chi2 plot of residues of the P-glycoprotein structure of Onchocerca volvulus, *modeled using the 4F4C template. (a) Arg, (b) Asn, and (c) Asp.*

clustering is at −5.66 kcal/mol. This shows that convergence to the best pose can be achieved through consecutive dockings with more iterations. **Figure 8(b)** depicts the binding site on the receptor, and **Figure 8(c)** shows the interacting residues in the benznidazole-P-glycoprotein (*Leishmania major*) docked complex viewed through RasMol 2.1.

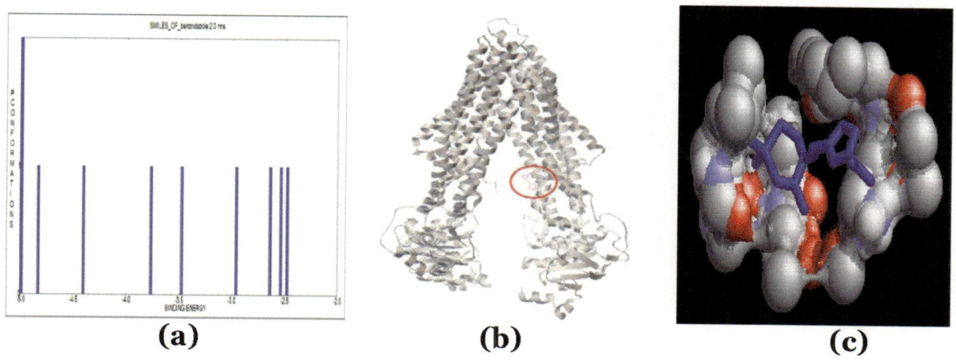

Figure 8.
*(a) Clustering analysis of the benznidazole- P-glycoprotein docked complex. (b) Location of the binding site on the receptor (P-glycoprotein [*Leishmania major*]). (c) The interacting residues in the benznidazole-P-glycoprotein (*Leishmania major*) docked complex is viewed using RasMol 2.1.*

Rank of complex	Free binding energy (kcal/mol)
1	−5.29
2	−5.01
3	−4.78
4	−5.14
5	−5.08
6	−5.02
7	−4.59
8	−4.53
9	−4.42
10	34.78

Table 8.
*Interaction of the drug niclosamide with P-glycoprotein (*Onchocerca volvulus*).*

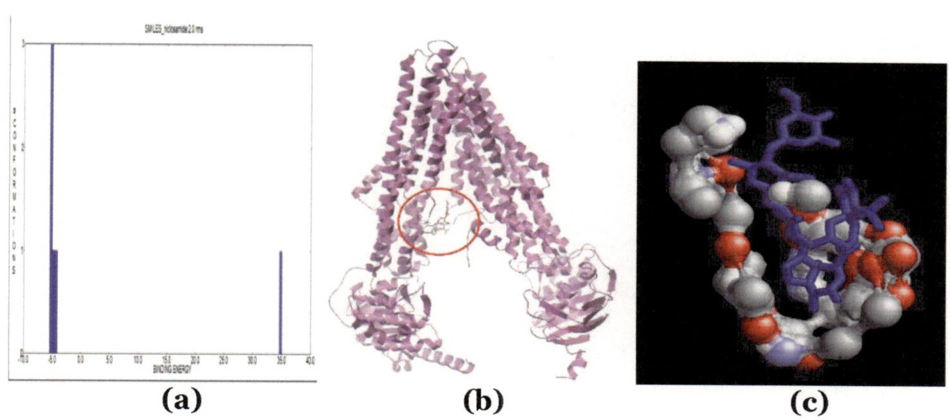

Figure 9.
*(a) Clustering analysis of the niclosamide- P-glycoprotein docked complex. (b) Location of the binding site on the receptor (P-glycoprotein [*Onchocerca volvulus*]). (c) The interacting residues in the niclosamide- P-glycoprotein (*Onchocerca volvulus*) docked complex is viewed using RasMol 2.1.*

Rank of complex	Free binding energy (kcal/mol)
1	−5.83
2	−5.51
3	−5.21
4	−4.71
5	−4.47
6	−4.15
7	−3.57
8	9.34
9	29.83
10	36.47

Table 9.
*Interaction of the drug Praziquantel with P-glycoprotein (*Schistosoma mansoni*).*

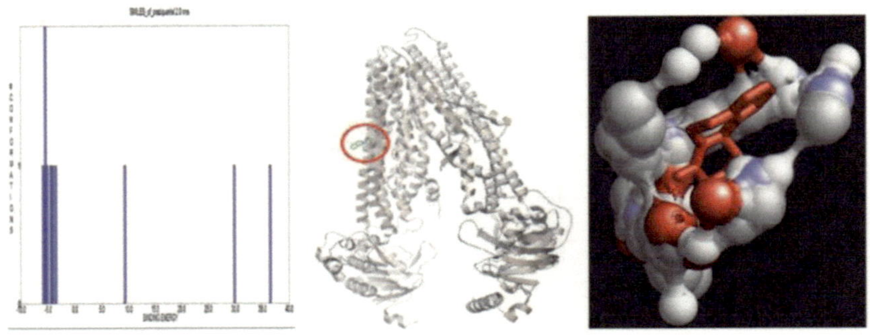

(a) (b) (c)

Figure 10.
*(a) Clustering analysis of the Praziquantel-P-glycoprotein docked complex. (b) Location of the binding site on the receptor (P-glycoprotein [*Schistosoma mansoni*]). (c) The interacting residues in the Praziquantel-P-glycoprotein [*Schistosoma mansoni*] docked complex are viewed using RasMol 2.1.*

Rank of complex	Free binding energy (kcal/mol)
1	−6.23
2	−6.04
3	−5.67
4	−4.81
5	−5.92
6	−5.25
7	−4.89
8	−4.83
9	−4.15
10	−3.34

Table 10.
*Interaction of the drug cladosporin with P-glycoprotein (*Trypanosoma cruzi*).*

3.6.2 Molecular docking results of niclosamide with P-glycoprotein (Onchocerca volvulus)

The best pose has a free binding energy of −5.29 kcal/mol (**Table 8**). The clustering figure shows the most number of conformations at −1.30 kcal/mol (**Figure 9**). **Figure 9(b)** depicts the binding site on the receptor, and **Figure 9(c)** shows the interacting residues in the niclosamide-P-glycoprotein (*Onchocerca volvulus*) docked complex viewed through RasMol 2.1.

3.6.3 Molecular docking results of praziquantel with P-glycoprotein (Schistosoma mansoni)

The best pose has a free binding energy of −5.83 kcal/mol (**Table 9**). The clustering figure (**Figure 10**) shows the most number of conformations at −5.0 kcal/mol. **Figure 10(b)** depicts the binding site on the receptor, and **Figure 10(c)** shows the interacting residues in the Praziquantel-P-glycoprotein (*Schistosoma mansoni*) docked complex viewed through RasMol 2.1.

3.6.4 Molecular docking results of cladosporin with P-glycoprotein (Trypanosoma cruzi)

The best pose has a free binding energy of −6.23 kcal/mol (**Table 10**). The clustering figure (**Figure 11**) shows the most number of conformations at −5.0 kcal/mol. **Figure 11(b)** depicts the binding site on the receptor, and **Figure 11(c)** shows the interacting residues in the the cladosporin-P-glycoprotein (*Trypanosoma cruzi*) docked complex viewed through RasMol 2.1.

These steps were carried out for each receptor-ligand complex, and the least free binding energy of each docked complex was determined. These results are summarized in **Table 11**.

3.7 Calculation of differential ligand binding affinity

The differential affinity of the potential drug for a given efflux pump protein relative to the known drug is estimated as the difference between the binding energies of the known and potential drugs.

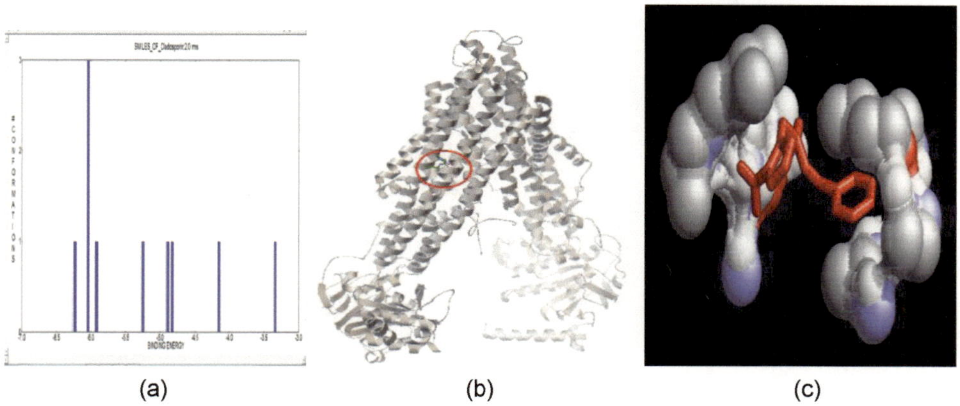

Figure 11.
*(a) Clustering analysis of the cladosporin-P-glycoprotein docked complex. (b) Location of the binding site on the receptor (P-glycoprotein [*Trypanosoma cruzi*]). (c) The interacting residues in the cladosporin-P-glycoprotein [*Trypanosoma cruzi*] docked complex are viewed using RasMol 2.1.*

	Known drugs	Free binding energy	Investigational drugs	Free binding energy
L. major	Amphotericin B	−6.44	Cladosporin	−6.42
	Fluconazole	−3.12	Jaspamide	−5.98
	Pentamidine	−2.67	Nifurtimox	−5.66
	Miltefosine	1.21	Praziquantel	−5.59
			Dapsone	−5.48
			Benznidazole	−5
			Tipifarnib	−4.54
			Flubendazole	−4.36
			Terbinafine	−4.25
			Sodium stibogluconate	−3.7
			Paromomycin	−3.07
			Meglumine antimoniate	
T. cruzi	Nifurtimox	−5.22	Cladosporin	−6.23
	Benznidazole	−5.02	Tipifarnib	−5.87
			Jaspamide	−5.82
			Fexinidazole	−4.62
			Suramin	−4.25
			Ravuconazole	−3.69
			Posaconazole	−2.52
			AN2690	
O. volvulus	Mebendazole	−5.36	Praziquantel	−6.16
	Albendazole	−5.22	Moxidectin	−5.53
	Suramin	−4.79	Niclosamide	−5.29
	Diethylcarbamazine	−3.76	Flubendazole	−4.58
	Ivermectin	−1.35	Thiabendazole	−4.35
			Metrifonate	−2.09
			Emodepside	−1.92
S. mansoni	Praziquantel	−5.83	Cladosporin	−6.07
	Mebendazole	−5.1	Jaspamide	−6.06
	Oxamniquine	−4.4	Niclosamide	−5.72
	Albendazole	−3.83	Nifurtimox	−5.62
			Artesunate	−5.12
			Benznidazole	−4.41
			Tipifarnib	−4.38
			Imatinib	−4.29
			Furozan	−4.27
			Suramin	−3.85
			Metrifonate	−3.54

Table 11.
Free binding energy of all known and investigational drugs, including repurposed antibiotics.

Leishmaniasis	**ΔG unknown**	**ΔΔG amphotericin B**	**ΔΔG fluconazole**	**ΔΔG pentamidine**	**ΔΔG miltefosine**
Cladosporin	−6.42	0.02	−3.3	−3.75	−7.63
Jaspamide	−5.98	0.46	−2.86	−3.31	−7.19
Nifurtimox	−5.66	0.78	−2.54	−2.99	−6.87
Praziquantel	−5.59	0.85	−2.47	−2.92	−6.8
Dapsone	−5.48	0.96	−2.36	−2.81	−6.69
Benznidazole	−5	1.44	−1.88	−2.33	−6.21
Tipifarnib	−4.54	1.9	−1.42	−1.87	−5.75
Flubendazole	−4.36	2.08	−1.24	−1.69	−5.57
Terbinafine	−4.25	2.19	−1.13	−1.58	−5.46
Sodium stibogluconate	−3.7	2.74	−0.58	−1.03	−4.91
Paromomycin	−3.07	3.37	0.05	−0.4	−4.28

Trypanosomiasis	**ΔG unknown**	**ΔΔG Benznidazole**	**ΔΔG Nifurtimox**
Cladosporin	−6.23	−1.21	−1.01
Tipifarnib	−5.87	−0.85	−0.65
Jaspamide	−5.82	−0.8	−0.6
Fexinidazole	−4.62	0.4	0.6
Suramin	−4.25	0.77	0.97
Ravuconazole	−3.69	1.33	1.53
Posaconazole	−2.52	2.5	2.7

Onchocerciasis	**ΔG unknown**	**ΔΔG Mebendazole**	**ΔΔG Albendazole**	**ΔΔG Suramin**	**ΔΔG Diethylcarbamazine**	**ΔΔG Ivermectin**
Praziquantel	−6.16	−0.8	−0.94	−1.37	−2.4	−4.81
Moxidectin	−5.53	−0.17	−0.31	−0.74	−1.77	−4.18
Niclosamide	−5.29	0.07	−0.07	−0.5	−1.53	−3.94
Flubendazole	−4.58	0.78	0.64	0.21	−0.82	−3.23
Thiabendazole	−4.35	1.01	0.87	0.44	−0.59	−3
Metrifonate	−2.09	3.27	3.13	2.7	1.67	−0.74
Emodepside	−1.92	3.44	3.3	2.87	1.84	−0.57

Schistosomiasis	**ΔG unknown**	**ΔΔG Praziquantel**	**ΔΔG Mebendazole**	**ΔΔG Oxamniquine**	**ΔΔG Albendazole**
Cladosporin	−6.07	−0.24	−0.97	−1.67	−2.24
Jaspamide	−6.06	−0.23	−0.96	−1.66	−2.23
Niclosamide	−5.72	0.11	−0.62	−1.32	−1.89
Nifurtimox	−5.62	0.21	−0.52	−1.22	−1.79
Artesunate	−5.12	0.71	−0.02	−0.72	−1.29
Benznidazole	−4.41	1.42	0.69	−0.01	−0.58
Tipifarnib	−4.38	1.45	0.72	0.02	−0.55
Imatinib	−4.29	1.54	0.81	0.11	−0.46
Furozan	−4.27	1.56	0.83	0.13	−0.44
Suramin	−3.85	1.98	1.25	0.55	−0.02
Metrifonate	−3.54	2.29	1.56	0.86	0.29

Table 12.
Differential ligand binding affinity for each known-potential drug pair.

$$\Delta\Delta G_{invest.} = \Delta G_{bind,potential} - \Delta G_{bind,known}$$

where $\Delta\Delta G_{invest.}$ = differential ligand affinity, kcal/mol;ΔG_{bind} = free energy of binding, kcal/mol.

For each disease, the differential ligand binding affinity is calculated for every known-potential drug pair. The $\Delta\Delta G_{investiational}$ values are given in **Table 12**. All values are expressed in kcal/mol. The drugs having $\Delta\Delta G_{invest}$ values greater than the $\Delta G_{invest.}$ values may have better antihelminthic activity.

All values are expressed in kcal/mol. It can be inferred from these results that many of the repurposed antiparasitic drugs show promise for treatment against other helminths. The results shown in **Table 12** serve as an indicator of which drugs may be promising antihelminthics:

1. Leishmaniasis: Cladosporin (−7.63 kcal/mol), Jaspamide (−7.19 kcal/mol), and Nifurtimox (−6.87 kcal/mol).

2. Trypanosomiasis: Cladosporin (−1.21 kcal/mol) and Tipifarnib (−0.85 kcal/mol)

Receptor	Drug	Interacting residues
4M1M	Amphotericin B	Thr172, Asp173, Ser176, Ala683, Asp687, Ser876, Ala879, Leu880, Lys883, Lys884, Glu887, Lys996
	Fluconazole	Val129, Cys133, Ala136, Asn179, Glu180, Gly181, Gly183, Asp184, Lys185, Met188, Leu875, Asp882, Lys930, Phe934
	Pentamidine	Glu239, Leu240, Ala242, Tyr243, Ala244, Gly247, Ala248, Glu251, Arg785, Thr811, Ala815, Asn816, Ala819, Gln820
	Miltefosine	Asp173, Ser176, Lys177, Glu180, Lys185, Leu875, Ala879, Leu880, Lys883
	Cladosporin	Gln434, Leu437,Leu439, Val468, Ser470, Glu472, Val474,Asn899, Arg901, Thr902, Ser905
	Jaspamide	Leu254, Ala255, Ala256, Ile257,Arg258, Thr259,Phe800, Asn805, Thr806, Thr807, Gly808, Leu810, Glu1115, Ile1117
	Nifurtimox	Ala288, Asn292, Gln769, Gly770, Phe773, Gly774, Glu778, Ala819, Gln820, Lys822, Gly823, Ser827, Phe990, Pro992
	Praziquantel	Leu254, Ala255, Ala256, Ile257,Arg258, Thr259, Phe800, Asn805, Thr806, Thr807, Leu810, Ser1113
	Dapsone	Val474, Leu475, Phe476, Ala477, Gly521, Glu522, Lys523, Lys891, Thr894, Glu895, Glu898, Asn899, His1003, Arg1006, Ile1007, Lys1010
	Benznidazole	Asp685, Val688, Pro689, Trp799, Asp802, Lys804, Asn805, Arg813, His1003, Arg1006, Ile1007, Lys1010
	Tipifarnib	Ala244, Gly247, Ala248, Val249, Glu251, Glu252, Asp1120, Gly1166, Asp1167, Lys1168
	Flubendazole	Phe159, Asp160, His162, Asp163, Val164, Ser470, Glu472, Val474, Ile897, Gly898, Asn899, Phe900, Arg901, Thr902
	Terbinafine	Phe159, Val164, Gln434, Gln437, Leu439, Val468, Ser470, Glu472, Val474, Ile897, Glu898, Asn899, Phe900, Arg901, Thr902, Ser905
	Sodium stibogluconate	Ser470, Glu472, Pro473, Val474, Leu475, Ala477, Glu522, Lys532, Glu895, Glu898, Asn899, Arg901, Thr902
	Paromomycin	Val164, Glu472, Pro473, Val474, Glu522, Glu898, Asn899

Table 13.
Interacting residues between the P-glycoprotein of Leishmania major *and the chosen drugs.*

Receptor	Drug	Interacting residues
4F4C	Mebendazole	Glu267, Thr268, Tyr271, Ala272, Gly275, Lys276, Lys315, Arg830, Ala860, Thr861, Pro864, Arg867
	Albendazole	Glu36, Gly37, Asp38, Ile40, Glu267, Thr268, Tyr271, Val305, Ala308, Lys309, Glu823, Thr826, Arg827, Arg830, Ala860, Thr861, Pro864, Arg867
	Suramin	Lys720, Leu723, Ser724, Lys727, Lys923, Val925, Lys936
	Diethylcarbamazine	Arg918, Arg919, Phe920, Gly922, Lys923, Asn924, Gln979
	Ivermectin	Arg172, Thr197, Phe200, Asp201, Glu204, Lys720, Lys923, Asn924, Val925, Ala928, Phe931, Ala932, Gly935, Lys936, Ile939
	Praziquantel	Leu11, Glu165, Lys207, Asp212, Arg918, Phe920, Lys923, Asn924, Ser927, Phe931, Ala972, Glu975, Gln979
	Moxidectin	Leu11, Arg12, Asp15, Lys26, Lys30, Glu33, Pro374, Gln913, Tyr914, Arg916, Gly917, Gly1032, Phe1033, Thr1035, Ser1036, Pro1039
	Niclosamide	Asn4, Gly5, Ser6, Leu7, Ile48, Thr49, Val56, Lys59, Gly380, Thr381, Gln383, Gly384
	Flubendazole	Glu36, Gly37, Ser42, Thr268, Tyr271, Ala272, Gly275, Arg830, Ala860, Thr861, Pro864, Asn865, Arg867, Lys1043
	Thiabendazole	Phe504, Asn505, Cys506, Asp933, Lys936, Ile937, Glu940, Phe957, Asn960, Lys964
	Metrifonate	Gln840, His841, Gly843, Phe844, Ser847, Gln849, Asn850, Lys1057, Ile1058, Lys1060
	Emodepside	Arg172, Thr197, Phe200, Asp201, Glu204, Asp550, Val925, Ser929, Phe931, Ala932, Gly935, Lys936, Ile939

Table 14.
Interacting residues between the P-glycoprotein of Onchocerca volvulus *and the chosen drugs.*

Receptor	Drug	Interacting residues
4F4C	Nifurtimox	Asn733, Asn734, Gln849, Asn850, Arg1056, Lys1057, Ile1058
	Benznidazole	Asn4, Gly5, Ser6, Leu7, Thr49, Glu55, Val56, Arg205,Thr381,Gly384, Ala385
	Cladosporin	Tyr35, Glu36, Ile40, Glu267, Thr268, Tyr271, Ala272, Gly275, Lys315, Arg830, Ala860, Thr861, Arg867
	Tipifarnib	Gly373, Asp38, Ile40, Asp41, Ser42, Asn43, Glu267, Thr268, Tyr271, Ala272, Gly275, Arg830, Ser856, Thr857, Ala860, Thr861, Arg867
	Jaspamide	Gln913, Tyr914, Arg916, Gly917, Arg918, Arg919, Lys923, Gly1032, Ph1033, Thr1035, Ser1036, Phe1038, pro1039
	Fexinidazole	Gly37, Asp38, Ile40, Asp41, Ser42, Asn43, Thr268, Tyr271, Ala272, Gly275, Arg830, Ser856, Ala860, Thr861, Pro864, Arg867
	Suramin	,Lys727, Lys923, Val925
	Ravuconazole	Ala910, Gln913, Tyr914, Gly917, Arg919, Gly1032, Phe1033, Thr1035, Ser1036, Pro1039
	Posaconazole	Glu33, Leu161, Gly917, Arg918, Arg919, Phe920, Gly922, Lys923, Asn924, Glu975, Ala976, Gln979, Phe1033, Thr1035, Ser1036, Pro1039

Table 15.
Interacting residues between the P-glycoprotein of Schistosoma mansoni *and the chosen drugs.*

3. Schistosomiasis: Cladosporin (−2.24 kcal/mol) and Jaspamide (−2.23 kcal/mol)

4. Onchocerciasis: Praziquantel (−4.81 kcal/mol) and Moxidectin (−4.18 kcal/mol)

3.8 Analysis of interacting residues in each docked complex

The best pose of each docked complex was viewed using RasMol 2.1, and all interacting residues within a radius of 4.5 Å of the ligand were restricted and analyzed. The results are summarized in **Tables 13–16**.

The interacting residues are shown and the binding pockets found in each protein sequence with respect to different drugs are highlighted. Analysis of the interacting residues showed certain binding pockets in each efflux pump protein studied. Certain residues were found to be preferred over others, for drug binding. These preferred binding pockets are:

Receptor	Drug	Interacting residues
4F4C	Praziquantel	Asn505, Arg551, Asp933, Lys936, Ile937, Ile939, Glu940, Glu943, Asn944, Lys964
	Mebendazole	Arg172, Thr197, Phe200, Asp201, Glu204, Lys207, Asn924, Ala928, Phe931, Ala932, Gly935, Ile939
	Oxamniquine	Ile40, Asp41, Ser42, Thr268, Tyr271, Ala272,Gly275, Arg830, Ser856, Ala860, Pro864, Arg867
	Albendazole	Ser42, Glu267, Thr268, Tyr271, Ala272, Gly275, Lys276, Lys315, Arg830, Ala860, Arg867
	Cladosporin	Tyr35, Glu36, Gly37, Ile40, Phe263, Ala264, Ile265, Glu267, Thr268, Tyr271, Lys315, Arg830, Thr861, Asn865, Arg867, Thr868, Glu1040, Lys1043
	Jaspamide	Leu161, Lys207, Glu208, Gly211, Asp212, Lys213, Gly917, Arg918, Arg919, Phe920, Gly922, Lys923, Asn924, Gln979
	Niclosamide	Lys26 Lys30, Ala910, Gln913, Tyr914, Arg916, Gly917, Leu1031, Gly1032, Phe1033,Thr1035
	Nifurtimox	Tyr35, Glu36, Gly37, Ile40, Phe263, Ala264, Glu267, Thr268, Lys315, Pro864,Asn865, Arg867, Thr868, Glu1040, Lys1043
	Artesunate	Arg8, Leu11, Arg12,Asp15, Lys26, Lys30, Leu371, Pro374, Arg916, Phe1033, Thr1035
	Benznidazole	Asn505, Arg551, Asp933, Lys936, Ile937, Glu940, Lys964
	Tipifarnib	Leu161, Glu204, Lys207, Glu208, Gly211, Asp212, Lys213, Val378, Asn924
	Imatinib	Ser42, Asn43, Glu267, Tr268, Tyr271, Ala272, Gly275, Lys276, Arg830, Ala860, Thr861, Arg867,
	Furozan	Ala910, Gln913, Tyr914, Arg916, Gly917, Arg919, Lys923, Gly1032, Thr1035, Ser1036, Phe1038, Pro1039
	Suramin	Asn4, Arg8, Asp51, Glu55, Thr194, Asp201, Asn202, Arg205, Glu716, Gly719, Lys720, Asp721
	Metrifonate	Ile40, Phe263, Ala264, Glu267, Thr268, Tyr271, Ala308, Lys315, Arg830, Ala860, Pro864, Arg867

Table 16.
Interacting residues between the P-glycoprotein of Trypanosoma cruzi *and the chosen drugs.*

1. P-glycoprotein (*Leishmania major*): **(Ser470, Glu472, Val474, Ile897, Glu898, Asn899, Phe900, Arg901, Thr902, Ser905)**

2. P-glycoprotein (*Onchocerca volvulus*): **(Arg830, Ala860, Thr861, Pro864, Arg867)**

3. P-glycoprotein (*Schistosoma mansoni*): **(Glu267, Thr268, Tyr271, Ala272, Gly275, Lys276)**

4. P-glycoprotein (*Trypanosoma cruzi*): **(Arg830, Ala860, Thr861, Arg867); (Gly917, Arg918, Arg919, Phe920, Gly922, Lys923); (Phe1033, Thr1035, Ser1036, Pro1039)**

3.9 P-glycoprotein in *E. coli*

A PSI-BLAST was performed to search for P-glycoprotein homologs in *E. coli* using hPGP as the query. The top BLAST hits showed low percentage identity (< 30%) and low score and were not annotated as bacterial P-glycoprotein. Though we could not reliably ascertain P-glycoprotein homologs in *E. coli*, there exist other mechanisms that could potentially lead to multidrug resistance phenotypes in *E. coli*. Multidrug efflux systems are of five types, namely the super-families ATP Binding Cassete (ABC) and Major Facilitador Super-family (MFS), Small Multidrug Resistance (SMR), Resistance, Nodulation, Division (RND) and Multidrug and Toxic Compound Extrusion (MATE). In *E.coli*, the examples for various systems include: MFS system Bcr, EmrB and EmrD; SMR family EmrE; RND family AcrB; and Mate family YdhE9 [39]. *E. coli* contains five putative ABC-type MDR-like transporters. These systems were all cloned and expressed in a drug-sensitive *E. coli* strain, and the drug resistance phenotypes were investigated. None of these systems provided an appreciable drug resistance to *E. coli*, except for YbjYZ, which conferred resistance to erythromycin [40]. The AcrAB-TolC system of *E. coli* is one of the best-characterized MDR transporters that is responsible for the acquisition of multiple antimicrobial resistance of the *mar* mutants, including resistance to tetracycline, chloramphenicol, ampicillin, nalidixic, and rifampicin [41, 42]. *E. coli* infections could modulate the pharmacokinetics of the drug enrofloxacin by altering the expression of intestinal P-glycoprotein in broilers [43].

4. Conclusions

The study of the human P-glycoprotein homologs, namely the P-glycoproteins of *Leishmania major*, *Onchocerca volvulus*, *Schistosoma mansoni*, and *Trypanosoma cruzi* has provided an insight into their drug resistance mechanisms. The investigational drugs such as cladosporin, jaspamide, nifurtimox, and tipifarnib are strong contenders for novel antihelminthic treatment. Known drugs such as praziquantel and moxidectin have shown great promise for use as treatment against other helminthic diseases.

Author details

Nivedita Jaishankar[1], Sangeetha Muthamilselvan[2] and Ashok Palaniappan[2*]

1 Department of Biotechnology, Sri Venkateswara College of Engineering, Sriperumbudur, India

2 Department of Bioinformatics, School of Chemical and BioTechnology, SASTRA Deemed University, Thanjavur, India

*Address all correspondence to: apalania@scbt.sastra.edu

References

[1] Nikaido H. Multidrug resistance in bacteria. Annual Review of Biochemistry. 2009;**78**:119-146. DOI: 10.1146/annurev.biochem.78.082907.145923

[2] Handzlik J, Matys A, Kieć-Kononowicz K. Recent advances in multi-drug resistance (MDR) efflux pump inhibitors of gram-positive bacteria S. aureus. Antibiotics. 2013; **2**(1):28-45. DOI: 10.3390/antibiotics2010028

[3] Sharma A, Gupta VK, Pathania R. Efflux pump inhibitors for bacterial pathogens: From bench to bedside. The Indian Journal of Medical Research. 2019;**149**(2):129-145. DOI: 10.4103/ijmr.IJMR_2079_17

[4] Higgins CF, Callaghan R, Linton KJ, Rosenberg MF, Ford RC. Seminars in cancer biology. Structure of the multidrug resistance P-glycoprotein. 1997;**8**:135-142

[5] Amin ML. P-glycoprotein inhibition for optimal drug delivery. Drug Target Insights. 2013;**7**:27-34. DOI: 10.4137/DTI.S12519

[6] Jin MS, Oldham ML, Zhang Q, Chen J. Crystal structure of the multidrug transporter P-glycoprotein from *C. elegans*. Nature. 2012; **490**(7421):566-569. DOI: 10.1038/nature11448

[7] Sauna ZE, Muller MM, Kerr KM, Ambudkar SV. The mechanism of action of multidrug-resistance- linked P-glycoprotein. Journal of Bioenergetics and Biomembranes. 2001;**33**(6): 481-491. DOI: 10.1023/A:1012875105006

[8] Fortuna A, ALves G, Falcao A. In vitro and in vivo relevance of the P-glycorotein probe substrates in drug discovery and development: Focus on rhodamine 123, digoxin and talinolol. Journal of Bioequivalence and Bioavailablity. 2012;**01**(02):22. DOI: 10.4172/jbb.S2-001

[9] Sheps JA, Ralph S, Zhao Z, et al. The ABC transporter gene family of *Caenorhabditis elegans* has implications for the evolutionary dynamics of multidrug resistance in eukaryotes. Genome Genome Biology. 2004;**5**:R15. DOI: 10.1186/gb-2004-5-3-r15

[10] Laing N, Speicher LA, Smith CD, Tew KD. P-glycoprotein binding and modulation of the multidrug resistance phenotype by Estramustine. Journal of the National Cancer Institute. 1994; **86**(9):688-694. DOI: 10.1093/jnci/86.9.688

[11] Bourguinat C, Ardelli BF, Pion SDS, Kamgno J, Gardon J, Duke BOL, et al. P-glycoprotein-like protein, a possible genetic marker for ivermectin resistance selection in *Onchocerca volvulus*. Molecular and Biochemical Parasitology. 2008;**158**(2):101-111. DOI: 10.1016/j.molbiopara.2007.11.017

[12] Gamarro F, Chiquero MJ, Amador MV, Castanys S. P-glycoprotein overexpression in methotrexate-resistant *Leishmania*. Biochemical Pharmacology. 1994;**47**(11):1939-1947. DOI: 10.1016/0006-2952(94)90067-1

[13] Piscopo TV, Mallia AC. Leishmaniasis. Postgraduate Medical Journal. 2007;**83**(976):649-657. DOI: 10.1136/pgmj.2006.047340corr1

[14] Barrett MP, Croft SL. Management of trypanosomiasis and leishmaniasis. British Medical Bulletin. 2012;**104**: 175-196. DOI: 10.1093/bmb/lds031

[15] Wyllie S, Cunningham ML, Fairlamb AH. Dual action of antimonial drugs on thiol redox metabolism in the human pathogen *Leishmania donovani*.

The Journal of Biological Chemistry. 2004;**279**(38):39925-39932. DOI: 10.1074/jbc.M405635200

[16] Soares ROA, Echevarria A, Bellieny MSS, Pinho RT, de Leo RMM, Seguins WS, et al. Evaluation of thiosemicarbazones and semicarbazones as potential agents anti-*Trypanosoma cruzi*. Experimental Parasitology. 2011; **129**(4):381-387. DOI: 110.1016/j.exppara.2011.08.019

[17] de Silva NR, Brooker S, Hotez PJ, Montresor A, Engels D, Savioli L. Soil-transmitted helminth infections: Updating the global picture. Trends in Parasitology. 2003;**19**(12):547-551. DOI: 10.1016/j.pt.2003.10.002

[18] Colley DG, Bustinduy AL, Secor WE, King CH. Human schistosomiasis. Lancet. 2014; **383**(9936):2253-2264. DOI: 10.1016/S0140-6736(13)61949-2

[19] Pinto-Almeida A, Mendes T, Armada A, et al. The role of efflux pumps in *Schistosoma mansoni* Praziquantel resistant phenotype. PLoS One. 2015;**10**(10):e0140147. DOI: 10.1371/journal.pone.0140147

[20] Cobo F. 10—Trypanosomiasis. In: Cobo F, editor. Imported Infectious Diseases. Cambridge: Woodhead Publishing; 2014. pp. 137-153. ISBN: 9781907568572. DOI: 10.1533/9781908818737.137

[21] Maya JD, Cassels BK, Iturriaga-Vásquez P. Mode of action of natural and synthetic drugs against *Trypanosoma cruzi* and their interaction with the mammalian host. Comparative Biochemistry and Physiology A Molecular and Integrative Physiology. 2007;**146**(4):601-620. DOI: 10.1016/j.cbpa.2006.03.004

[22] Bahia MT, de Andrade IM, Martins TA, et al. Fexinidazole: A potential new drug candidate for Chagas disease. PLoS Neglected Tropical Diseases. 2012;**6**(11):e1870. DOI: 10.1371/journal.pntd.0001870

[23] Liu J, Hajibeigi A, Ren G, Lin M, Siyambalapitiyage W, Liu Z, et al. Retention of the radiotracers 64Cu-ATSM and 64Cu-PTSM in human and murine tumors is influenced by MDR1 protein expression. Journal of Nuclear Medicine. 2009;**50**(8):1332-1339. DOI: 10.2967/jnumed.109.061879

[24] Rappa G, Lorico A, Liu MC, Kruh GD, Cory AH, Cory JG, et al. Overexpression of the multidrug resistance genes mdr1, mdr3 and mrp in L1210 leukemia cells resistant to inhibitors of ribonucleotide reductase. Biochemical Pharmacology. 1997;**54**: 649-655. DOI: 10.1016/s0006-2952(97)00210-4

[25] Campos MCO, Castro-Pinto DB, Ribeiro GA, Berredo-Pinho MM, Gomes LHF, Bellieny MS, et al. P-glycoprotein efflux pump plays an important role in *Trypanosoma cruzi* drug resistance. Parasitology Research. 2010;**112**:2341-2351. DOI: 10.1007/s00436-010-1988-6

[26] Kumar A, Muthamilselvan S, Palaniappan A. Computational studies of drug repurposing targeting P-glycoprotein-mediated multidrug resistance phenotypes in priority infectious agents. In: Creatinine—a Comprehensive Update. Rijeka: Intechopen; 2020. DOI: 10.5772/intechopen.90745

[27] Bhagwat M, Aravind L. PSI-BLAST tutorial. In: Bergman NH, editor. Comparative Genomics: Volumes 1 and 2. Totowa (NJ): Humana Press; 2007. Available from: https://www.ncbi.nlm.nih.gov/books/NBK2590/.ch 10

[28] Chenna R, Sugawara H, Koike T, Lopez R, Gibson TJ, Higgins DG, et al. Multiple sequence alignment with the Clustal series of programs. Nucleic

Acids Research. 2013;**31**(13):3497-3500. DOI: 10.1093/nar/gkg500

[29] Biasini M, Bienert S, Waterhouse A, Arnold K, Studer G, Schmidt T, et al. SWISS-MODEL: Modelling protein tertiary and quaternary structure using evolutionary information. Nucleic Acids Research. 2014;**42**(Web Server Issue): W252-W258. DOI: 10.1093/nar/gku340

[30] Schwede T, Kopp J, Guex N, Peitsch MC. SWISS-MODEL: An automated protein homology-modeling server. Nucleic Acids Research. 2003; **31**(13):3381-3385. DOI: 10.1093/nar/gkg520

[31] Weininger D, Weininger A, Weininger JL. SMILES 2. Algorithm for generation of unique SMILES notation. Journal of Chemical Information and Computer Sciences. 1989;**29**(2):97-101. DOI: 10.1021/ci00062a008

[32] Morris GM, Huey R, Lindstrom W, Sanner MF, Belew RK, Goodsell DS, et al. Autodock4 and AutoDockTools4: Automated docking with selective receptor flexiblity. Journal of Computational Chemistry. 2009;**30**(16): 2785-2791. DOI: 10.1002/jcc.21256

[33] Sayle RA, Milner-White EJ. RASMOL: Biomolecular graphics for all. Trends in Biochemical Sciences. 1995; **20**(9):374. DOI: 10.1016/s0968-0004(00)89080-5

[34] Messerli SM, Kasinathan RS, Morgan W, Spranger S, Greenberg RM. *Schistosoma mansoni* P-glycoprotein levels increase in response to praziquantel exposure and correlate with reduced praziquantel susceptibility. Molecular And Biochemical Parasitology. 2009;**167**(1): 54-59. DOI: 10.1016/j.molbiopara.2009.04.007

[35] Buckner FS, Waters NC, Avery VM. Recent highlights in anti-protozoan drug development and resistance research. International Journal of Parasitology Drugs and Drug Resistance. 2012;**2**:230-235. DOI: 10.1016/j.ijpddr.2012.05.002

[36] García MT, Lara-Corrales I, Kovarik CL, Pope E, Arenas R. Tropical skin diseases in children: A review-part II. Pediatric Dermatology. 2016;**33**: 264-274. DOI: 10.1111/pde.12778

[37] Kappagoda S, Singh SMU, Blackburn BG. Antiparasitic therapy. Mayo Clinic Proceedings. 2011;**86**(6): 561-583. DOI: 10.4065/mcp.2011.0203

[38] Varghese S, Palaniappan A. Computational pharmacogenetics of P-glycoprotein mediated antiepileptic drug resistance. The Open Bioinformatics Journal. 2018;**11**:197-207. DOI: 10.2174/1875036201811010197

[39] Moreira MAS, de Souza EC, de Moraes CA. Multidrug efflux systems in gram-negative bacteria. Brazilian Journal of Microbiology. 2004;**35**(1):2. DOI: 10.1590/S1517-83822004000100003

[40] Nishino K, Yamaguchi A. Analysis of a complete library of putative drug transporter genes in *Escherichia coli*. Journal of Bacteriology. 2001;**183**: 5803-5812

[41] Ma D, Cook DN, Alberti M, Pon NG, Nikaido H, Hearst JE. Molecular cloning and characterization of acrA and acrE genes of *Escherichia coli*. Journal of Bacteriology. 1993;**175**: 6299-6313

[42] Ma D, Cook DN, Alberti M, Pon NG, Nikaido H, Hearst JE. Genes acrA and acrB encode a stress-induced efflux system of *Escherichia coli*. Molecular Microbiology. 1995;**16**:45-55

[43] Lubelski J, Konings WN, Driessen AJM. Distribution and physiology of ABC-type transporters contributing to multidrug resistance in bacteria. Microbiology and Molecular Biology Reviews. 2007;**71**(3):463-476. DOI: 10.1128/MMBR.00001-07

Section 3

Disinfection and Antibiotic Resistance

Chapter 8

Effects of UV-LED Irradiation on *E. coli* in Water Disinfection

Paul Onkundi Nyangaresi, Baoping Zhang and Liang Shen

Abstract

Ultraviolet light-emitting diode (UV-LED) is a newly emerging UV light source with a potential of replacing the conventional chemical methods, mercury UV lamps and xenon lamps in water disinfection applications. In this chapter, we will first give a general description on the status of *E. coli* disinfection in water by UV-LEDs. Then the main text will concentrate on our experimental studies. We will discuss the effects of single and combined UV-LED irradiation on *E. coli* in water, including the inactivation efficiency, the recover percentage after the UV-LED irradiation, the optimal wavelength for low energy consumption, differences in pulsed and continuous operations of UV-LEDs, effect of UVA-LED followed by UVC-LED irradiation and vice versa, and finally the effect of TiO_2-assisted photocatalytic disinfection.

Keywords: UV-LED, disinfection, *E. coli*, water

1. Introduction

Millions of people including children die every year from infectious diseases caused by various waterborne pathogens [1]. Among the pathogens, a group of bacteria called *Escherichia coli* (*E. coli*) is one of the known carrier of the diseases such as diarrhea, urinary tract infections, respiratory illness and pneumonia [2]. Since *E. coli* are typically found in the environment, foods and intestines of humans and animals, they have been widely used as fecal indicator bacteria in water quality analysis [3]. Numerous countries and world organizations put a limit count of zero per 100 ml *E. coli* for drinking water. Passing this limit, it is an indication of the presence of faecally related pathogens in water, and hence a potential risk of high level of microbial waterborne disease outbreak [4]. Therefore, different water disinfection methods have been employed using *E. coli* as an inactivation target either in laboratory tests or in water disinfection plants. Among the different methods, the conventional use of chemicals such as chlorine can lead to introduction of disinfectant-resistance to bacteria [5], change of water taste and production of odor [6] and harmful disinfection by-products (DBPs) such as trihalomethane (THM) compounds, and haloacetic acids (HAAs) that are carcinogenic, mutagenic and reproductive toxicants [7]. Ozone is reported as an effective alternative disinfectant to chlorine due to its ability of reducing microbiological challenge to downstream disinfection. However, the ozone is also known in forming DBPs, particularly bromate [4], that can cause irreversible effects on humans such as renal failure and deafness [8]. The latest water disinfection method employs the use of ultraviolet (UV) light irradiation whose wavelength ranges from 100 to 400 nm. The UV light irradiation is currently attracting extensive attention in water and wastewater disinfection because of it is DBPs-free, and no need of chemicals that can cause ecological problems [9].

UV light is usually divided into four regions: vacuum (V) UV (100–200 nm), UV-C (200–280 nm), UV-B (280–315 nm) and UV-A (315–400 nm) [10]. Note that, water and air absorb all wavelengths below 190 nm. Therefore, only the wavelengths between 190 and 380 nm can cause biological effects [11]. Absorption of UV light by deoxyribonucleic acid (DNA)/ and ribonucleic acid (RNA) of a microorganism or virus inhibits its normal replication leading to cell death [12]. However, the UV damaged DNA of some microorganisms particularly *E. coli* is known to undergo repair by mechanisms such as photo-repair that requires light in the wavelength range of 300-500 nm to activate a photolyase enzyme and dark repair that is light independent [13, 14]. This can greatly decrease the UV light disinfection efficiency hence posing a great chance of health risks of infection. The common UV light sources include: the sun, mercury pressure lamps, xenon lamps and newly emerging UV-light emitting diodes (UV-LEDs). Although the sun gives a cheap and green natural source of light, it is mostly unreliable and only UVA, and approximately 10% of UVB light reaches the earth's surface [15]. Mercury pressure lamps which exists in two types: low pressure (LP) and medium pressure (MP) mercury lamps emitting a monochromatic light at a wavelength of 254 nm and polychromatic light at a broad range of 185–600 nm respectively [12], are the commonly used UV light sources in the current water disinfection systems [16]. However, these lamps are usually characterized with fixed wavelengths and limitations like short bulb lifetime, low energy efficiency, high operating temperatures and environmental pollution due to mercury [17]. On the other hand, xenon lamps are characterized by a broad range of wavelength (200–1100 nm), with 40% being UV consisting of UVC, UVB and UVA in ratio about 20%, 8% and 12%, respectively [18]. Therefore, the xenon lamp can exhibit both photochemical effect due to the effect of the UV light, photophysical and photothermal effects due to its high intense pulses [19]. The three multi-target effects can lead to complete destruction of the cell wall and the nucleic acid structure of a microorganism [20]. In addition, the xenon lamps have high penetration, high energy conversion, no pre-heating is needed, faster start-up and no ozone generation [21]. Although the xenon lamp exhibits the above mentioned advantages over the sun and mercury pressure lamps, they have a high energy demand which is un-preferable especially in developing countries. The lamps are also limited in adjusting the duty rates and pulse frequency due to overheating that will affect disinfection efficiency [22]. The newly emerging UV-LEDs are characterized with diversity in wavelengths within the UV range and have advantages such as environmental friendly (no mercury), compact and durable, faster start-up, potential to minimize energy consumption, longer lifetime, and a high frequency switching [23–25].

Therefore, due to their characteristics and advantages, the UV-LEDs have arisen as a very promising UV light sources in water disinfection applications as demonstrated in literature [23, 26–28]. Especially, UV-LED reactors can be utilized in small scale since they can be photovoltaic powered, which is convenient in remote areas since they can be photovoltaic powered [29–31]. Although the wall plug efficiency (WPE) of UV mercury lamps (15–35%) is higher than that of UV-LEDs (<10%), the latter is expected to be improved significantly, being similar to the case seen in visible LEDs whose WPE is currently around 80% [32, 33]. In water disinfection, the UV-LED irradiation can be applied in two modes: (i) pulsed light

(PL) and (ii) continuous wave (CW) mode. Whereas PL irradiation is a fast non-thermal technology for decontamination based on the application of short pulses of high intensity of light [22], CW application on the other hand is based on the application of low light intensity [34]. Furthermore, the diverse nature of the UV-LED wavelengths allows for tailored irradiation in which the wavelengths can be irradiated at the same time (simultaneous) or one after the other (sequential). During the disinfection applications, the mechanism of the two irradiation modes can either be photolytic or photocatalytic. In photolytic disinfection, only UV light is involved such that the absorbed photons inactivate the pathogen [12]. Meanwhile, photocatalytic disinfection involves combining UV light and a photocatalyst such as TiO_2, that has the ability to absorb UV light of appropriate photon energy (Eq. (1)), and in an air-saturated or water environment, radicals such as OH• and $•O_2^-$ that are highly destructive towards microorganisms are produced [35]. Therefore, this chapter discusses effects of UV-LED irradiation on *E. coli* in water, including inactivation efficiency, recover percentage after the UV-LED irradiation, and energy consumption, in terms of single and combined wavelength, PL and CW operations, simultaneous and sequential modes, and finally the effect of TiO_2-assisted photocatalysis.

$$E = h\frac{c}{\lambda} \tag{1}$$

where E is the photon energy, h is the plank's constant =6.63×10^{-34}J s, c is the speed of light in a vacuum = 3.0×10^8 m/s and λ is the wavelength of the UV light (m).

2. Indices of inactivation and repair performance for UV-LED disinfection

2.1 Evaluation of inactivation

2.1.1 Inactivation efficiencies

The inactivation efficiency of *E. coli* was analyzed by calculating log inactivation using Eq. (2).

$$\text{Log inactivation} = \text{Log}\left(\frac{N_0}{N}\right) \tag{2}$$

where N_0 and N are the colony count (CFU/mL) before and immediately after inactivation, respectively.

2.1.2 Synergistic inactivation efficiencies

Synergistic effect of combined wavelengths on the *E. coli* inactivation is compared from the results of log inactivation by combined UV-LEDs and the results from the sum of log inactivation by individual UV-LEDs. Therefore, the synergy values were calculated using the relation:

Synergy (Log units) = Log inactivation by combined UV-LEDs – Sum of log inactivation by individual UV-LEDs.

2.2 Evaluation of repair

2.2.1 Repair efficiencies

The percentage of repair either due to photo-repair or dark repair was quantified using Eq. (3) [36].

$$\text{Percentage of repair } (\%) = \frac{N_t - N}{N_0 - N} \cdot 100\% \tag{3}$$

where N_0 is the cell number before UV irradiation (CFU/mL), N is the immediate cell number after UV irradiation (CFU/mL), N_t is the cell number after repair for a period of time, t (CFU/mL).

In addition, the repair can be expressed as a function of the survival ratio (Eq. (4)) in respect of the initial microorganism concentration before the inactivation process [37].

$$S = \frac{N_t}{N_0} \cdot 100\% \tag{4}$$

where S is the survival ratio at time t (%); N_0 and N_t have the same meaning as above.

2.2.2 Repair kinetics

2.2.2.1 Modeling photo-repair

A non-linear regression model was used to model photo-repair (Eq. (5)) [38, 39].

$$S = \frac{S_m}{1 + \left(\frac{S_m}{S_0} - 1\right) \cdot e^{-k_2 \cdot S_m \cdot t}} \tag{5}$$

where S_m is the maximum limit of the microorganisms' survival by repair and S_0 is the survival ratio immediately after UV irradiation, k_2 is the growth second-order repair rate constant.

Note that k_2 is not a pure repair rate constant, it is rather a model parameter that is adjusted to predict the experimental data whose physical meaning is related to the time required to reach S_m and then the stabilization phase [38, 39]. Therefore, a pure repair rate constant, K (Eq. (6)) can be obtained from the derivatives of Eq. (5) and its maximum value, K_{max} (Eq. (7)) is obtained when S reaches half of S_m [40].

$$K = \frac{ds}{dt} = k_2(S_m - S) \cdot S \tag{6}$$

$$K_{max} = \frac{k_2(S_m)^2}{4} \tag{7}$$

2.2.2.2 Modeling dark repair

A model that considers a low and brief repair period and a decay phase was used in modeling dark repair (Eq. (8)) [38, 39].

$$S = \frac{S_m}{1 + \left(\frac{S_m}{S_0} - 1\right) \cdot e^{-k_2 \cdot S_m \cdot t}} - M \cdot t \tag{8}$$

where M is the mortality, a zero-order decay rate constant, while the other parameters have the same meaning as in Eq. (5). Note that, S, S_m, S_0, k_2, M and t in Eqs. (5) and (8) have a clear physical significance.

2.3 UV-LED technical parameters

2.3.1 Emission spectrum and optical power

The action spectrum of a microorganism is directly related to the LED emission spectrum i.e., the wavelength and the full width at half maximum [41–44]. Therefore, the determination of the LED emission spectrum before any experimental study is crucial. In this chapter, UV-LEDs with emissions at 265, 280, 310 and 365 nm, optical power of 1.8, 1.6, 1.3, 100 mW respectively at current of 20, 20, 20, 350 mA achieved at voltages of 6.0, 4.0, 6.0 and 4.0 V respectively (Great Bright Company, China) were used. The optical power was measured by an integrating sphere. Meanwhile the emission spectra measured with Spectro 320 Optical Scanning Spectrometer exhibited peak wavelengths at 267, 275, 310 and 370 nm with full widths at half-maximum of about 12, 10, 9 and 8 nm respectively (**Figure 1**).

2.3.2 Fluence measurement

The log inactivation of most pathogens is proportional to the applied UV light fluence as given in Eq. (9), where k is the inactivation rate constant that varies from one microorganism to another.

$$\text{Log inactivation} = k \cdot \text{Fluence} \tag{9}$$

Therefore, determining of fluence is critical for UV-LED disinfection applications. The common UV fluence determination methods include: Radiometry and chemical actinometry (iodide-iodate (KI) and ferrioxalate (FeO_x) actinometry). For UV-LEDs, fluence determination protocol employing the two methods for pathogen

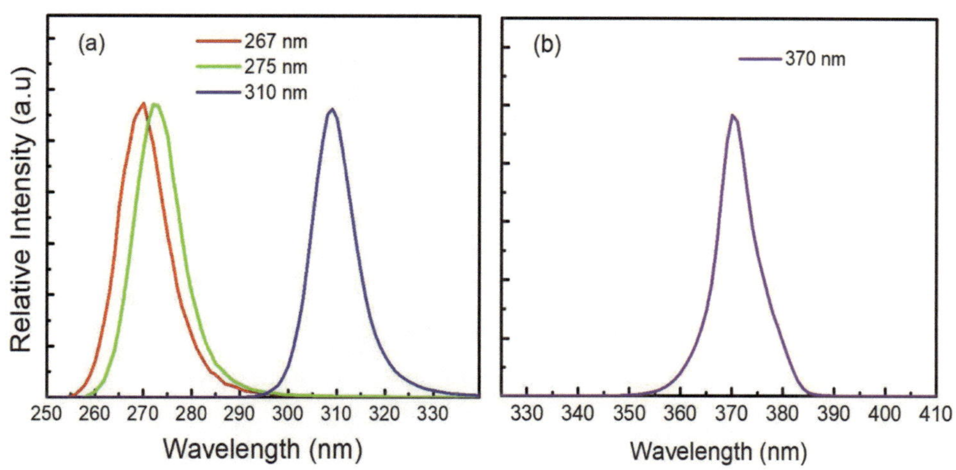

Figure 1.
Emission spectra from the (a) 267, 275, 310 and (b) 370 nm UV-LEDs.

inactivation is well described in [45]. Therefore, this chapter employed radiometry only in the UV-LED fluence determination.

Average incident irradiance, $\overline{E_0}$ (mW/cm^2) was first determined using IL-1700 radiometer with a SED 270 detector (International Light, USA), at the same distance as the water surface level from the UV-LEDs (L cm). Average fluence $\overline{F_0}$ (mJ/cm^2) inside the Petri dish was then calculated using Eq. (10) [45].

$$\overline{F_0} = \frac{\overline{E_0} \cdot PF \cdot DF \cdot WF \cdot RF \cdot t}{CF} \tag{10}$$

where PF the petri factor, DF the divergence factor calculated using Eq. (11), WF the water factor calculated using Eq. (12), RF the reflection factor taken to be 0.975 [46], t (s) the exposure time and CF is the collimation factor which was taken to be 1.

$$\mathrm{DF} = \frac{\mathrm{L}}{\mathrm{L} + \mathrm{D}} \tag{11}$$

where L (cm) is the distance between microbial suspension surface and the UV-LED and D (cm) the microbial suspension depth (**Figure 2**).

$$\mathrm{WF} = \frac{\mathrm{I}_\lambda \cdot \left(1 - 10^{-\alpha_\lambda \cdot \mathrm{D}}\right)}{\mathrm{I} \cdot \alpha_\lambda \cdot \mathrm{D} \cdot \ln(10)} \tag{12}$$

where I (mW/cm^2) and I_λ (mW/cm^2/nm) are the total radiant power of the UV-LED and the radiant power at λ of the UV-LED, respectively, α_λ (cm^{-1}) is the decadic absorption coefficient of the microbial suspension at λ, and D (cm) is the microbial suspension depth. The decadic absorption coefficient is the absorbance for 1 cm path length.

2.3.3 Electrical energy determination

The electrical energy ($E_{E,N}$) for a specific N-log inactivation of microorganisms can be determined using Eq. (13).

$$\mathrm{E}_{\mathrm{E,N}} = \frac{\pi\left(\frac{d}{2}\right)^2 \cdot \mathrm{F_N}}{3.6 \cdot 10^3 \cdot \mathrm{V} \cdot \mathrm{C} \cdot \mathrm{WF}} \tag{13}$$

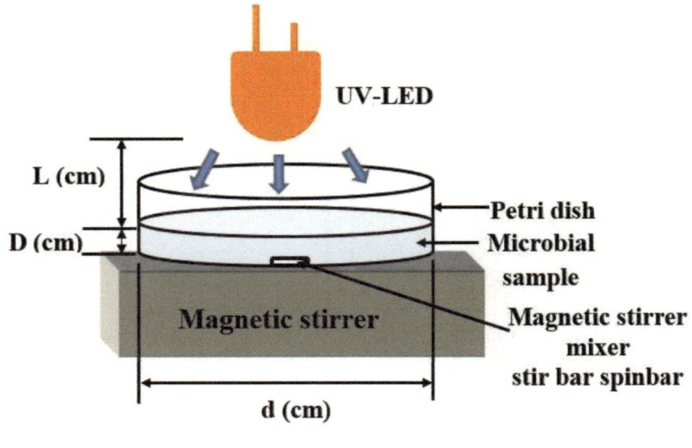

Figure 2.
Set-up of a batch disinfection reactor.

where $E_{E,N}$ is the electrical energy for a specific *N*-log reduction of each sample, (in kWh/m^3), d (cm) is the internal diameter of the Petri dish (**Figure 2**) and F_N is the fluence required for N-log inactivation (mJ/cm^2). The value of 3.6×10^3 is a unit conversion constant for W and kW, s and h, mL and m^3. V is the volume of the sample (mL). C is the wall plug efficiency calculated using Eq. (14) [47] and WF is the water factor calculated using Eq. (12) [46].

$$C = \frac{P_{output}}{P_{input}} = \frac{F_A}{I_A \cdot V_A} \quad (14)$$

where P_{output} is optical power (mW) of the UV-LEDs, P_{input} is the applied electrical power (mW), I_A is the applied current (mA), V_A is the applied voltage (V), and F_A is the radiant flux (mW).

3. Disinfection performance of UV-LED

The UV-LED disinfection efficiencies were discussed in four parts: (i) inactivation; (ii) repair; (iii) synergistic effect; and (iv) electrical energy efficiency.

3.1 *E. coli* inactivation efficiency

Comparative experiments with or without TiO_2 confirmed that, after 40 min of stirring in the dark, no inactivation occurred (data not shown). This indicates that UV light is the key requirement in both photolytic and photocatalyic inactivation. In both the photolytic and photocatalytic experiments, lower wavelengths were found to have a higher inactivation efficiency than longer wavelengths (267 > 275 > 310 > 370 nm) [48]. Specifically, in photolytic inactivation, an average fluence of 5, 7, 800 and 900 mJ/cm^2 was required by the 267, 275, 310 and 370 nm UV-LEDs, respectively per order of log inactivation. Note that, a 4-log inactivation is required especially in Austria and Germany [12] in the inactivation of most microorganisms. Therefore, the 267 and 275 nm UV-LEDs required an average fluence of 12 and 15 mJ/cm^2, respectively for the 4-log to be achieved in *E. coli* inactivation. Meanwhile the other UV-LEDs required a relatively higher fluence for the same 4-log inactivation to be achieved [48, 49]. This finding indicated that UVC wavelengths have a higher germicidal effect in the inactivation of *E. coli* as also confirmed by their relatively higher average inactivation rate constant (k) of 0.4 and 0.3 for the 267 and 275 nm UV-LEDs, respectively compared to insignificant <0.03 for the 310 and 365 nm UV-LEDs. The finding was also consisted with the other studies in literature as reviewed in Ref. [50]. The DNA of most microorganisms is believed to have an absorption maximum of light between 260 and 270 nm [51], hence confirming the findings.

In photocatalytic disinfection, addition of TiO_2 (1.0 g/L) resulted an interesting finding. Whereas the inactivation efficiency was increased in both the 310 and 370 nm UV-LEDs by the addition of TiO_2, that for the 267 and 275 nm UV-LEDs was drastically decreased [48]. Note that, anatase phase of TiO_2 that was used in our work has a bandgap of around 3.20 eV [52]. Therefore, in an air saturated or water environment, UV photon energy, $E \sim 5.12 \times 10^{-19}$ J is required to induce the generation of the reactive OH• radicals from the TiO_2 surface. The photon energy from the 267, 275, 310 and 370 UV-LEDs was calculated and found to be 7.45×10^{-19}, 6.87×10^{-19}, 6.42×10^{-19}, and 5.11×10^{-19} J, respectively. This indicates that, UVA wavelength is the most appropriate in photocatalytic disinfection as was confirmed by a significant enhanced inactivation efficiency by the 370 nm UV-LED when anatase phase of TiO_2 was added in the *E. coli* suspension [48]. The enhanced inactivation efficiency by the 370 nm UV-LED with TiO_2 is therefore attributable to the huddle effect of the UV photons and $OH^{\bullet}$ radicals. Other than their lower capability of radical production from the TiO_2 surface due to UV photon energy not within the optimum, the inactivation efficiency by the 267 and 275 nm UV-LEDs decreased with addition of the TiO_2 due to a screening effect by the TiO_2 which protected the *E. coli* against the strong UV photon of the UV-LEDs [53].

In another experiment, PL and CW UV-LED irradiation showed similar inactivation efficiency at equivalent average fluence [54]. Meanwhile, 267 nm UV-LED still had a slightly higher inactivation efficiency than the 275 nm UV-LED (**Figure 3**), which is in agreement with previous findings explained in the preceding paragraphs and also confirmed by reports in Ref. [51]. Although different UV-LEDs were employed, similar findings were also reported in other studies reported in literature [55–57]. However, an enhanced inactivation efficiency by PL over CW UV-LED irradiation is reported [58–61]. These discrepancies could be attributed mainly to unequal fluences between the PL and CW UV-LEDs, which is key in microbial inactivation. PL from xenon lamps is reported to cause enhanced inactivation efficiency than CW UV irradiation by mercury lamps [62]. The finding is due to xenon lamps' broad-spectrum UV content, short duration intense pulses and the high peak power which can lead to three multi-target mechanisms (photochemical, photophysical and photothermal) [63]. It should be noted that, the PL irradiation produced by xenon lamps is much different from that of the UV-LEDs in terms of emission spectrum, intensity, frequency switching. Therefore, the inactivation mechanisms of the PL xenon lamp may not apply to the UV-LEDs whose wavelengths are just within 200–400 nm and if a single UV-LED is used, almost a monochromatic wavelength is obtained compared with the broad range (200–1100 nm) from the xenon lamp. In addition, the current peak power of the UV-LEDs is still low (mW) which requires more improvements [64], compared to that of xenon lamps which is relatively high (kW) [65]. Unless the optical power is significantly improved, the *E. coli* inactivation efficiency by PL and CW UV-LED will still be equivalent. The only significant advantage of PL over CW UV-LED is its ability to suppress the heat generated during the UV-LED operation [54, 56]. This is due to the PL irradiation's ability to generate heat only during the short pulse and a

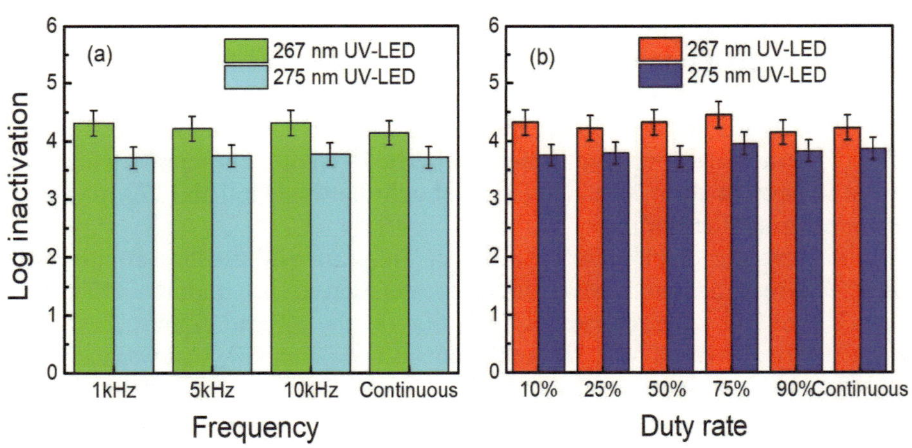

Figure 3.
Log inactivation at equivalent fluence of 17.3 mJ/cm² on E. coli *inactivation by the PL and CW UVC-LED irradiation after (a) varying frequency at 50% pulse rate and (b) varying duty rate at frequency = 1 kHz. Error bars represent standard deviation from triplicate experimental data.*

cooling period can occur between each pulse. This ability was clearly observed when the PL showed a lower solder temperature as compared to the CW UV-LED at similar driving currents and ambient temperature (**Figure 4**).

Diversity of UV-LED wavelengths allows UV-LED for tailored irradiation like involving 2 or more wavelengths either in simultaneous or sequential manner. It is well known that, irradiation by UVC and UVB can induce lesion formation in the genomic DNA of a microorganism [66–68]. Meanwhile, irradiation by UVA causes formation of active substances such as reactive oxygen species that have lethal effects to a microorganism [69]. Due to their different inactivation mechanisms, this part of the chapter therefore concentrated only on simultaneous and sequential irradiation involving a combination of UVC(or UVB) and UVA wavelengths. Note that, "UVC(B)" used here and henceforth in this chapter stands for UVC or UVB. Compared to sum of corresponding single wavelength, simultaneous irradiation of 267, 275 or 310 with 370 nm UV-LED led to lower log inactivation values of 1.27, 1.23 and 0.64, respectively. Similarly, lower log inactivation of 0.92, 0.90 and 0.63 was also obtained in sequential irradiation of 267, 275 and 310 nm followed by the 370 nm UV-LED, respectively (**Figure 5**). These results indicate that the 370 nm UV-LED irradiation could have functioned in repairing the already UV damaged DNA, rather than further damaging it [70, 71]. This assumption could be possible since the 370 nm is within the range of photo-repair light, 300–480 nm [13, 14]. On the other hand, higher log inactivation of 2.15 and 2.13 were achieved in sequential irradiation of 370 nm followed by 267 or 275 nm UV-LEDs, respectively. This log inactivation was also higher than that from the sum of corresponding single wavelength UV-LED irradiations, except for sequential irradiation of 370 nm followed by 310 nm UV-LEDs which achieved 0.98 log inactivation (**Figure 5**). Although the 370 nm (UVA) radiation can repair an already UV damaged DNA, the radiation on the other hand has an adverse effect when irradiated on un UV damaged DNA [72]. This phenomenon is known as concomitant photo-repair phenomenon in which inactivating light itself has the potential to photo-repair the UV-injured DNA [66]. Note also that, the 310 nm (UVB) is within the photo-repair light (300–480 nm), the 310 nm could have a concomitant photo-repair phenomenon similar to the 370 nm wavelength. While, 310 nm is still found to produce lesions in DNA that damage microorganisms. These findings are consistent with the other studies in literature [71, 73].

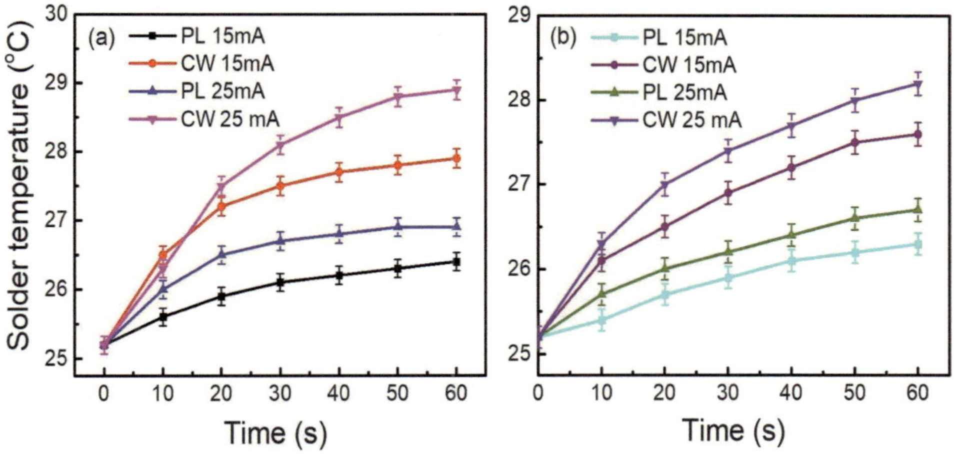

Figure 4.
Solder temperature as a function of operation period of the UVC-LEDs when operating in PL and CW mode; an ambient temperature of ~25°C, 50% duty rate at a frequency = 1 kHz for 267 nm (a) and 275 nm (b). Error bars represent standard deviation from triplicate experimental data.

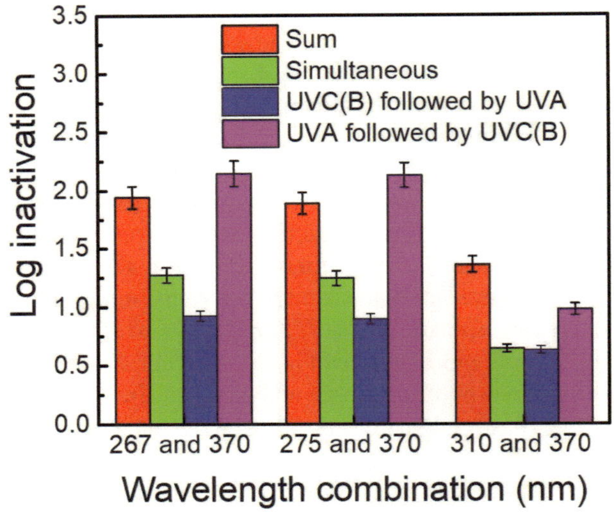

Figure 5.
E. coli *inactivation by combined wavelengths from different UV-LEDs. The 267, 275, 310 and 370 nm UV-LEDs provided an average fluence of 2.6, 2.6, 511.3 and 539.6 mJ/cm², respectively. Error bars represent standard deviation from triplicate experimental data.*

3.2 *E. coli* repair efficiency

As mentioned earlier in the introduction section, *E. coli* has the ability to undergo repair after damage from UV light irradiation. In all the experiments conducted, photo-repair was more dominant with an average of above 5% of photo-repair and negligible or no dark repair occurred [48, 49], demonstrating that photo-effect is the dominant mechanism of *E. coli* repair. The dominance of photo-effect in *E. coli* repair was also reported in other studies in Refs. [40, 74, 75]. Considering the 267, 275 nm UV-LED, the same observation was confirmed by the highest rate of photo-repair constant, $K_{max} > 4\%\ h^{-1}$ compared to that of dark repair, $K_{max} < 0.02\%\ h^{-1}$ [49]. By analyzing the photo-repair after photolytic inactivation, 275 and 370 nm wavelengths were found to be appropriate in suppressing the photo-repair. In addition, when the same wavelengths were applied, 275 nm followed after the 370 nm UV-LED irradiation has a much lower percentage of photo-repair compared to the simultaneous irradiation of 275 nm and 370 nm. This observation is attributed mostly to the damage of *E. coli*'s membrane at 370 nm [76], and as well as both DNA and proteins at 275 nm [77]. Note that, no significant difference was observed in the percentage of photo-repair for PL and CW UV-LED irradiation [54]. However, the addition of TiO_2 led to an insignificant % of *E. coli* photo-repair (<1%) and for dark repair, mortality was registered [48]. The observation is attributed to the concomitant effect of the photons from the UV-LEDs and the $OH^•$ radicals generated from the surface of UV irradiated TiO_2 that led to more damage to the *E. coli*. In addition, the mortality in the dark repair is attributed to a residual disinfecting effect of the $OH^•$ [78].

3.3 Synergistic effect

During the *E. coli* inactivation, different wavelengths were combined and their synergistic effect was evaluated. The irradiations were performed in both simultaneous and sequential manner. From the results obtained, simultaneous irradiation involving 267/275, 267/310 and 275/310 wavelength combinations from the UV-LEDs did not yield synergy in *E. coli* inactivation [49]. Note that, the 267 and

275 nm belong to the UVC, meanwhile the 310 nm belong to the UVB. The UVC and UVB have similar inactivation mechanism [66–68], which explains the absence of synergy in this case. Although UVC(B) and UVA wavelengths are reported to have different disinfection mechanisms as highlighted in the introduction section, interesting findings were found both in simultaneous and sequential irradiation on *E. coli* inactivation. Simultaneous irradiation of 267, 275, 310 nm and their combination with 370 nm UV-LED led to lower log inactivation compared to the sum of log inactivation of the corresponding single wavelengths. Similarly, lower log inactivation was achieved for 267, 275 and 310 nm followed by 370 nm UV-LED irradiation (**Figure 6**). These findings highlighted the concomitant photo-repair phenomenon of the 370 nm UV-LED. It should be noted that, the 370 nm is within the range of photo-repair light (300–500 nm). Therefore, other than damaging the *E. coli* bacteria, the 370 nm light could have performed the role of photo-repair as also discussed in previous studies [71, 72]. No synergy was found for 370 nm followed by 310 nm UV-LED irradiation. However, synergistic effect was found for 370 nm followed by 267 or 275 nm UV-LED (**Figure 6**). Because the 370 nm light can cause cell membrane damage, when irradiated with UVA first then followed by the UVC wavelengths, more damage was realized, leading to the synergy ultimately. However, irradiating 310 nm UV-LED after the 370 nm could have resulted to the repair of the *E. coli* since the 310 nm UV-LE is within the photo-repair light, hence absence of synergy in that case.

3.4 Electrical energy efficiency

To make a viable decision in choosing an appropriate UV-LED to be applied in disinfection applications, it is necessary to determine the electrical energy efficiency ($E_{E,N}$) of the UV-LEDs for microorganism inactivation in water. For the combined wavelengths, the sequential irradiation of UVA followed by UVC-LED showed higher inactivation and repair repression efficiencies of *E. coli* compared to the other combinations. Therefore, the electrical energy efficiency per order of magnitude ($E_{E,O}$) was determined only for single wavelength irradiation (in both photolytic and photocatalytic) and UVA followed by UVC(B)-LED irradiation on the

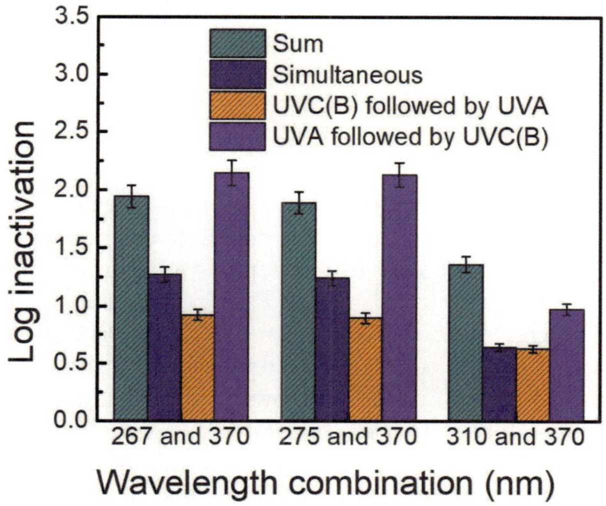

Figure 6.
Synergy from the combined UV-LEDs. The 267, 275, 310 and 370 nm UV-LEDs provided an average fluence of 2.6, 2.6, 511.3 and 539.6 mJ/cm², respectively. Error bars represent standard deviation from 3 experimental data.

Mode of irradiation	Photolytic/photocatalytic inactivation	UV-LED wavelength (nm)	$E_{E,O}$ (kWh/m³)
Single wavelength	Photolytic	267	0.4
		275	0.3
		310	17.2
		365	4.0
	Photocatalytic	267	0.6
		275	0.4
		310	16.0
		365	2.0
Combined wavelength (UVA followed by UVC or UVB	Photolytic	370 followed by 267	0.7
		370 followed by 275	0.5
		370 followed by 310	1.7

Table 1.
Average values of $E_{E,O}$ *for different wavelength irradiations in* E. coli *inactivation in water.*

E. coli in water (**Table 1**). In both photolytic and photocatalytic disinfection, the 275 nm UV-LED required lower $E_{E,O}$. Although the addition of TiO_2 to the *E. coli* suspension led to an increase in the $E_{E,O}$ for the 267 and 275 nm UV-LEDs, that for the 310 and 370 nm UV-LEDs decreased. Meanwhile, for the 370 nm followed by 275 nm UV-LED irradiation, it required lower $E_{E,O}$ than the other combination manners. The lower $E_{E,O}$ for the 275 nm UV-LED, and 370 nm followed by 275 nm UV-LED irradiation is mainly attributed to the higher wall plug efficiencies of these two kinds of UV-LEDs [48, 49]. A similar finding has also been reported in Ref. [79]. Note that, the decrease in $E_{E,O}$ for mostly the 370 nm UV-LED in photocatalytic disinfection is attributed two things: (i) its higher wall plug efficiency; and (ii) its photon energy being within the required to induce radicals on TiO_2 surface.

4. Conclusions

In this chapter, recent achievements about *E. coli* disinfection in water by UV-LEDs has been highlighted, as well as a general description on UV-LEDs. The main text concentrated more on our experimental studies in which the effects of single and combined UV-LED irradiation on *E. coli* in water, including the inactivation efficiency, the recover percentage after the UV-LED irradiation, the best wavelength for low energy consumption, differences in PL and CW operations of UV-LEDs, combination with UVA-LED followed by UVC-LED irradiation and vice versa, and finally the effect of TiO_2 photo-catalyst, were discussed. Whereas the 267 nm UV-LED showed higher inactivation efficiency, the 275 nm UV-LED was more competitive with comprehensive consideration of higher repressive ability on *E. coli* repair and higher electrical energy efficiency. For photocatalytic disinfection, the 370 nm UV-LED was the most appropriate. Although PL UV-LED was found to be effective in suppressing temperature rising than CW operation, the two modes showed insignificant difference in *E. coli* inactivation and repair efficiency. For the combined wavelengths, UVA (370 nm) followed by UVC (275 nm) irradiation was effective in all aspects of inactivation, repair and electrical energy efficiencies.

Acknowledgements

We acknowledge the support from the National Key R&D Program of China (no. 2016YFB0400803), and the China Government Scholarship Council, no. 2017GXZ023553.

Conflict of interest

None.

Author details

Paul Onkundi Nyangaresi[1], Baoping Zhang[1*] and Liang Shen[2]

1 Department of Electronic Engineering, Laboratory of Micro/Nano-Optoelectronics, Xiamen University, Xiamen, Fujian, China

2 Department of Chemical and Biochemical Engineering, College of Chemistry and Chemical Engineering, Xiamen University, Xiamen, Fujian, China

*Address all correspondence to: bzhang@xmu.edu.cn

References

[1] Griffiths JK. Waterborne diseases. In: Quah SR, editor. International Encyclopedia of Public Health. 2nd ed. Oxford: Academic Press; 2017. pp. 388-401. DOI: 10.1016/B978-0-12-803678-5.00490-2

[2] Woodall CJ. Waterborne diseases—What are the primary killers? Desalination. 2009;**248**(1):616-621. DOI: 10.1016/j.desal.2008.05.110

[3] Jackson PJ, Dillon GR, Irving TE, Stanfield G. Manual on treatment for small water supply systems. In: Final Report to the Department of the Environment, Transport and the Regions. DETR/DWI 4936/1. Henley Road, Medmenham, Marlow, Buckinghamshire SL7 2HD; 2001

[4] EPA. Water Treatment Manual: Disinfection. 2011. pp. 3-4. Available from: https://www.epa.ie/pubs/advice/drinkingwater/Disinfection2_web.pdf [Accessed: 08 October 2019]

[5] Puzari M, Sharma M, Chetia P. Emergence of antibiotic resistant Shigella species: A matter of concern. Journal of Infection and Public Health. 2018;**11**(4):451-454. DOI: 10.1016/j.jiph.2017.09.025

[6] Wang AQ, Lin YL, Xu B, Hu CY, Gao ZC, Liu Z, et al. Factors affecting the water odor caused by chloramines during drinking water disinfection. Science of the Total Environment. 2018;**639**:687-694. DOI: 10.1016/j.scitotenv.2018.05.188

[7] Mian HR, Hu G, Hewage K, Rodriguez MJ, Sadiq R. Prioritization of unregulated disinfection by-products in drinking water distribution systems for human health risk mitigation: A critical review. Water Research. 2018;**147**:112-131. DOI: 10.1016/j.watres.2018.09.054

[8] WHO. Bromate in Drinking-water. 2005. pp. 10-11. Available from: https://www.who.int/water_sanitation_health/dwq/chemicals/bromate260505.pdf [Accessed: 14 October 2019]

[9] Kowalski W. Ultraviolet Germicidal Irradiation Handbook. Berlin: Springer; 2009. pp. 368-371. DOI: 10.1007/978-3-642-01999-9

[10] Meulemans CCE. The basic principles of UV-disinfection of water. Ozone: Science & Engineering. 1987;**9**(4):299-313. DOI: 10.1080/01919518708552146

[11] Harm W. Biological Effects of Ultraviolet Radiation. New York: Cambridge University Press; 1980. pp. 31-39. Available from: http://assets.cambridge.org/97805212/93624/frontmatter/9780521293624_frontmatter.pdf

[12] Bolton JR, Cotton CAC. Mechanism of UV disinfection. In: The Ultraviolet Disinfection Handbook. New York, USA: American Water Works Association; 2008. pp. 25-40

[13] Friedberg EC. DNA damage and repair.pdf. Nature. 2003;**421**(23):436-440. DOI: 10.1038/nature01408

[14] Oguma K, Katayama H, Ohgaki S. Photoreactivation of *Escherichia coli* after low- or medium-pressure UV disinfection determined by an endonuclease sensitive site assay. Applied and Environmental Microbiology. 2002;**68**(12):6029-6035. DOI: 10.1128/aem.68.12.6029-6035.2002

[15] Roy CR, Gies HP, Lugg DJ, Toomey S, Tomlinson DW. The measurement of solar ultraviolet radiation. Mutation Research. 1998;**422**(1):7-14. DOI: 10.1016/S0027-5107(98)00180-8

[16] Bolton JR, Cotton CA. UV disinfection equipment. In: The Ultraviolet Disinfection Handbook. New York, USA: American Water Works Association; 2008. pp. 49-62

[17] Vilhunen S, Särkkä H, Sillanpää M. Ultraviolet light-emitting diodes in water disinfection. Environmental Science and Pollution Research. 2009; **16**(4):439-442. DOI: 10.1007/s11356-009-0103-y

[18] Wekhof A. Disinfection with flash lamps. PDA Journal of Pharmaceutical Science and Technology. 2000;**54**(3): 264-276. http://www.steribeam.com/info/PDA-2000.pdf

[19] Kramer B, Muranyi P. Effect of pulsed light on structural and physiological properties of listeria innocua and *Escherichia coli*. Journal of Applied Microbiology. 2014;**116**(3): 596-611. DOI: 10.1111/jam.12394

[20] Dharmadhikari AK, Bharambe H, Dharmadhikari JA, D'Souza JS, Mathur D. DNA damage by OH radicals produced using intense, ultrashort, long wavelength laser pulses. Physical Review Letters. 2014;**112**(13):138105. DOI: 10.1103/PhysRevLett.112.138105

[21] Ling J, Zhang X, Zheng W, Xing J. Evaluation of biological inactivation efficacy for in-duct pulsed xenon lamp. Building and Environment. 2018;**143**: 178-185. DOI: 10.1016/j.buildenv.2018.07.015

[22] Gómez-López VM, Ragaert P, Debevere J, Devlieghere F. Pulsed light for food decontamination: A review. Trends in Food Science & Technology. 2007;**18**(9):464-473. DOI: 10.1016/j.tifs.2007.03.010

[23] Muramoto Y, Kimura M, Nouda S. Development and future of ultraviolet light-emitting diodes: UV-LED will replace the UV lamp. Semiconductor Science and Technology. 2014;**29**(8): 084004. DOI: 10.1088/0268-1242/29/8/084004

[24] Würtele MA, Kolbe T, Lipsz M, Külberg A, Weyers M, Kneissl M, et al. Application of GaN-based ultraviolet-C light emitting diodes—UV LEDs—For water disinfection. Water Research. 2011;**45**(3):1481-1489. DOI: 10.1016/j.watres.2010.11.015

[25] McDermott SL, Walsh JE, Howard RG. A comparison of the emission characteristics of UV-LEDs and fluorescent lamps for polymerisation applications. Optics & Laser Technology. 2008;**40**(3):487-493. DOI: 10.1016/j.optlastec.2007.07.013

[26] Chen J, Loeb S, Kim J-H. LED revolution: Fundamentals and prospects for UV disinfection applications. Environmental Science: Water Research & Technology. 2017;**3**(2):188-202. DOI: 10.1039/c6ew00241b

[27] Ibrahim MAS, MacAdam J, Autin O, Jefferson B. Evaluating the impact of LED bulb development on the economic viability of ultraviolet technology for disinfection. Environmental Technology. 2014;**35**(4):400-406. DOI: 10.1080/09593330.2013.829858

[28] Lawal O, Cosman J, Pagan J. UV-C LED Devices and Systems: Current and Future State. 2018. pp. 1-7. Available from: https://www.iuvanews.com/stories/pdf/IUVA_2018_Quarter1_Lawal-article_hyperlinks.pdf [Accessed: 12 July 2019]

[29] Crawford MH, Banas MA, Ruby DS, Ross MP, Nelson JS, Allerman AA, et al. Final LDRD Report: Ultraviolet Water Purification Systems for Rural Environments and Mobile Applications. SAND2005-7245. Albuquerque, New Mexico 87185 and Livermore, California 94550: Sandia National Laboratories; 2005

[30] Lui GY, Roser D, Corkish R, Ashbolt N, Jagals P, Stuetz R.

Photovoltaic powered ultraviolet and visible light-emitting diodes for sustainable point-of-use disinfection of drinking waters. Science of the Total Environment. 2014;**493**:185-196. DOI: 10.1016/j.scitotenv.2014.05.104

[31] Matsumoto T, Tatsuno I, Hasegawa T. Instantaneous water purification by deep ultraviolet light in water waveguide: *Escherichia coli* bacteria disinfection. Water. 2019;**11**: 968. DOI: 10.3390/w11050968

[32] Kuritzky LY, Espenlaub AC, Yonkee BP, Pynn CD, DenBaars SP, Nakamura S, et al. High wall-plug efficiency blue III-nitride LEDs designed for low current density operation. Optics Express. 2017;**25**(24): 30696-30707. DOI: 10.1364/OE.25.030696

[33] CREE. Cree Launches the Industry's Brightest and Most Efficient Royal Blue LED. Available from: https://www.cree.com/news-events/news/article/cree-launches-the-industrys-brightest-and-most-efficient-royal-blue-led [Accessed: 12 July 2019]

[34] Gómez-López VM, Bolton JR. An approach to standardize methods for fluence determination in bench-scale pulsed light experiments. Food and Bioprocess Technology. 2016;**9**(6): 1040-1048. DOI: 10.1007/s11947-016-1696-z

[35] Wang W, Huang G, Yu JC, Wong PK. Advances in photocatalytic disinfection of bacteria: Development of photocatalysts and mechanisms. Journal of Environmental Sciences (China). 2015;**34**:232-247. DOI: 10.1016/j.jes.2015.05.003

[36] Lindenauer KG, Darby JL. Ultraviolet disinfection of wastewater: Effect of dose on subsequent photoreactivation. Water Research. 1994;**28**(4):805-817. DOI: 10.1016/0043-1354(94)90087-6

[37] Kashimada K, Kamiko N, Yamamoto K, Ohgaki S. Assessment of photoreactivation following ultraviolet light disinfection. Water Science and Technology. 1996;**33**(10):261-269. DOI: 10.1016/0273-1223(96)00428-3

[38] Nebot Sanz E, Salcedo Davila I, Andrade Balao JA, Quiroga Alonso JM. Modelling of reactivation after UV disinfection: Effect of UV-C dose on subsequent photoreactivation and dark repair. Water Research. 2007;**41**(14): 3141-3151. DOI: 10.1016/j.watres.2007.04.008

[39] Salcedo I, Andrade JA, Quiroga JM, Nebot E. Photoreactivation and dark repair in UV-treated microorganisms: Effect of temperature. Applied and Environmental Microbiology. 2007;**73**(5): 1594-1600. DOI: 10.1128/AEM.02145-06

[40] Li G-Q, Wang W-L, Huo Z-Y, Lu Y, Hu H-Y. Comparison of UV-LED and low pressure UV for water disinfection: Photoreactivation and dark repair of *Escherichia coli*. Water Research. 2017; **126**:134-143. DOI: 10.1016/j.watres.2017.09.030

[41] Beck SE, Rodriguez RA, Linden KG, Hargy TM, Larason TC, Wright HB. Wavelength dependent UV inactivation and DNA damage of adenovirus as measured by cell culture infectivity and long range quantitative PCR. Environmental Science & Technology. 2014;**48**(1):591-598. DOI: 10.1021/es403850b

[42] Chen RZ, Craik SA, Bolton JR. Comparison of the action spectra and relative DNA absorbance spectra of microorganisms: Information important for the determination of germicidal fluence (UV dose) in an ultraviolet disinfection of water. Water Research. 2009;**43**(20):5087-5096. DOI: 10.1016/j.watres.2009.08.032

[43] Jagger J. Introduction to Research in Ultraviolet Photobiology. Englewood Cliffs, N.J.: Prentice-Hall; 1967

[44] Mamane-Gravetz H, Linden KG, Cabaj A, Sommer R. Spectral sensitivity of bacillus subtilis spores and MS2 coliphage for validation testing of ultraviolet reactors for water disinfection. Environmental Science & Technology. 2005;**39**(20):7845-7852. DOI: 10.1021/es048446t

[45] Kheyrandish A, Mohseni M, Taghipour F. Protocol for determining ultraviolet light emitting diode (UV-LED) fluence for microbial inactivation studies. Environmental Science & Technology. 2018;**52**(13):7390-7398. DOI: 10.1021/acs.est.7b05797

[46] Bolton JR, Linden KG. Standardization of methods for fluence (UV dose) determination in bench-scale UV experiments. Journal of Environmental Engineering. 2003; **129**(3):209-215. DOI: 10.1061//asce/0733-9372/2003/129:3/209

[47] Zurfi A, Zhang J. Model identification and wall-plug efficiency measurement of white LED modules. American Journal of Engineering and Applied Sciences. 2016;**9**(2):412-419. DOI: 10.3844/ajeassp.2016.412.419

[48] Nyangaresi PO, Qin Y, Chen G, Zhang B, Lu Y, Shen L. Comparison of UV-LED photolytic and UV-LED/TiO2 photocatalytic disinfection for *Escherichia coli* in water. Catalysis Today. 2019;**335**:200-207. DOI: 10.1016/j.cattod.2018.11.015

[49] Nyangaresi PO, Qin Y, Chen G, Zhang B, Lu Y, Shen L. Effects of single and combined UV-LEDs on inactivation and subsequent reactivation of *E. coli* in water disinfection. Water Research. 2018;**147**:331-341. DOI: 10.1016/j.watres.2018.10.014

[50] Li X, Cai M, Wang L, Niu F, Yang D, Zhang G. Evaluation survey of microbial disinfection methods in UV-LED water treatment systems. Science of the Total Environment. 2019; **659**:1415-1427. DOI: 10.1016/j.scitotenv.2018.12.344

[51] Gates FL. A study of the bactericidal action of ultra violet light: III. The absorption of ultraviolet light by bacteria. Journal of General Physiology. 1930;**14**(1):31-42. Available from: https://www.ncbi.nlm.nih.gov/pmc/articles/PMC2141090/pdf/31.pdf

[52] Belver C, Bedia J, Gómez-Avilés A, Peñas-Garzón M, Rodriguez JJ. Chapter 22—Semiconductor photocatalysis for water purification. In: Thomas S, Pasquini D, Leu S-Y, Gopakumar DA, editors. Nanoscale Materials in Water Purification. Elsevier; 2019. pp. 581-651. DOI: 10.1016/B978-0-12-813926-4.00028-8

[53] Herrmann JM. Heterogeneous photocatalysis: Fundamentals and applications to the removal of various types of aqueous pollutants. Catalysis Today. 1999;**53**:115-129. DOI: 10.1016/S0920-5861(99)00107-8

[54] Nyangaresi PO, Qin Y, Chen G, Zhang B, Lu Y, Shen L. Comparison of the performance of pulsed and continuous UVC-LED irradiation in the inactivation of bacteria. Water Research. 2019;**157**:218-227. DOI: 10.1016/j.watres.2019.03.080

[55] Gillespie JB, Maclean M, Given MJ, Wilson MP, Judd MD, Timoshkin IV, et al. Efficacy of pulsed 405-nm light-emitting diodes for antimicrobial photodynamic inactivation: Effects of intensity, frequency, and duty cycle. Photomedicine and Laser Surgery. 2017; **35**(3):150-156. DOI: 10.1089/pho.2016.4179

[56] Song K, Taghipour F, Mohseni M. Microorganisms inactivation by continuous and pulsed irradiation of ultraviolet light-emitting diodes (UV-LEDs). Chemical Engineering Journal. 2018;**343**:362-370. DOI: 10.1016/j.cej.2018.03.020

[57] Sholtes K, Linden KG. Pulsed and continuous light UV LED: Microbial inactivation, electrical, and time efficiency. Water Research. 2019;**165**: 114965. DOI: 10.1016/j.watres.2019. 114965

[58] Li J, Hirota K, Yumoto H, Matsuo T, Miyake Y, Ichikawa T. Enhanced germicidal effects of pulsed UV-LED irradiation on biofilms. Journal of Applied Microbiology. 2010;**109**(6): 2183-2190. DOI: 10.1111/ j.1365-2672.2010.04850.x

[59] Tran T, Racz L, Grimaila MR, Miller M, Harper WF Jr. Comparison of continuous versus pulsed ultraviolet light emitting diode use for the inactivation of *Bacillus globigii* spores. Water Science and Technology. 2014; **70**(9):1473-1480. DOI: 10.2166/ wst.2014.395

[60] Wengraitis S, McCubbin P, Wade MM, Biggs TD, Hall S, Williams LI, et al. Pulsed UV-C disinfection of *Escherichia coli* with light-emitting diodes, emitted at various repetition rates and duty cycles. Photochemistry and Photobiology. 2013; **89**(1):127-131. DOI: 10.1111/ j.1751-1097.2012.01203.x

[61] Zou XY, Lin YL, Xu B, Cao TC, Tang YL, Pan Y, et al. Enhanced inactivation of *E. coli* by pulsed UV-LED irradiation during water disinfection. Science of the Total Environment. 2019; **650**(Pt 1):210-215. DOI: 10.1016/j. scitotenv.2018.08.367

[62] Bohrerova Z, Shemer H, Lantis R, Impellitteri CA, Linden KG. Comparative disinfection efficiency of pulsed and continuous-wave UV irradiation technologies. Water Research. 2008;**42**(12):2975-2982. DOI: 10.1016/j.watres.2008.04.001

[63] Wekhof A, Trompeter FJ, Franken O. Pulsed UV-disintegration (PUVD): A new sterilisation mechanism for packaging and broad medical-hospital applications. In: Proceedings of the First International Conference on Ultraviolet Technologies. Washington, DC, USA; 2001. pp. 1-15

[64] Kneissl M, Seong T-Y, Han J, Amano H. The emergence and prospects of deep-ultraviolet light-emitting diode technologies. Nature Photonics. 2019;**13**(4):233-244. DOI: 10.1038/ s41566-019-0359-9

[65] Takeshita K. Damage of yeast cells induced by pulsed light irradiation. International Journal of Food Microbiology. 2003;**85**(1–2):151-158. DOI: 10.1016/s0168-1605(02)00509-3

[66] Jagger J. Near-UV radiation effects on microorganisms. Photochemistry and Photobiology. 1981;**34**(6):761-768

[67] Ravanat J-L, Douki T, Cadet J. Direct and indirect effects of UV radiation on DNA and its components. Journal of Photochemistry and Photobiology B: Biology. 2001;**63**(1): 88-102. DOI: 10.1016/S1011-1344(01) 00206-8

[68] Sinha RP, Häder D-P. UV-induced DNA damage and repair: A review. Photochemical & Photobiological Sciences. 2002;**1**(4):225-236. DOI: 10.1039/b201230h

[69] Oppezzo OJ, Pizarro RA. Sublethal effects of ultraviolet A radiation on Enterobacter cloacae. Journal of Photochemistry and Photobiology B: Biology. 2001;**62**(3):158-165. DOI: 10.1016/S1011-1344(01)00180-4

[70] Song K, Mohseni M, Taghipour F. Mechanisms investigation on bacterial inactivation through combinations of UV wavelengths. Water Research. 2019; **163**:114875. DOI: 10.1016/j.watres.2019. 114875

[71] Xiao Y, Chu XN, He M, Liu XC, Hu JY. Impact of UVA pre-radiation on

UVC disinfection performance: Inactivation, repair and mechanism study. Water Research. 2018;**141**: 279-288. DOI: 10.1016/j.watres. 2018.05.021

[72] Pizarro RA, Orce LV. Membrane damage and recovery associated with growth delay induced by near-UV radiation in *Escherichia coli* K-12. Photochemistry and Photobiology. 1988;**47**:391-397. DOI: 10.1111/j.1751-1097.1988. tb02742.x.

[73] Song K, Taghipour F, Mohseni M. Microorganisms inactivation by wavelength combinations of ultraviolet light-emitting diodes (UV-LEDs). Science of the Total Environment. 2019;**665**:1103-1110. DOI: 10.1016/j. scitotenv.2019.02.041

[74] Quek PH, Hu J. Indicators for photoreactivation and dark repair studies following ultraviolet disinfection. Journal of Industrial Microbiology & Biotechnology. 2008;**35**(6):533-541. DOI: 10.1007/ s10295-008-0314-0

[75] Quek PH, Hu J. Influence of photoreactivating light intensity and incubation temperature on photoreactivation of Escherichia coli following LP and MP UV disinfection. Journal of Applied Microbiology. 2008;**105**(1):124-133. DOI: 10.1111/ j.1365-2672.2008.03723.x

[76] Pizarro RA. UV-A oxidative damage modified by environmental conditions in Escherichia coli. International Journal of Radiation Biology. 1995;**68**(3): 293-299. DOI: 10.1080/ 09553009514551221

[77] Schmid FX. Biological macromolecules: UV-visible spectrophotometry. In: Encyclopedia of Life Sciences. Chichester: eLS. John Wiley & Sons Ltd.; 2001. DOI: 10.1038/npg.els.0003142

[78] Xiong P, Hu J. Inactivation/ reactivation of antibiotic-resistant bacteria by a novel UVA/LED/TiO2 system. Water Research. 2013;**47**(13): 4547-4555. DOI: 10.1016/j. watres.2013.04.056

[79] Beck SE, Ryu H, Boczek LA, Cashdollar JL, Jeanis KM, Rosenblum JS, et al. Evaluating UV-C LED disinfection performance and investigating potential dual-wavelength synergy. Water Research. 2017;**109**:207-216. DOI: 10.1016/j.watres.2016.11.024

Chapter 9

Antimicrobial Resistance in *Escherichia coli*

Mario Galindo-Méndez

Abstract

In the last decades, antimicrobial resistance has become a global threat to public health systems worldwide. Among those bacteria that pose the greatest threat to human health because of its growing resistance to antibiotics are the members of the *Enterobacteriaceae* family, particularly *Escherichia coli* and *Klebsiella* spp. Among the different antibiotic-resistant mechanisms developed by bacteria, the ones found in *Enterobacteriaceae* are more diverse than those in other families and include resistance to different antibiotic groups, advantages that partially explain why these microorganisms are among the most common causes of antibiotic-resistant bacterial infections in humans. Due to the continuously increasing number of infections caused by multidrug-resistant *E. coli* due to its ease of transmission via the fecal-oral route among humans and from environmental sources, the understanding of the epidemiology of these strains and their mechanisms of resistance are key components in the fight against these infections.

Keywords: antimicrobial resistance, multidrug resistance, antibiotics, *Escherichia coli*, *Enterobacteriaceae*

1. Introduction

Escherichia coli is one of the most studied bacteria in the world and is arguably the best understood of all model microorganisms [1]. In the context of human and animal ecology, this microorganism participates as both a commensal of the gut, being one the first bacterial species to colonize it right after birth [2], and one of the most important human and animal pathogens, being able to cause intestinal and extra-intestinal infections. In humans, *E. coli* is the most frequent cause of urinary tract infections and has been identified as the causative agent of disease in practically every anatomical site of the human body, causing appendicitis, pneumonia, bloodstream, gastrointestinal infections, skin abscesses, intra-amniotic and puerperal infection in pregnant women, meningitis and endocarditis. Furthermore, *E. coli* can cause both community-acquired infections and health care-related infections, and is able to cause disease in all age groups.

Since the introduction of penicillin in the 1940s, which started the era of antibiotics, these agents have been recognized as one of the greatest advances in modern medicine and a turning point in human history. In 1900, infectious disease was a leading cause of death; in 2000, infectious diseases were responsible for only a small percentage of deaths in developed nations [3]. Unfortunately for humans, bacteria have evolved different mechanisms that have rendered them resistant to

IntechOpen

antibiotics, to the point that since not long ago antimicrobial resistance has become a global threat to public health systems worldwide.

The ability of bacteria to develop resistance against antibiotics began soon after their introduction, as penicillin resistance by *S. aureus* was identified just a few years after its introduction in hospitalized patients [4]. In the case of *E. coli*, resistance against antibiotics has been steadily increasing since the first reported cases and, due to its impact in human health, is now included, along with the rest of the *Enterobacteriaceae* family, in the World Health Organization's (WHO) list of the 12 families of bacteria that pose the greatest threat to human health [5].

The contribution of *E. coli* to the antimicrobial resistance phenomenon should be analyzed under two different, but complementary, contexts that at some point meet in one common issue: a broad impact on human health. These two perspectives include the increasing number of infections worldwide caused by multidrug-resistant *E. coli* strains *per se* and the ability of this bacterium to transmit its genetic-resistant traits to other bacteria. *E. coli* has evolved these two attributes that have made this microorganism such a key player in the antibiotic resistance pandemic due to its ease of transmission among humans and from animals to humans via the fecal-oral route. Secondly, the microorganism's ability to colonize the gut of humans and animals allow it to be in close interaction with an abounding number of different bacteria, interaction that grant *E. coli* the duality to behave as a donor of genetic material to other bacteria and the ability to acquire resistance genes from other microorganisms.

This chapter describes the human actions that have contributed to the development of *E. coli* resistance to antibiotics, including the major impact of hygiene on the transmission and maintenance of its multidrug-resistant strains, and the known mechanisms developed by this organism to resist the actions of commonly used antibiotics.

2. Onset and spread of *E. coli* resistance to antibiotics

The emergence of antibacterial resistance in *E. coli* and other bacteria is multifactorial, but has paralleled the incorporation of these agents into the therapeutic arsenal in human and veterinary medicine. Data show that *E. coli* present the highest rates of resistance against those antibiotics that have been in use the longest time [6], as is evidenced by the high resistance rate worldwide against sulfonamides [7], whose use in humans started in the 1930s and its first *E. coli*-resistant clones were identified as early as 1950 [6]. Additionally, it is of no coincidence that those regions of the world with the highest consumption of antibiotics are low- to mid-income countries (**Table 1**), whose antibiotic-resistant rates are higher than those found in high-income nations.

Antibiotic resistance (AR) is largely believed to be the sole result of human activity and antibiotic chemotherapy; however, genomic studies of human bacterial commensals and environmental bacteria have revealed the presence of considerable numbers of resistance determinants within their genomes [9] that were not acquired from horizontal transmission and predated the clinical introduction of antibiotics. This type of AR is known as intrinsic resistance and provides a selective benefit for the producing strains by inhibiting or eliminating other bacteria competing for resources. Intrinsic resistance differentiates from the newly developed extrinsic antibiotic resistance in that in the former there is no contribution of human activities and the latter is mainly driven by antibiotic selection pressure [10]. In the current era of increasing AR and lack of new antibacterial agents, the study of intrinsic resistance becomes highly attractive as a new mechanism to counteract

Country	Daily doses per 1000 inhabitants/day (% of total) [8]
Mongolia	64.4
Iran	38.8
Turkey	38.2
Sudan	35.3
Serbia	31.6
Montenegro	29.3
Romania	28.5

Table 1.
Countries with the highest antibiotic consumption in the world.

bacterial resistance, as inhibition of elements that comprise the intrinsic resistome renders bacteria hyper-susceptible to antibiotics [11]. In the case of Gram-negative bacteria, like *E. coli*, two major contributors to the bacterium intrinsic resistance are its outer membrane, which is impermeable to many molecules, and its expression of numerous efflux pumps, that effectively reduce the intracellular concentration of certain antibiotics [12].

The acquired, or extrinsic, and continuously increasing resistance of *E. coli* to antibiotics is already considered a major public health problem around the world. In 2018, more than half of the *Escherichia coli* isolates reported to the European Centre for Disease Prevention and Control were resistant to at least one antimicrobial group under surveillance, and combined resistance to several antimicrobial groups was frequent [13]; in the United States in 2017, the national prevalence of extended spectrum β-lactamases (ESBL)-producing *E. coli* strains isolated from urinary tract infections (UTI) was 15.7%, whereas levofloxacin and trimethoprim-sulfamethoxazole-resistant rates were ≥24% among all isolates [14]. In developing countries the situation worsens, as reported by national surveillance data from Mexico, China and Turkey, where *E. coli*-resistant strains has been shown to have a prevalence >40% to cephalosporins, quinolones and trimethoprim/sulfamethoxazole (TSX), drugs widely used around the world to empirically treat bacterial infections (**Table 2**).

During 1945, just a few years after the introduction into clinical practice of penicillin, Alexander Fleming warned the world about antibiotic overuse, warning that became reality a few years later when the first *S. aureus* strain was reported to be resistant to penicillin. Several human activities have been identified as key drivers of the current AR crisis, but it has been demonstrated that the overuse of antibiotics clearly influences the evolution of resistance [18]. The reported actions that have led to the overuse of antibiotics are multifactorial and include different players in different industries such as the health, the livestock and the pharmaceutical industries. Examples of these actions comprise inappropriate prescription of antibiotics by healthcare providers, extensive use of antibiotics in livestock and fish farming, patients not following antibiotic treatment regimes, poor hygiene, bacterial mutations and lack of new antibiotics developed [19].

2.1 Overuse/inappropriate prescribing

One of the most significant factors that have contributed to the current antibacterial resistance crisis is the rapid evolution of bacteria under selective antibiotic pressure, since a continuous interaction between any given antibiotic and bacteria is an important aspect for the increase in multidrug-resistant strains [20].

Country	Resistance rates (%)		
	Cephalosporins	Quinolones	TSX
Mexico [15]	54.4	59.0	62.1
China [16]	52.4	69.8	ND
Turkey [17]	40.7	47.2	58.0

ND: not done.

Table 2.
Escherichia coli *antibiotic resistance rates to different antibiotics.*

Unfortunately, overuse and inappropriate prescription of these drugs are two large contributors to such issue. In any given antibiotic treatment against a bacterial infection, susceptible bacteria will be killed; if properly targeted, the pathogenic microorganism will be eradicated; however, along infecting bacteria, those members of the individual's microbiota, sensitive to the antibiotic in use, will also be wiped out. In case resistant microorganisms exist, either belonging to the normal microbiota or the pathogenic microorganisms being targeted, these survivors will replicate and will become the prevailing strain within the respective anatomical site.

The discovery and use of antibiotics have revolutionized the field of medicine and saved millions of lives each year; unfortunately, seen as the "miracle drug," healthcare providers and patients around the world have abused their use. Despite the marked increase of infections caused by multidrug-resistant bacteria around the world, the global response to this crisis has been inadequate, as people not only continue to misuse antibiotics but have continuously increased their abuse. Using a global database of antibiotic sales, Klein et al. [21] found that the antibiotic consumption rate around the world increased dramatically from 11.3 daily doses/1000 inhabitants per day to 15.7, an increase of 39%, between 2000 and 2015. In this same study, it was reported that the mean antibiotic consumption rate was primarily driven by the consumption in low- and mid-income countries, as no coincidence present the highest prevalence of multi drug-resistant bacteria-related infections. To make matters worse, the consumption of last-resort antibiotics such as carbapenems and colistin is also on the rise [21], situation that is consistent with the appearance of *E. coli*-resistant strains to these agents. To date, resistance of this organism to carbapenems is rare, with its prevalence depending on the area of the world under study, but not exceeding 3% [22]. However, in the future, an increase of resistance to this agent might be seen in *E. coli*, as the enzymes responsible for its hydrolysis, and thus inactivation, carbapenemases, are encoded mainly on plasmids, and are highly transmissible [23].

A key contributor to the increasing selective pressure of antibiotics is their overprescription. Recent data indicates that over 70% of prescribed antibiotics by primary care providers in the United States are inappropriate, the majority of which are for acute respiratory tract infections [24]; unfortunately, this rate of antibiotic misuse is probably a situation found in most countries. Coincidently, ciprofloxacin, one of the two most likely antibiotic to be prescribed inappropriately [24] is one to which *E. coli* present the highest rates of resistance around the world [15–17].

In addition to the contribution of the abuse of antibiotics to the selection of resistance, Zhang et al. [25] found epidemiological evidence that antibiotic resistance and *E. coli* diarrheagenic virulence phenotypes might be partially linked. They found that subjects with diarrhea had more frequent use of antibiotics before their onset of symptoms, linkage that might be explained as antibiotics might disrupt the intestinal microbiota, allowing overgrowth of resistant pathogens [25].

2.2 Use of antibiotics in livestock

Antibiotics are used in livestock to treat clinical disease, to prevent and control common disease events, and to enhance animal growth [26]. Unfortunately, this use of antibiotics has favored spread and persistence of resistant bacteria in humans by means of two different mechanisms: (a) human ingestion of antibiotics by means of the antibiotic-contaminated meat that enters the body and induces selective pressure on the host's microbiota and (b) resistant bacteria found in the gut of food animals are transmitted to humans via contaminated meat.

When livestock are treated or are provided with antibiotics, these agents exercise the same selective pressure on their microbiota as when humans ingest these drugs; thus, overuse of antibiotics on food animals has led to a high colonization rate of intestinal bacteria, including members of the *Enterobacteriaceae* family, such as *E. coli* and *Klebsiella* spp., that become resistant to different antimicrobials. Different studies around the world have shown that ready-to-eat animal products are contaminated with *E. coli* strains resistant to different kinds of antibiotics, mainly to β-lactams by means of the bacterial production of extended spectrum β-lactamases (ESBL) [27, 28]. These studies show that animal meat contaminated with *E. coli*-resistant strains is far more prevalent in developing than in developed nations, probably due to different hygiene habits; German studies have reported a prevalence of ESBL-contaminated meat of 24.1% [27], whereas in Mexico this prevalence has been reported to be above 60.0% [28]. *E. coli* strains isolated in meat have also shown resistance to other antibiotics, including to last-resort ones such as carbapenems [29] and colistin [30]. If this contaminated meat is ingested undercooked by humans, gut colonization is likely, establishing a reservoir for future antibiotic-resistant infections, as Ruppé et al. have shown that people with high gut colonization rates of ESBL-producing *E. coli* strains present higher risk to develop urinary tract infections with these clones than patients with no ESBL gut colonization [31].

Figure 1 shows a resumed representation of the main reservoirs, including livestock, of antibiotic-resistant *E. coli* and their interaction with humans.

2.3 Hygiene/fecal colonization

Higher consumption of antibiotics in unprivileged areas of the world plays a key role in the emergence and maintenance of antimicrobial resistance due to selective pressure by these agents on resident microbiota. However, studies have shown that inhabitants of these areas can be highly colonized with antibiotic-resistant *E. coli* strains despite not being in contact with antibiotics for 3–6 months [28], indicating that additional factors play important roles in the increased prevalence of AR worldwide. Global evidence suggests that elements in people's environment such as poor waste, non-potable drinking water, housing overcrowding and lack of hygiene facilitate the development and transmission of resistant bacteria [32].

The ability of *E. coli* to colonize different environments, including the gut of humans and animals, has provided this organism with the evolutionary advantage to acquire antibiotic resistance traits from other bacteria within its environment, as well as to be easily transmitted via the fecal-oral route. The gut microbiota of humans can harbor more than 1000 different antibiotic-resistant genes [33] and transmission of these traits among gut commensals is a constant phenomenon. Major examples of the transference of resistance genes between environmental bacteria, including gut commensals, and human pathogens, are the bla*CTX-M* genes, which is the most prevalent ESBL gene in *E. coli* and *Klebsiella* spp., and

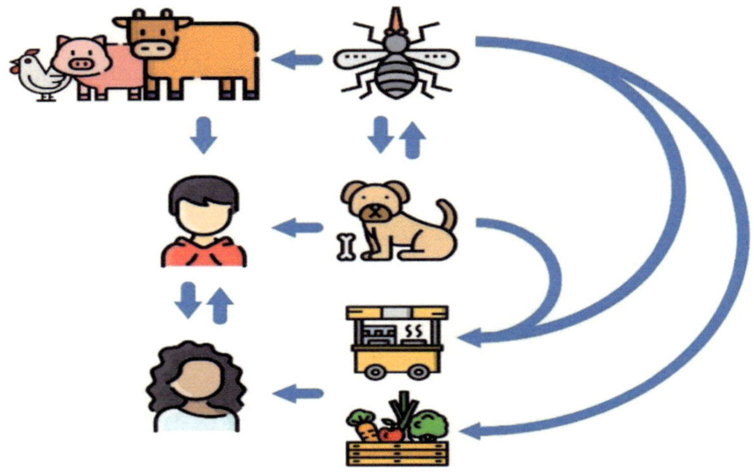

Figure 1.
Reservoirs of antibiotic-resistant E. coli *and their interaction with humans. Arrows show the* E. coli *flux from the different reservoirs.*

the OXA-48-type carbapenem-hydrolyzing β-lactamase genes, which are increasingly reported in *Enterobacteriaceae* around the globe. The potential origin of the bla*CTX-M* genes was identified in the chromosomal DNA of various environmental *Kluyvera* species [34], whereas that of OXA-48 was found to originate from the waterborne, environmental *Shewanella* species [35].

As many antibiotic resistance genes are associated with elements such as plasmids or transposons, and while the transfer of these elements may also occur through transformation or transduction, conjugation is often considered as the most likely responsible mechanism for the transmission of these traits [36]. The aforementioned ESBL and carbapenemase genes are primary examples of resistant genes with high impact on human health that have spread between bacteria via plasmid conjugation. Studies in China [37] have demonstrated that transmission via conjugation of ESBL genes in *E. coli* do occur even in the food chain, situation that partially explain the high fecal prevalence of ESBL-producing *E. coli* around the world.

The gut of humans and animals is a major reservoir of antibiotic-resistant *E. coli* and shedding of these strains through the feces of colonized individuals, livestock and domestic animals allows them to reach humans via contaminated water and food (see **Figure 1**). Human fecal colonization by antibiotic-resistant *E. coli* strains present the highest rates in deprived areas of the world, situation that begins since birth. Whereas in high-income nations the prevalence of *E. coli* strains resistant to antibiotics colonizing the gastrointestinal system of neonates is low [38], in low-income countries the prevalence of *E. coli* strains resistant to antibiotics such as tetracycline, ampicillin and trimethoprim/sulfamethoxazole exceeds 50% [39]. Fecal colonization of humans by resistant *E. coli* is on the rise around the world since the mid-2000s and the situation has worsened as fecal colonization by strains resistant to last-resource antibiotics, such as colistin, has been recently reported in different countries [40, 41]. As the prevalence of fecal colonization by these *E. coli* strains increase, so will the number of human infections caused by them, as it has been previously shown that fecal colonization with resistant microorganisms increases the risk factor of developing urinary tract infections by a factor of 13.0 [31].

Antibiotic consumption has contributed to the selection of resistance and is largely accepted as one of the major drivers of AR development; however, the high prevalence of antibiotic resistance around the world, especially in low- and mid-income countries, can be more likely attributed to the dissemination and maintenance of resistant clones via poor sanitation and lack of hygiene habits [32]. Ingestion of contaminated food and water, close contact with colonized animals and household members and abundance of flies are factors that contribute to the transmission of kl *E. coli* strains. As these conditions are considerably less frequent in developed areas of the world, this situation partially explains the reduced prevalence of these strains in these nations. However, due to the current globalization, resistant strains can easily be transmitted from one country to another. In a large cohort study of Dutch travelers to regions of the world with high prevalence of ESBL-producing bacteria, 34.3% subjects who were ESBL negative before travel had acquired these clones during their time abroad, with the highest number of acquisitions being among those who traveled to southern Asia, and remained colonized at 12 months after return [42]. Additionally, this same study showed that the estimated probability of onward transmission within households was 12%. Similar results were reported in a study in Spain, in which up to 66% of the isolates from patients with ESBL-producing *E. coli* infections were indistinguishable from those isolated from fecal samples from their household members [43]. These results indicate that acquisition of *E. coli*-resistant clones during travel is high and that transmission between household members can maintain such clones in the community for long periods of time.

As anthropogenic activities largely shape the resistome of different environments, transmission of resistant genes between bacteria in a community can be influenced by its contamination with human and animal feces and its impact is largely driven not by the presence of resistant bacteria but rather from the presence of human-related mobile resistance genes [44]. If poor sanitation, manifested by fecal contamination, of a given community is the key to transmit and maintain resistant clones, the reduction of antibiotic consumption will not be sufficient to control antimicrobial resistance. Thus, strategies to control the AR pandemic should also include improving sanitation conditions in all parts of the world.

3. Antibiotic resistance mechanisms

Few microorganisms have shown the ability to develop resistance to as many classes of antibiotics as the *Enterobacteriaceae.* Of the large list of bacterial genus that belong to this family, *E. coli* is only surpassed by *Klebsiella* in the number of human infections associated to multidrug-resistant bacteria [15–17, 27] and the past two decades have witnessed major increases in the emergence and spread of *E. coli* resistance strains to major classes of antibiotics such as β-lactams, quinolones, aminoglycosides, sulfonamides and fosfomycin. Unfortunately, this resistance has spread to last resource antibiotic classes such as the polymyxins and carbapenems. The following sections will briefly described the resistance mechanisms developed by *E. coli* against one of the major antibiotic groups currently used in the treatment against this organism: the β-lactams.

3.1 Resistance to β-lactams

Antibiotics belonging to the β-lactams class share a common feature: a three-carbon and one-nitrogen ring (beta-lactam ring), which is the molecular constituent responsible for the bacteriolytic mechanism of action of these agents against bacteria. β-Lactams act by inhibiting the bacterial synthesis of peptidoglycan, a vital constituent of the microorganism cell wall. The targets for the actions of beta-lactam antibiotics are known as penicillin-binding proteins (PBPs).

Bacteria have evolved different mechanisms of resistance against β-lactams: (a) Inactivation of these agents by the production of beta-lactamases; (b) decreased penetration of the antibiotic to the target site; (c) alteration of target site PBPs; and (d) efflux from the periplasmic space through specific pumping mechanism. However, in the case of *E. coli*, resistance to these antibiotics is mediated by the production of a group of enzymes referred as the "β-lactamases." These enzymes are ancient compounds, currently exceeding 2800 unique proteins, which emerged from environmental sources [45].

To date, β-lactamases are usually classified based on functional or structural criteria. Currently, the most widely used classification for these enzymes is the Ambler structural classification, which is based on sequence similarity, and separates these proteins into four classes: the classes A, C, and D of serine-β-lactamases and the class B of metallo-β-lactamases [46].

Gram-negative bacteria have evolved the production of different β-lactamases; in the case of *E. coli*, the most important ones from the medical point of view are the extended spectrum β-lactamases (ESBL), AmpC β-lactamases (AmpC) and the carbapenemases. Each of these groups of enzymes presents different spectrum of hydrolytic activity, thus presenting resistance to different types of β-lactams, as shown in **Table 3**.

3.1.1 Extended spectrum β-lactamases (ESBL)

Among the β-lactamases, ESBL are worthy of the attention of the scientific and medical community over the last decades because of their increasing prevalence as cause of antibiotic-resistant infections around the world. These enzymes can be produced by any member of the *Enterobacteriaceae*, but *Klebsiella* spp. and *E coli* are the predominant ESBL-producing genus.

ESBL belong mostly to class A of the Ambler classification, are generally plasmid encoded and confer resistance to those bacteria that produce them to penicillins, first-, second-, and third-generation cephalosporins and monobactams (e.g., aztreonam), but cannot hydrolyze cephamycins (cefoxitin) or carbapenems (imipenem, meropenem), and are inhibited by β-lactamase inhibitors such as clavulanic acid, tazobactam and sulbactam [47].

β-Lactamase	**Spectrum of activity**	**Inhibition by β-lactamase inhibitors**	**Activity against broad-spectrum cephalosporins**
ESBL	Penicillins First to third generation cephalosporins Monobactams	Yes	No
AmpC	Penicillins First to third generation cephalosporins Monobactams	No	Yes
Carbapenemases New Delhi metallo-β-lactamase	All β-lactams except aztreonam	No	Yes
Carbapenem-hydrolyzing oxacillinase-48	All β-lactams except broad spectrum cephalosporins	No	Weak

Table 3.
Spectrum of activity of the major types of β-lactamases produced by Escherichia coli.

When ESBL were first identified, most ESLB-related infections were caused by strains producing the TEM and SHV types. However, since then, ESBL CTX-M has emerged as the predominant type, both in humans and animals, in commensal organisms and in pathogenic strains and in community and healthcare-associated infections. Since the first isolation of SHV- and TEM-producing strains, more than 100 different variants of each type have been described and all have arisen from the original strains; contrary to the SHV and TEM types, CTX-M groups seem to have originated from the chromosomally encoded ESBL genes from different *Kluyvera* species [48].

HSV, TEM and CTX-M show different hydrolytic activities against different β-lactams. When first identified, SHV β-lactamases proved its activity against penicillins and first generation cephalosporins; as of today, the three sub-groups used to classify this group of enzymes present different antibiotic resistance phenotypes: (a) subgroup 2b hydrolyze penicillins and early cephalosporins (cephaloridine and cephalothin) and are strongly inhibited by clavulanic acid and tazobactam; (b) subgroup 2br are broad-spectrum β-lactamases that acquired resistance to clavulanic acid; and (c) subgroup 2be comprises ESBL that can also hydrolyze one or more oxyimino β-lactams (cefotaxime, ceftazidime, and aztreonam) [49]. In the case of TEM β-lactamases, the bacteria carrying these genes are able to hydrolyze penicillin and first generation cephalosporins such as cephaloridine; furthermore, TEM-1 is able to hydrolyze ampicillin at a greater rate than carbenicillin, oxacillin, or cephalothin, and has negligible activity against extended-spectrum cephalosporins [50]. Finally, CTX-M enzymes have the property of having potent hydrolytic activity against cefotaxime, with CTX-M-producing microorganisms showing cefotaxime MICs in the resistant range (>64 μg/ml), while ceftazidime MICs are usually in the apparently susceptible range (2 to 8 μg/ml); however, some CTX-M-type ESBLs may actually hydrolyze ceftazidime and confer resistance to this cephalosporin; aztreonam MICs are variable. CTX-M-type β-lactamases hydrolyze cefepime with high efficiency [50].

The exponential global increase in the number of infections caused by ESBL-producing strains has coincided with the appearance of the *CTX-M* genes. When originally reported, these strains were predominantly found in three geographic areas: South America, the Far East, and Eastern Europe. However, due to the extremely transferable plasmids which harbor bla*CTX-M* genes [49], with a frequency of transmission from 10^{-7} to 10^{-2} per donor cell [48], these strains are now increasingly reported as cause of human infections in every continent, to the point that it could be speculated that CTX-M-type ESBLs are now the most frequent ESBL type worldwide [50]. An additional factor that has been suggested as a key contributor to the dissemination of these clones is the frequent co-existence of bla*CTX-M* with genes conferring resistance to other classes of antibiotics like fluoroquinolones and aminoglycosides, situation that might lead to high rates of co-selection [51].

To date, over 150 CTX-M types have been identified and described (https://www.lahey.org/studies/other.asp). These ESBL types have been grouped into five clusters (CTX-M-1, CTX-M-2, CTX-M-8, CTX-M-9, and CTX-M-25) [52], with each cluster presenting different variants, and more variants constantly being described, as shown by the recent discovery of two novel ones, named $bla_{\text{CTX-M-14.2}}$ and $bla_{\text{CTX-M-15.2}}$ [53]. Out of the different CTX-M variants, different reports in different continents indicate that $bla_{\text{CTX-M-15}}$, belonging to cluster CTX-M-1, is now the most prevalent ESBL in *E. coli* around the world [54]. The increasing predominance of the $bla_{\text{CTX-M-15}}$ allele might be due to the powerful ability of this enzyme to hydrolyze different β-lactams, which probably offers the producing bacteria a selective advantage, especially when multiple antibiotics are concomitantly or consecutively prescribed [53].

One of the key players in the global dissemination of CTX-M-15-producing *E. coli* strains is clone ST131. A study performed in *E. coli* ST131 strains isolated between 2002 and 2004, before de ESBL pandemic, showed that only 2% of those strains carried the *CTX-M-15* gene [55]; almost two decades later, ST131 is one of the main clones isolated in the worldwide spread of ESBL-producing *E. coli* [56], particularly subclone H30Rx [57]. How this *E. coli* clone went from being a non-factor in the global ESBL transmission to a key player is probably multifactorial. Although ST131 strains are not considered hypervirulent, most of them show the presence of fluoroquinolone-resistant genes, they have the ability to be persistent gut colonizers even in the absence of antibiotic exposure, a condition that precedes some infections such as those in the urinary tract, and can be easily transmitted between people of all ages [58]. All of these factors have allowed this clone to be successful human pathogen, even before the spread of the ESBL genes; however, the acquisition by ST131 strains of the CTX-M-15 plasmid has made this *E. coli* lineage an even more successful pathogen and has probably exasperated the spread of such clone [59] and the rapid global spread of CTX-M-15-producing *E. coli*.

3.1.2 AmpC β-lactamases (AmpC)

Although the production of class A extended spectrum β-lactamases is the most common mechanism of resistance in *E. coli* against β-lactam agents, class C β-lactamases, or AmpC, can also confer those strains that produce them the ability to inactivate some of these compounds. Similar to ESBL, AmpC-producing organisms hydrolyze amino- and ureidopenicillins, oxyimino-β-lactams such as ceftazidime, ceftiofur, and aztreonam, but contrary to the former enzymes, AmpC also inactivates broad and extended-spectrum cephalosporins such as cephamycins (cefoxitin) and are not inhibited by β-lactamase inhibitors such as clavulanic acid. Neither ESBL nor AmpC confer bacteria resistance to carbapenems.

Originally, AmpC were described as chromosomally encoded enzymes and were detected in a few bacterial species such as *Enterobacter cloacae*, *Citrobacter freundii*, *Serratia marcescens, Acinetobacter* spp., *Aeromonas* spp. and *Pseudomonas aeruginosa* [60]. As the use of β-lactamase inhibitors increased among the population, dissemination of AmpC genes among bacterial species began by means of horizontal travel through plasmids, phenomenon that led to the appearance of AmpC-resistant traits in bacteria that previously lacked such genes or expressed them at low levels, such as *E. coli*, *Klebsiella* spp. and *Shigella* spp. [60].

In *E. coli*, the subject of this chapter, resistance by AmpC can be plasmid encoded or due to the overexpression of the chromosomal AmpC genes. Contrary to the AmpC enzymes of other members of the *Enterobacteriaceae*, such as *Enterobacter* spp. and *Citrobacter freundii*, that of *E. coli* exhibits a non-inducible phenotype that is constitutive and its production depends on either the strength of the *ampC* promoter [61], the presence of >1 copy of the *ampC* gene, the incorporation of a stronger promoter sequence as part of an insertion element or by the acquisition of a strong promoter of other bacterial species [62]. As stated before, this organism can carry *ampC* genes either chromosomally or in plasmids; however, the latter is being recognized as the major threat since plasmid-encoded AmpC are easily transferable between bacterial species, can cause nosocomial outbreaks, is associated with multidrug resistance and, in combination with porin loss, may lead to resistance to carbapenems [63].

Bacterial resistance to β-lactams is a major public health problem around the world. Although ESBL production clearly exceeds AmpC production as the major cause of β-lactam resistance, the later enzymes are now being recognized as a growing problem in different members of the *Enterobacteriaceae*, including *E. coli*, as evidenced by the increasing number of these strains being reported across the globe. Sources of AmpC-producing *E. coli* strains include livestock [64], the environment [65], as colonizers of the human gut [66] and as cause of human infections. The prevalence of these strains isolated as causative agents of human infections varies, ranging from 2.0% reported in a Portuguese hospital [67] to 16.7% from three university hospitals in Iran [68] to 29.0% from five referral hospitals in Sudan [69].

When comparing the epidemiology of today's AmpC-producing *E. coli* to that of ESBL-producing bacteria of two decades ago, they present several common features: high gut colonization in both animals and humans, reduced prevalence as cause of human infections, environmental contamination by these multidrug-resistant strains, higher isolation of both types of β-lactamase-producing strains in developing countries and their ability to be transmitted via plasmids among different bacterial species. As these two types of β-lactamase-producing strains behave similarly, it would be of no surprise to witness in the near future a booming increase of reports of infections caused by AmpC-producing strains, as witness two decades ago with ESBL. To make matter worse, infectious disease specialists are starting to see an increase of cases of *E. coli* strains that co-express ESBL and AmpC genes, complicating antimicrobial treatment even further. Different reports in India [70, 71] have shown that co-expression of bla*ESBL* and bla*AmpC* genes by *E. coli* strains isolated from different human infections is not uncommon, thus continuous monitoring of these resistance patterns is a necessity that will help prevent the further spread of these multidrug-resistant microorganisms.

3.1.3 Carbapenemases

Since ESBL- and AmpC-producing *E. coli* are increasingly being reported as cause of severe infections, carbapenems represent in many cases the last option for effective treatment against these infections. Nevertheless, with an increasing consumption of these agents, carbapenem-resistant strains, particularly *Klebsiella* spp. and in a lesser degree *E. coli*, have become a public health concern, particularly in the hospital setting. Carbapenems bind to penicillin-binding proteins and induce spheroplast formation and cell lysis without filament formation. The carbapenems include four agents: imipenem, meropenem, ertapenem and doripenem.

As in the case of ESBL- and AmpC-producing *Enterobacteriaceae*, reports from different countries show that resistance to carbapenems has been constantly increasing in the last few years, becoming a public health problem. In Europe, 11 countries have reported an increase in the number of infections caused by carbapenemase-producing *Enterobacteriaceae* in the period from 2015 to 2018 [72] and in China, Tian et al. [73] have reported an increase in the prevalence of carbapenemase-producing *E. coli* from 0% in 2011 to 1.9% in 2017.

The reported carbapenemases in *E. coli* primarily include *Klebsiella pneumoniae* carbapenemases (KPC), metallo-β-lactamases (MBL), including the VIM, IMP, GIM and NDM type, and oxacillin-hydrolyzing metallo-β-lactamases (OXA) [74]; however, different reports around the world have shown that the predominant types in *E. coli* are of the New Delhi metallo-β-lactamase (NDM-1) and carbapenem-hydrolyzing oxacillinase-48 (OXA-48) types [73, 75, 76].

3.1.3.1 New Delhi metallo-β-lactamase

The New Delhi metallo-β-lactamase (NDM-1) and closely related enzymes are a group of zinc-requiring metallo-β-lactamases capable of hydrolyzing a broad range of β-lactams including all penicillins, cephalosporins and carbapenems, just sparing monobactams, and are among the most recently identified carbapenemases. The gene encoding these enzymes, *bla*$_{NDM}$, has been identified on bacterial chromosomes and plasmids [77]; however, in the case of *E. coli*, *bla*$_{NDM}$ is mainly plasmid encoded with only few strains carrying it chromosomally [78].

NDM-1 was first identified in 2008 in India, a country that has been pointed out as the primary reservoir of NDM strains [77], followed by the Balkan states [79] and the Middle East [80]. From these three spots, *bla*$_{NDM-1}$-carrying bacterial strains have spread around the world, mainly due to the ability of the carrying microorganisms to horizontally transfer the carbapenemase resistance trait via plasmids. An additional factor that has contributed to the worldwide dissemination of NDM-1-producing strains is the frequent co-existence of the *bla*$_{NDM-1}$ gene on plasmids carrying additional antibiotic resistance genes, situation that has allowed the plasmid-carrying strains to thrive under environments of antibiotic selective pressure.

Since the first report of NDM-1, over 20 NDM variants have been reported; however, in *E. coli*, NDM-1, followed by NDM-5, are the predominant variants in human infections in different parts of the world [81, 82]. Surprisingly, in a study by Shen et al. [83] published in 2018, the highest prevalence in the human gut and livestock was of the NMD-5 variant, suggesting a possible shift from NDM-1 to NDM-5 in the community in China. An additional, and important finding of this study, was the identification, albeit small, of NDM-5 *E. coli* strains that co-express colistin resistance genes, *mcr*-1, in the gut of healthy individuals, situation that if not properly controlled might contribute to the future dissemination of *E. coli* strains that are resistant to last resource antibiotics.

3.1.3.2 Carbapenem-hydrolyzing oxacillinase-48 (OXA-48)

As with any other β-lactamase, OXA-48 hydrolyzes β-lactam antibiotics, including carbapenemases, but paradoxically spares broad-spectrum cephalosporins. OXA-48 genes were originally traced to the aquatic bacterium *Shewanella oneidensis*, but further studies now trace its origin to *Shewanella xiamenensis* [84]. Since the first description in Europe of OXA-48-carrying *Enterobacteriaceae*, several variants have been reported, including OXA-162, OXA-163, OXA-181, OXA-204, OXA-232, OXA244 and OXA-245.

Mainly found in *Klebsiella* species, reports on the detection of *bla*$_{oxa}$-carrying *E. coli* have increased in the last 3 years in different parts of the world, being reported in studies in Myanmar [85], the United States [86] and Thailand [87]. In all three studies, the isolated strains were co-expressing *bla*$_{OXA-48}$ or its variants and *bla*$_{NDM5}$. Oxa-48-carrying *E. coli* strains have also been isolated in Europe, between January and October 2019, 134 cases of *E. coli* strains carrying the OXA-48 variant OXA-244 were isolated from clinical samples in Germany; this same variant was further identified in 119 *E. coli* strains isolated from other European countries [88]. The source and route of transmission of these strains is currently unclear.

As carbapenems are considering in many clinical instances as a last resource antibiotic, worldwide monitoring on the prevalence of *E. coli* carrying resistant traits against these agents should be continuously performed in order to prevent the spread of these strains, situation that can jeopardize even further the current antibiotic resistance crisis.

4. Conclusions

The ability of *Escherichia coli* to colonize the gut of humans and animals, thus facilitating its transmission via the fecal-oral route, and its ability to transmit and uptake antibiotic resistance genes via plasmids to and from other bacteria have made this organism a key target in the fight against antimicrobial resistance. As discussed in this chapter, *E. coli* has evolved different mechanisms to fight off the action of antibiotics, and in many cases a single strain can carry resistance genes to distinct classes of these agents, thus complicating treatment.

The emergence of antibiotic resistance has been shown to be multifactorial, but all elements coincide in a major topic: antibiotic over abuse, both in human and veterinary medicine. The establishment of antibiotic stewardship programs is a major necessity in all nations as a way to reduced antibiotic resistance. However, as the spread and maintenance of *E. coli*-resistant traits among humans and between animals and humans is driven by additional, and probably more difficult to tackle, social issues such as lack of hygiene, lack of drinking water and house overcrowding, these factors must be taken care of in order to truly impact antibiotic resistance.

Conflict of interest

The author declares no conflict of interest.

Author details

Mario Galindo-Méndez
Laboratorios Galindo SC, Universidad Anáhuac Oaxaca, Oaxaca, Mexico

*Address all correspondence to: mgm@laboratoriosgalindo.com

References

[1] Riley M, Abe T, Arnaud MB, Berlyn MK, Blattner FR, Chaudhuri RR, et al. *Escherichia coli* K-12: A cooperatively developed annotation snapshot—2005. Nucleic Acids Research. 2006;**34**:1-9. DOI: 10.1093/nar/gkj405

[2] Hewitt JH, Rigby J. Effect of various milk feeds on numbers of *Escherichia coli* and *Bifidobacterium* in the stools of new-born infants. Journal of Hygiene (London). 1976;**77**(1):129-139. DOI: 10.1017/s0022172400055601

[3] Mohr K. History of antibiotics research. Current Topics in Microbiology and Immunology. 2016;**398**:237-272. DOI: 10.1007/82_2016_499

[4] Lowy FD. Antimicrobial resistance: The example of *Staphylococcus aureus*. The Journal of Clinical Investigation. 2003;**111**(9):1265-1273. DOI: 10.1172/JCI18535

[5] World Health Organization. WHO Publishes List of Bacteria for Which New Antibiotics Are Urgently Needed. 2017. Available from: https://www.who.int/news-room/detail/27-02-2017-who-publishes-list-of-bacteria-for-which-new-antibiotics-are-urgently-needed [Accessed: 09 April 2020]

[6] Tadesse DA, Zhao S, Tong E, Ayers S, Singh A, Bartholomew MJ. Antimicrobial drug resistance in *Escherichia coli* from humans and food animals, United States, 1950-2002. Emerging Infectious Diseases. 2012;**18**(5):741-749. DOI: 10.3201/eid1805.111153

[7] Ny S, Edquist P, Dumpis U, Gröndahl-Yli-Hannuksela K, Hermes J, Kling AM, et al. Antimicrobial resistance of *Escherichia coli* isolates from outpatient urinary tract infections in women in six European countries including Russia. Journal of Global Antimicrobial Resistance. 2019;**17**:25-34. DOI: 10.1016/j.jgar.2018.11.004

[8] World Health Organization. WHO Report on Surveillance of Antibiotic Consumption. 2018. Available from: https://apps.who.int/iris/bitstream/handle/10665/277359/9789241514880-eng.pdf [Accessed: 09 April 2020]

[9] Sommer MO, Dantas G, Church GM. Functional characterization of the antibiotic resistance reservoir in the human microflora. Science. 2009;**325**(5944):1128-1131. DOI: 10.1126/science.1176950

[10] Cox G, Wright GD. Intrinsic antibiotic resistance: Mechanisms, origins, challenges and solutions. International Journal of Medical Microbiology. 2013;**303**(6-7):287-292. DOI: 10.1016/j.ijmm.2013.02.009

[11] Liu A, Tran L, Becket E, Lee K, Chinn L, Park E, et al. Antibiotic sensitivity profiles determined with an *Escherichia coli* gene knockout collection: Generating an antibiotic bar code. Antimicrobial Agents and Chemotherapy. 2010;**54**:1393-1403. DOI: 10.1128/AAC.00906-09

[12] Nikaido H, Rosenberg EY, Foulds J. Porin channels in *Escherichia coli*: Studies with beta-lactams in intact cells. Journal of Bacteriology. 1983;**153**(1):232-240. DOI: 10.1128/JB.153.1.232-240.1983

[13] European Centre for Disease Prevention and Control. Surveillance of Antimicrobial Resistance in Europe. 2018. Available from: https://www.ecdc.europa.eu/en/publications-data/surveillance-antimicrobial-resistance-europe-2018 [Accessed: 10 April 2020]

[14] Critchley I, Cotroneo N, Pucci M, Mendes R. The burden in antimicrobial resistance among urinary tract isolates

in *Escherichia coli* in the United States in 2017. PLoS One. 2019;**14**(12):e0220265. DOI: 10.1371/journal.pone.0220265

[15] Garza-Gonzalez E, Morfın-Otero R, Mendoza-Olazara S, Bocanegra-Ibarias S, Flores-Treviño S, Rodrıguez-Noriega E, et al. A snapshot of antimicrobial resistance in Mexico. Results from 47 centers from 20 states during a six-month period. PLoS One. 2019;**14**(3):e0209865. DOI: 10.1371/journal.pone.0209865

[16] Huan Y, Oguto JO, Gu J, Ding F, You Y, Huo Y, et al. Comparative analysis of quinolone resistance in clinical isolates of *Klebsiella pneumoniae* and *Escherichia coli* from Chinese children and adults. BioMed Research International Journal. 2015;**168292**. DOI: 10.1155/2015/168292

[17] Cöplü N, Simsek H, Gür D, Gözalan A, Hasdemir U, Gülay Z. The first results of national antimicrobial resistance surveillance system in Turkey. Türk Hijiyen ve Tecrübi Biyoloji Dergisi. 2018;**75**(4):333-344. DOI: 10.5505/TurkHijyen.2018.68878

[18] Read AF, Woods RJ. Antibiotic resistance management. Evolution, Medicine, and Public Health. 2014;**2014**(1):147. DOI: 10.1093/emph/eou024

[19] World Health Organization. Antimicrobial Resistance. Available from: https://www.who.int/health-topics/antimicrobial-resistance [Accessed: 10 April 2020]

[20] Kolar M, Urbánek K, Látal T. Antibiotic selective pressure and development of bacterial resistance. International Journal of Antimicrobial Agents. 2001;**17**(5):357-363. DOI: 10.1016/s0924-8579(01)00317-x

[21] Klein EY, Van Boeckel TP, Martinez EM, Pant S, Gandra S, Levin SA, et al. Global increase and geographic convergence in antibiotic consumption between 2000 and 2015. Proceedings of the National Academy of Sciences of the United States of America. 2018;**115**(15):E3463-E3470. DOI: 10.1073/pnas.1717295115

[22] Nordmann P, Poirel L. Epidemiology and diagnostics of carbapenem resistance in Gram-negative bacteria. Clinical Infectious Diseases. 2019;**69**(7):S521-S528. DOI: 10.1093/cid/ciz824

[23] Nordmann P, Poirel L. Strategies for identification of carbapenemase-producing *Enterobacteriaceae*. The Journal of Antimicrobial Chemotherapy. 2013;**68**:487-489. DOI: 10.1093/jac/dks426

[24] Shively NR, Buehrle DJ, Clancy CJ, Decker BK. Prevalence of inappropriate antibiotic prescribing in primary care clinics within a veterans affairs health care system. Antimicrobial Agents and Chemotherapy. 2018;**62**(8):e00337-18. DOI: 10.1128/AAC.00337-18

[25] Zhang L, Levy K, Trueba G, Cevallos W, Trostle J, Foxman B. Effects of selection pressure and genetic association on the relationship between antibiotic resistance and virulence in *Escherichia coli*. Antimicrobial Agents and Chemotherapy. 2015;**59**(11):6733-6740. DOI: 10.1128/AAC.01094-15

[26] McEwen SA, Fedorka-Cray PJ. Antimicrobial use and resistance in animals. Clinical Infectious Diseases. 2002;**34**(Suppl 3):S93-S106. DOI: 10.1086/340246

[27] Kaesbohrer A, Bakran-Lebl K, Irrgang A, Fischer J, Kämpf P, Schiffmann A. Diversity in prevalence and characteristics of ESBL/pAmpC producing *E. coli* in food in Germany. Veterinary Microbiology. 2019;**233**:52-60. DOI: 10.1016/j.vetmic.2019.03.025

[28] Galindo-Méndez M. Reservoirs of CTX-M extended spectrum

β-lactamase-producing *Enterobacteriaceae* in Oaxaca, Mexico. Journal of Microbiology & Experimentation. 2019;**7**(1):43-47. DOI: 10.15406/jmen.2019.07.00239

[29] Köck R, Daniels-Haardt I, Becker K, Mellmann A, Friedrich AW, Mevius D. Carbapenemase-resistant *Enterobacteriaceae* in wildlife, food-producing, and companion animals: A systematic review. Clinical Microbiology and Infection. 2018;**24**(12):1241-1250. DOI: 10.1016/j.cmi.2018.04.004

[30] Joshi PR, Thummeepak R, Leugtongkam U, Pooarlai R, Paudel S, Acharya M, et al. The emergence of colistin-resistant *Escherichia coli* in chicken meats in Nepal. FEMS Microbiology Letters. 2019;**366**(20):fnz237. DOI: 10.1093/femsle/fnz237

[31] Ruppé E, Lixandru B, Cojocaru R, Buke C, Paramythiotou E, Angebault C, et al. Relative fecal abundance of extended-spectrum-β-lactamase-producing *Escherichia coli* strains and their occurrence in urinary tract infections in women. Antimicrobial Agents and Chemotherapy. 2013;**57**(9):4512-4517. DOI: 10.1128/AAC.00238-13

[32] Collignon P, Beggs JJ, Walsh TR, Gandra S, Laxminarayan R. Anthropological and socioeconomic factors contributing to global antimicrobial resistance: A univariate and multivariable analysis. The Lancet Planetary Health. 2018;**2**:e398-e405. DOI: 10.1016/S2542-5196(18)30186-4

[33] Hu Y, Yang X, Qin J, Lu N, Cheng G, Wu N. Metagenome-wide analysis of antibiotic resistance genes in a large cohort of human gut microbiota. Nature Communications. 2013;**4**:2151. DOI: 10.1038/ncomms3151

[34] Canton R, Coque TM. The CTX-M β-lactamase pandemic. Current Opinion in Microbiology. 2006;**9**(5):466-475. DOI: 10.1016/j.mib.2006.08.011

[35] Poirel L, Potron A, Nordmann P. OXA-48-like carbapenemases: The phantom menace. The Journal of Antimicrobial Chemotherapy. 2012;**67**(7):1597-1606. DOI: 10.1093/jac/dks121

[36] von Wintersdorff CJH, Penders J, van Niekerk JM, Mills ND, Majumder S, van Alphen LV. Dissemination of antimicrobial resistance in microbial ecosystems through horizontal gene transfer. Frontiers in Microbiology. 2016;**7**:173. DOI: 10.3389/fmicb.2016.00173

[37] Ye Q, Wu Q, Zhang S, Zhang J, Yang G, Wang J, et al. Characterization of extended-Spectrum β-lactamase-producing *Enterobacteriaceae* from retail food in China. Frontiers in Microbiology. 2018;**9**:1709. DOI: 10.3389/fmicb.2018.01709

[38] Prelog M, Grif K, Decristoforo C, Wurzner R, Kiechl-Kohlendorfer U, Brunner A, et al. Tetracycline-resistant *Escherichia coli* strains are inherited from parents and persist in the infant's intestines in the absence of selective pressure. European Journal of Pediatrics. 2009;**168**(10):1181-1187. DOI: 10.1007/s00431-008-0901-0

[39] Bryce A, Costelloe C, Hawcroft C, Wootton M, Hay AD. Faecal carriage of antibiotic resistant *Escherichia coli* in asymptomatic children and associations with primary care antibiotic prescribing: A systematic review and meta-analysis. BMC Infectious Diseases. 2016;**16**:359. DOI: 10.1186/s12879-016-1697-6

[40] Bourrel AS, Poirel L, Royer G, Darty M, Vuillemin X, Kieffer N, et al. Colistin resistance in Parisian inpatient faecal *Escherichia coli* as the result of two distinct evolutionary pathways. The Journal of Antimicrobial Chemotherapy.

2019;**74**(6):1521-1530. DOI: 10.1093/jac/dkz090

[41] Merida-Vieyra J, De Colsa-Ranero A, Arzate-Barbosa P, Arias-de la Garza E, Méndez-Tenorio A, Murcia-Garzón J. First clinical isolate of *Escherichia coli* harboring mcr-1 gene in Mexico. PLoS One. 2019;**14**(4):e0214648. DOI: 10.1371/journal.pone.0214648. eCollection 2019

[42] Arcilla MS, van Hattem JM, Haverkate MR, Bootsma MCJ, van Genderen PJJ, Goorhuis A. Import and spread of extended-spectrum β-lactamase-producing *Enterobacteriaceae* by international travellers (COMBAT study): A prospective, multicentre cohort study. The Lancet Infectious Diseases. 2017;**17**(1):78-85. DOI: 10.1016/S1473-3099(16)30319-X

[43] Valverde A, Grill F, Coque TM, Pintado V, Baquero F, Canton R, et al. High rate of intestinal colonization with extended spectrum-beta-lactamase-producing organisms in household contacts of infected community patients. Journal of Clinical Microbiology. 2008;**46**(8):2796-2799. DOI: 10.1128/JCM.01008-08

[44] Lee K, Kim DW, Lee DH, Kim YS, Bu JH, Cha JH. Mobile resistome of human gut and pathogen drives anthropogenic bloom of antibiotic resistance. Microbiome. 2020;**8**(1):2. DOI: 10.1186/s40168-019-0774-7

[45] Bush K. Past and present perspectives on β-lactamases. Antimicrobial Agents and Chemotherapy. 2018;**62**(10):e01076-18. DOI: 10.1128/AAC.01076-18

[46] Bush K, Jacoby G. Updated functional classification of β lactamases. Antimicrobial Agents and Chemotherapy. 2010;**54**(3):969-976. DOI: 10.1128/AAC.01009-09

[47] Philippon A, Labia R, Jacoby G. Extended-spectrum beta-lactamases. Antimicrobial Agents and Chemotherapy. 1989;**33**:1131-1136. DOI: 10.1128/aac.33.8.1131

[48] Bonnet R. Growing group of extended-spectrum β-lactamases: The CTX-M enzymes. Antimicrobial Agents and Chemotherapy. 2004;**48**(1):1-14. DOI: 10.1128/AAC.48.1.1-14.2004

[49] Liahopoulus A, Mevius D, Ceccarelli D. A review of SHV extended-spectrum β-lactamases: Neglected yet ubiquitous. Frontiers in Microbiology. 2016;7:1374. DOI: 10.3389/fmicb.2016.01374

[50] Paterson DL, Bonomo RA. Extended-spectrum β-lactamases: A clinical update. Clinical Microbiology Reviews. 2005;**18**(4):657-686. DOI: 10.1128/CMR.18.4.657-686.2005

[51] Cantón R, Ruiz-Garbajosa P. Co-resistance: An opportunity for the bacteria and resistance genes. Current Opinion in Pharmacology. 2011;**11**:477-485. DOI: 10.1016/j.coph.2011.07.007

[52] Boyd DA, Tyler S, Christianson S, McGeer A, Muller MP, Willey BM, et al. Complete nucleotide sequence of a 92-kilobase plasmid harboring the CTX-M-15 extended-spectrum beta-lactamase involved in an outbreak in long-term-care facilities in Toronto, Canada. Antimicrobial Agents and Chemotherapy. 2004;**48**(10):3758-3764. DOI: 10.1128/AAC.48.10.3758-3764.2004

[53] Ramadan A, Abdelaziz NA, Amin NA, Aziz RK. Novel *bla*CTX-M variants and genotype-phenotype correlations among clinical isolates of extended spectrum beta lactamase-producing *Escherichia coli*. Scientific Reports. 2019;**9**(1):4224. DOI: 10.1038/s41598-019-39730-0

[54] Bevan ER, Jones AM, Hawkey PM. Global epidemiology of

CTX-M β-lactamases: Temporal and geographical shifts in genotype. Journal of Antimicrobial Chemotherapy. 2017;**72**(8):2145-2155. DOI: 10.1093/jac/dkx146

[55] Johnson JR, Menard M, Johnston B, Kuskowski MA, Nichol K, Zhanel GG. Epidemic clonal groups of *Escherichia coli* as a cause of antimicrobial-resistant urinary tract infections in Canada, 2002 to 2004. Antimicrobial Agents and Chemotherapy. 2009;**53**:2733-2739. DOI: 10.1128/AAC.00297-09

[56] Rogers BA, Sidjabat HE, Paterson DL. *Escherichia coli* O25b-ST131: A pandemic, multiresistant, community-associated strain. The Journal of Antimicrobial Chemotherapy. 2011;**66**(1):1-14. DOI: 10.1093/jac/dkq415

[57] Nicolas-Chanoine MH, Bertrand X, Madec JY. *Escherichia coli* ST131, an intriguing clonal group. Clinical Microbiology Reviews. 2014;**27**(3):543-574. DOI: 10.1128/CMR.00125-13

[58] Whitmer GR, Moorthy G, Arshad M. The pandemic *Escherichia coli* sequence type 131 strain is acquired even in the absence of antibiotic exposure. PLoS Pathogens. 2019;**15**(12):e1008162. DOI: 10.1371/journal.ppat.1008162

[59] Peirano G, Pitout JDD. Molecular epidemiology of *Escherichia coli* producing CTX-M β-lactamases: The worldwide emergence of clone ST131 O25:H4. International Journal of Antimicrobial Agents. 2010;**35**(4):316-321. DOI: 10.1016/j.ijantimicag.2009.11.003

[60] Beceiro A, Bou G. Class β-lactamases: An increasing problem worldwide. Reviews in Medical Microbiology. 2004;**15**(4):141-152. DOI: 10.1097/00013542-200410000-00003

[61] Caroff N, Espaze E, Gautreau D, Richet H, Reynaud A. Analysis of the effects of -42 and -32 *ampC* promoter mutations in clinical isolates of *Escherichia coli* hyperproducing *AmpC*. The Journal of Antimicrobial Chemotherapy. 2000;**45**(6):783-788. DOI: 10.1093/jac/45.6.783

[62] Jaurin B, Normark S. Insertion of IS2 creates a novel ampC promoter in *Escherichia coli*. Cell. 1983;**32**(3):809-816. DOI: 10.1016/0092-8674(83)90067-3

[63] Matsumura Y, Tanaka M, Yamamoto M, Nagao M, Machida K, Ito Y, et al. High prevalence of carbapenem resistance among plasmid-mediated AmpC beta-lactamase-producing *Klebsiella pneumoniae* during outbreaks in liver transplantation units. International Journal of Antimicrobial Agents. 2015;**45**(1):33-40. DOI: 10.1016/j.ijantimicag.2014.08.015

[64] Vounba P, Arsenault J, Bada-Alambédji R, Fairbrother JM. Antimicrobial resistance and potential pathogenicity of *Escherichia coli* isolates from healthy broilers in Québec, Canada. Microbial Drug Resistance. 2019;**25**(7):1111-1121. DOI: 10.1089/mdr.2018.0403

[65] Sen K, Berglund T, Soares MA, Taheri M, Ma Y, Khalil L, et al. Antibiotic resistance of *E. coli* isolated from a constructed wetland dominated by a crow roost, with emphasis on ESBL and AmpC containing *E. coli*. Frontiers in Microbiology. 2019;**10**:1034. DOI: 10.3389/fmicb.2019.01034

[66] Nakayama T, Kumeda Y, Kawahara R, Yamamoto Y. Quantification and long-term carriage study of human extended spectrum/AmpC β-lactamase-producing *Escherichia coli* after international travel to Vietnam. Journal of Global Antimicrobial Resistance. 2019;**21**:229-234. DOI: 10.1016/j.jgar.2019.11.001

[67] Oliveira C, Amador P, Prudencio C, Tomaz CT, Tavarez-Ratado P, Fernander R. ESBL and AmpC β-lactamases in clinical strains of *Escherichia coli* from Serra da Estrela, Portugal. Medicina (Kaunas, Lithuania) 2019;**55**(6):pii: E272. DOI: 10.3390/medicina55060272

[68] Rizi KS, Mosavat A, Youssefi M, Jamehdar SA, Ghazyini K, Safdari H, et al. High prevalence of blaCMY AmpC Beta-lactamase in ESBL co-producing *Escherichia coli* and *Klebsiella* spp. clinical isolates in northeast of Iran. Journal of Global Antimicrobial Resistance. 2020;**S2213-7165**(20):30075-8. DOI: 10.1016/j.jgar.2020.03.011. [Published online ahead of print, 01 Apr 2020]

[69] Dirar M, Bilal N, Ibrahim ME, Hamid M. Resistance patterns and phenotypic detection of β-lactamase enzymes among *Enterobacteriaceae* isolates from referral hospitals in Khartoum State, Sudan. Cureus. 2020;**12**(3):e7260. DOI: 10.7759/cureus.7260

[70] Ghosh B, Mukherjee M. Emergence of co-production of plasmid mediated AmpC beta-lactamase and ESBL in cefoxitin-resistant uropathogenic *Escherichia coli*. European Journal of Clinical Microbiology & Infectious Diseases. 2016;**35**(9):1449-1454. DOI: 10.1007/s10096-016-2683-z

[71] Mirza S, Jadhav S, Misra RN, Das NK. Coexistence of β-lactamases in community-acquired infections in tertiary care hospital in India. International Journal of Microbiology. 2019;**2019**:7019578. DOI:10.1155/2019/7019578

[72] Brolund A, Lagerqvist N, Byfors S, Struelens MJ, Monnet DL, Albiger B, et al. Worsening epidemiological situation of carbapenemase-producing *Enterobacteriaceae* in Europe, assessment by national experts from 37 countries, July 2018. Journal of Euro Surveillance. 2019;**24**(9):1900123. DOI: 10.2807/1560-7917. ES.2019.24.9.1900123

[73] Tian X, Zheng X, Sun Y, Fang R, Zhang S, Zhang X, et al. Molecular mechanisms and epidemiology of carbapenem-resistant *Escherichia coli* isolated from Chinese patients during 2002-2017. Infection and Drug Resistance. 2020;**13**:501-512. DOI: 10.2147/IDR.S232010

[74] Nordmann P, Naas T, Poirel L. Global spread of carbapenemase-producing *Enterobacteriaceae*. Emerging Infectious Diseases. 2011;**17**:1791-1798. DOI: 10.3201/eid1210.110655

[75] Gauthier L, Dortet L, Cotellon G, Creton E, Cuzon G, Ponties V, et al. Diversity of carbapenemase-producing *Escherichia coli* isolates in France in 2012-2013. Antimicrobial Agents and Chemotherapy. 2018;**62**(8):e00266-18. DOI: 10.1128/AAC.00266-18

[76] Nordmann P, Poirel L. The difficult-to-control spread of carbapenemase producers among *Enterobacteriaceae* worldwide. Clinical Microbiology and Infection. 2014;**20**(9):821-830. DOI: 10.1111/1469-0691.12719

[77] Kumarasamy KK, Toleman MA, Walsh TR, Bagaria J, Butt F, Balakrishnan R, et al. Emergence of a new antibiotic resistance mechanism in India, Pakistan and the UK: A molecular, biological and epidemiological study. The Lancet Infectious Diseases. 2010;**10**(9):597-602. DOI: 10.1016/S1473-3099(10)70143-2

[78] Shen P, Yi M, Fu Y, Ruan Z, Du X, Yu Y, et al. Detection of an *Escherichia coli* sequence type 167 strain with two tandem copies of blaNDM-1 in the chromosome. Journal of Clinical Microbiology. 2016;**551**(1):199-205. DOI: 10.1128/JCM.01581-16

[79] Zarfel G, Hoenigl M, Würstl B. Emergence of carbapenem-resistant *Enterobacetriaceae* in Austria,

2001-2010. Clinical Microbiology and Infection. 2011;**17**(11):E5-E8. DOI: 10.1111/j.1469-0691.2011.03659.x

[80] Shibl A, Al-Agamy M, Memish Z, Senok A, Khader SA, Assiri A. The emergence of OXA-48 and NDM-1-positive *Klebsiella pneumonia* in Riyadh, Saudi Arabia. The Journal of Infectious Diseases. 2013;**17**(12):e1130-e1133. DOI: 10.1016/j.ijid.2013.06.016

[81] Ranjan A, Shaik S, Mondal A, Nandanwar N, Hussain A, Semmler T, et al. Molecular epidemiology and genome dynamics of New Delhi metallo-beta-lactamase-producing extraintestinal pathogenic *Escherichia coli* strains from India. Antimicrobial Agents and Chemotherapy. 2016;**60**(11):6795-6805. DOI: 10.1128/AAC.01345-16

[82] Hu X, Xu X, Wang X, Xue W, Zhou H, Zhang L, et al. Diversity of New Delhi metallo-beta-lactamase-producing bacteria in China. International Journal of Infectious Diseases. 2017;**55**:92-95. DOI: 10.1016/j.ijid.2017.01.011

[83] Shen Z, Hu Y, Sun Q, Hu F, Zhou H, Shu L, et al. Emerging carriage of NDM-5 and MCR-1 in *Escherichia coli* from healthy people in multiple regions in China: A cross sectional observational study. EClinicalMedicine. 2018;**6**:11-20. DOI: 10.1016/j.eclinm.2018.11.003

[84] Potron A, Poirel L, Nordmann P. Origin of OXA-181, an emerging carbapenem-hydrolyzing oxacillinase, as a chromosomal gene in *Shewanella xiamenensis*. Antimicrobial Agents and Chemotherapy. 2011;**55**(9):4405-4407. DOI: 10.1128/AAC.00681-11

[85] Aung MS, San N, Maw WW, San T, Urushibara N, Kawaguchiva M, et al. Prevalence of extended-spectrum beta-lactamase and carbapenemase genes in clinical isolates of *Escherichia coli* in Myanmar: Dominance of bla_{NMD5} and emergence of $bla_{OXA-181}$. Microbial Drug Resistance. 2018;**24**(9):1333-1344. DOI: 10.1089/mdr.2017.0387

[86] Hasassri ME, Boyce TG, Norgan SA, Cunningham PR, Jeraldo S, Weissman S, et al. An immunocompromised child with bloodstream infection caused by two *Escherichia coli* strains, one harboring NDM-5 and the other harboring OXA-48-like carbapenemase. Antimicrobial Agents and Chemotherapy. 2016;**60**:3270-3275. DOI: 10.1128/AAC.03118-15

[87] Lunha K, Chanawong A, Lulitanond C, Wilailuckana N, Charoensri L, Wonglakorn P. High level carbapenem-resistant OXA-48-producing *Klebsiella pneumoniae* with a novel OmpK36 variant and low-level, carbapenem-resistant, non-porin-deficient, OXA-181-producing *Escherichia coli* from Thailand. Diagnostic Microbiology and Infectious Disease. 2016;**85**:221-226. DOI: 10.1016/j.diagmicrobio.2016.03.009

[88] European Centre for Disease Prevention and Control. Increase in OXA-244-Producing *Escherichia coli* in the European Union/European Economic Area and the UK since 2013. 2020. Available from: https://www.ecdc.europa.eu/sites/default/files/documents/RRA-E-coli-OXA-244-producing-E-coli-EU-EEA-UK-since-2013.pdf [Accessed: 15 April 2020]

Chapter 10

Antibiotic Resistance among Iraqi Local *E. coli* Isolates

Hussein O.M. Al-Dahmoshi, Noor S.K. Al-Khafaji and Mohammed H.O. Al-Allak

Abstract

Escherichia coli is a famous Gram-negative bacillary bacterium that belongs to Enterobacteriaceae. It is either micro-biota innocent for human or pathogenic with arrays of diseases. The pathogenic *E. coli* can be assigned to intestinal (InPEC) or extraintestinal (ExPEC) with disease ranging from watery diarrhea to pulmonary distress. The most prevalent form of InPEC is enteropathogenic *E. coli* (EPEC), while the most prevalent ExPEC is uropathogenic *E. coli* (UPEC). The other InPEC includes Shiga toxin–producing (STEC), enteroaggregative (EAEC), enterotoxigenic (ETEC), enteroinvasive (EIEC), diffusely adherent (DAEC) and adherent invasive *E. coli* (AIEC). ExPEC was implicated in cystitis, pyelonephritis, sepsis, respiratory tract infection, cervicovaginal infection (CVEC), meningitis (NMEC), otitis media, cholecystitis and wound infection. Antibiotic resistance is the challenging in world nowadays. Multidrug-resistant (MDR) *Escherichia coli* has become challenging with existing antibiotic options. *E. coli* pathogens have various virulence factors that determine their pathogenesis and antimicrobial resistance mechanisms. The rapid and ongoing spread of antimicrobial-resistant organisms threatens our ability to successfully treat a growing number of infectious diseases. It is well established that antibiotic use is a significant, and modifiable, driver of antibiotic resistance. The most commonly used antibiotic classes for InPEC and ExPEC were third-generation cephalosporin, carbapenem, fluoroquinolone and aminoglycosides. Actually, the most effective prescribed medication is one of the following: cefotaxime, ceftriaxone, ciprofloxacin, amikacin, gentamycin, levofloxacin and imipenem. Generally, according to our review for more than 100 local Iraqi studies among Iraqi provinces, the results revealed the resistance rate from highest to lowest as follows: cefotaxime (76.5%), ceftriaxone (75.9%), gentamycin (41.65%), ciprofloxacin (32.13%), amikacin (17.3%), levofloxacin (15%) and imipenem (5.14%). The resistance mechanisms may include genes encoding antibiotic-modifying enzymes like those of extended-spectrum beta-lactamases gene comprising: *blaCTX-M*, *blaTEM*, *blaSHV*, *blaOXA*, *blaPER*, *blaVIM*, *blaIMP*, *blaNDM* and *blaAMPc*. Efflux pumping includes AcrAB, while resistance to quinolone may be mediated by mutation among qnrA, qnrB, qnrD and qnrS. Resistance to aminoglycosides includes encoding to aminoglycoside-modifying enzymes like *aac(6)-Ib*, *aph(3)-I*, *aph(3)-IIa*, *aph(3)-Ib*, *ant(3)-I*, *aac(3)-II* and *aac(3)-IV*.

Keywords: InPEC, ExPEC, CVEC, NMEC, DEC, *blaCTX-M*, *blaTEM*, *blaSHV*

1. Introduction

Escherichia coli is prominent Gammaproteobacteria, Gram-negative bacilli live facultatively. It is the principal non-pathogenic facultative flora of the human intestine with harmless effect in healthy individuals. The virulent pathotypes of *E. coli* strains have the capability to cause a collection of intestinal and extraintestinal diseases, especially in immune-compromised persons [1]. Intestinal disease includes diarrhea or dysentery caused by six pathotypes, while extraintestinal diseases consists of vaginosis, urinary tract infections, respiratory tract infection, otitis media and meningitis [2, 3]. The enteropathogenic or diarrheagenic *E. coli* is an imperative cause of diarrhea in the newborn, immunocompromised and travelers. It can be assigned to one of the seven pathotypes: enteropathogenic (EPEC), Shiga toxin–producing (STEC), enteroaggregative (EAEC), enterotoxigenic (ETEC), enteroinvasive (EIEC), diffusely adherent (DAEC) and adherent invasive *E. coli* (AIEC) [4–6].

Uropathogenic *Escherichia coli* (UPEC) strains are the most significant causative agents of UTIs in humans. The total prevalence of UTIs caused by the UPEC strains is about 30–70% [7]. UPEC is the most common cause of community- and hospital-acquired urinary tract infections (UTIs). Isolates from uncomplicated community-acquired UTIs express a variety of virulence traits that promote the efficient colonization of the urinary tract. In contrast, nosocomial UTIs can be caused by *E. coli* strains that differ in their virulence traits from the community-acquired UTI isolates. UPEC virulence markers are used to distinguish these facultative extraintestinal pathogens, which belong to the intestinal flora of many healthy individuals, from intestinal pathogenic *E. coli* (IPEC) [8, 9].

One of the important extraintestinal *E. coli* (ExPEC) infections is nosocomial ventilator-associated pneumonia (VAP) with the mortality rate reaching 13%. Until the early 2000s, the ExPEC was not considered as a major pathogen responsible for ventilator-assisted pneumonia that may be due to focusing on other bacteria like *Staphylococcus aureus*, *Acinetobacter baumannii* and *Pseudomonas aeruginosa* [10–12]. Many studies stated the high frequency of ExPEC among VAP even more than *Staphylococcus aureus* and *Pseudomonas aeruginosa*. In developing countries, both hospital- and community-acquired respiratory tract infections (RTIs) are linked with emerging MDR *E. coli* [13–15].

Escherichia coli is most commonly associated with bloodstream infections and death due to sepsis. Sepsis is a life-threatening clinical condition affecting more than 40 million worldwide with mortality rate more than 15%. The incidence of sepsis caused by Gram-negative bacteria, such as *Escherichia coli* (*E. coli*), has been steadily increasing since the late 1990s. It ranks as the leading cause of death in intensive care units. *E. coli* accounts for approximately 14.1% of early onset neonatal sepsis and it is the second most common pathogen, along with coagulase negative Staphylococcus, after group B Streptococcus (GBS)2 [16–18].

Escherichia coli (especially K1) is one of the most common causative pathogens of neonatal meningitis, but the presence of *E coli* in an immunocompetent adult, causing meningitis, is rare with an annual incidence of less than one case per year. The penetration of *E. coli* through the blood-brain barrier is a key step of the meningitis pathogenesis. Diabetes mellitus, alcoholism, cirrhosis, HIV infection and malignancies are some of the risk factors to develop *E coli* meningitis. A distant source is usually identified, either from the urinary or the digestive tract [19–22].

According to the World Health Organization, *Enterobacteriaceae*, including *Escherichia coli*, are among the critical priority antibiotic-resistant bacteria. Multidrug-resistant (MDR) *Escherichia coli* has been listed as a priority pathogen by the World Health Organization (WHO) due to emerging antimicrobial resistance (AMR) [23, 24].

2. *Escherichia coli* diseases and antibiotic resistance

E. coli causes a wide range of diseases that can be assigned to either intestinal caused by intestinal *E. coli* (InPEC) or extraintestinal caused by extraintestinal *E. coli* (ExPEC) [25]. The most important diseases are as follows:

2.1 ExPEC-associated cystitis and antibiotic resistance

Cystitis is a common and expensive condition that impacts humans of different age groups from the neonate till geriatric age group. It is a pathogenic inflammation of the lower urinary tract. Women are more commonly afflicted with UTIs, and they are caused by common pathogens such as *Escherichia coli* (86%) [26]. Uropathogenic *Escherichia coli* (UPEC) is significantly associated with cystitis via sets of virulence factors (adhesins, siderophores, toxins, capsule production and protease) that assist its colonization, invasion, and survival within the host urinary system [27, 28]. High recurrence rates and increasing antimicrobial resistance among UPEC threaten to greatly increase the economic burden of these UTIs [5]. The resistance rate of Iraqi local UPEC to different antibiotic classes is summarized in **Table 1**. The resistance genes are listed in **Table 2**.

The average resistance rate is as follows: cefotaxime (77%), ceftriaxone (70%), ciprofloxacin (45.47%), amikacin (23.42%), gentamycin (45.69%) and imipenem (6.06%). Resistance to beta-lactams was attributed to many mechanisms, and one of them is to the modifying enzymes especially *blaTEM*, *blaSHV*, *blaCTX-M*, *blaOXA*, *blaPER* and *blaVIM*, while resistance to ciprofloxacin was interpreted

Antibiotic	No. of isolate/ study	Resistance %	Province	Reference
Cefotaxime	246	85.13	Babylon	[34, 40–43]
	76	88.9	Najaf	[30]
	62	76.9	Karbala	[44]
	91	75.8	Wasit	[36]
	50	82.5	Saladin	[45]
	90	89.15	Erbil	[46]
	106	52	Zakho	[47]
	234	70	Kirkuk	[48]
	381	70.8	Duhok	[49, 50]
	50	78	Sulemania	[51]
Ceftriaxone	176	81.21	Babylon	[34, 40, 41, 43]
	62	100	Karbala	[44]
	81	75	Missan	[31]
	91	74.7	Wasit	[36]
	7	73.15	Anbar	[33]
	90	52	Erbil	[46]
	106	70	Zakho	[47]
	234	67	Kirkuk	[48]
	381	71.3	Duhok	[49, 50]
	50	34.5	Sulemania	[51]

Antibiotic	No. of isolate/ study	Resistance %	Province	Reference
Ciprofloxacin	258	41.73	Babylon	[9, 34, 40, 41, 43, 52]
	60	68.3	Baghdad	[53]
	50	72.5	Saladin	[45]
	14	25	Tikrit	[54]
	381	53.25	Duhok	[49, 50]
	113	24.8	Anbar	[55]
	50	52.6	Sulemania	[51]
	50	52.6	Erbil	[39]
	81	23.4	Missan	[31]
	91	40.6	Wasit	[36]
Levofloxacin	102	38.5	Babylon	[42, 50]
	14	0	Tikrit	[54]
Amikacin	322	25.1	Babylon	[29, 34, 40, 41, 43, 51, 52]
	62	24	Sulemania	[44]
	50	95	Karbala	[45]
	90	25.15	Saladin	[46, 51]
	14	5	Erbil	[54]
	234	5.9	Tikrit	[48]
	331	0.85	Kirkuk	[48]
	50	46	Duhok	[49, 50]
	113	9.8	Anbar	[55]
	81	11.1	Missan	[31]
	91	9.8	Wasit	[36]
Gentamycin	286	37.48	Babylon	[29, 34, 40–43, 52]
	106	74.35	Sulemania	[30]
	76	22.2	Najaf	[45]
	50	75	Saladin	[46]
	90	64.5	Erbil	[54]
	14	25	Tikrit	[48]
	381	51.2	Duhok	[49, 50]
	113	50	Anbar	[55]
	81	11.1	Missan	[31]
	91	46.1	Wasit	[36]
Imipenem	42	11.9	Babylon	[41]
	7	25	Anbar	[33]
	91	0	Wasit	[36]
	106	0	Zakho	[47]
	234	1.5	Kirkuk	[48]
	381	2.25	Duhok	[49, 50]
	113	4.5	Anbar	[55]
	50	4.7	Sulemania	[51]
	50	4.7	Erbil	[54]

Table 1.
Distribution of antibiotic resistance among Iraqi local UPEC.

Antibiotic class	Genes	Province	Reference
Quinolones	*qnrA, qnrB, qnrD, qnrS*	Babylon	[9]
Beta-lactam	*blaTEM, blaCTX-M*	Sulemania	[29]
Beta-lactam	*blaTEM, blaCTX-M, blaSHV*	Najaf	[30–32]
Beta-lactam	*blaSHV*	Anbar	[33]
Beta-lactam	*blaTEM, blaCTX-M, blaSHV, blaOXA, AmpC*	Babylon	[34]
Beta-lactam	*blaCTX-M14, blaCTX-M15* *blaCTX-M24, blaCTX-M27*	Duhok	[35]
Beta-lactams	*blaTEM, blaCTX-M, blaSHV* *blaOXA*	Wasit	[36]
Beta-lactams	*blaCTX-M*	Karbala	[37]
Beta-lactams	*blaTEM, blaPER, blaVIM and blaCTX-M-2,* *blaTEM,*	Baghdad	[38]
Aminoglycosides	*aac(6)-Ib, aph(3)-I, aph(3)-IIa* *aph(3)-Ib, ant(3)-I, aac(3)-II* *aac(3)-IV*	Najaf	[31, 39]

Table 2.
Antibiotic resistance genes among Iraqi local UPEC.

due to the presence of *qnrA*, *qnrB*, *qnrD* and *qnrS* genes [9, 29–37]. Resistance to aminoglycosides among UPEC may be mediated by *aac(6)-Ib*, *aph(3)-I*, *aph(3)-IIa*, *aph(3)-Ib*, *ant(3)-I*, *aac(3)-II* and *aac(3)-IV* [31, 39].

2.2 ExPEC-associated sepsis and RTIs antibiotic resistance

Lower respiratory tract infections are a leading cause of morbidity and death worldwide. Optimizing the treatment of respiratory tract infections (RTIs) caused by multidrug-resistant (MDR) *Escherichia coli* has become challenging with existing antibiotic options. *E. coli* pathogens have various virulence factors that determine their pathogenesis and antimicrobial resistance (AMR) mechanisms [38]. The rapid and ongoing spread of antimicrobial-resistant organisms threatens our ability to successfully treat a growing number of infectious diseases. It is well established that antibiotic use is a significant, and modifiable, driver of antibiotic resistance [56, 57]. Physician visits for respiratory tract infections (RTI) commonly result in an antibiotic prescription, despite the fact that most upper RTIs are viral in nature. In these cases, antibiotics provide no benefit; thus, guidelines limit their recommended use to certain situations where the etiology is likely bacterial [58–60].

Over- and inappropriate prescribing of antibiotics is highly prevalent in the primary care setting, especially for upper respiratory tract infections (URTIs). In the outpatient setting, URTIs account for approximately 50–70% of total antibiotic prescriptions, even though most cases are of viral origin [61, 62]. The overuse of broad-spectrum antibiotics, such as third-generation cephalosporins, amoxicillin-clavulanate and fluoroquinolones, is strongly associated with the emergence of resistant strains, does not provide better clinical outcomes, and may lead to adverse events as well as unnecessary costs. Reducing unnecessary antibiotic prescriptions and the overuse of broad-spectrum agents may contain antimicrobial resistance and preserve the efficacy of existing antibiotics [63–65].

The Iraqi studies dealing with antibiotic resistance among sepsis-associated *E. coli* are summarized in **Table 3**. Most *E. coli* strains isolated from bloodstream

Antibiotic	No. of isolate/study	Resistance %	Province	Reference
Cefotaxime	7	92	Najaf	[66]
	19, 42, 2	94.8, 95.2, 100	Duhok	[67–69]
	9	41	Karbala	[70]
	41	51.1	Baghdad	[71]
Ceftriaxone	7	90	Najaf	[66]
	19, 42	94.8, 93	Duhok	[67, 68]
	9	29	Karbala	[70]
Ciprofloxacin	7	46	Najaf	[66]
	19	0.0	Duhok	[67, 68]
Amikacin	7	44	Najaf	[66]
	19, 42, 2	0.0, 35.7, 0.0	Duhok	[67–69]
	9	30	Karbala	[70]
	41	22	Baghdad	[71]
Gentamycin	7	46	Najaf	[66]
	19, 42, 2	78.5, 52.4, 40	Duhok	[67–69]
	9	38	Karbala	[70]
	41, 17	22, 29.4	Baghdad	[71, 72]
Imipenem	7	6	Najaf	[66]
	19, 42, 2	0.0, 9.5, 0.0	Duhok	[67–69]
	9	2	Karbala	[70]

Table 3.
Distribution of antibiotic resistance among Iraqi local sepsis-associated E. coli.

were resistant to most antimicrobials particularly β-lactam antibiotics and third-generation cephalosporins. It might be that long-term exposure to these antimicrobials by patients infected with bacteremia leads to horizontal transfer of plasmid-resistant antimicrobial genes between different strains of bacteria [66].

The average resistance rate is as follows: cefotaxime (79%), ceftriaxone (76.7%), ciprofloxacin (23%), amikacin (21.95%), gentamycin (43.7%) and imipenem (3.5%). The third-generation cephalosporins were the most commonly prescribed antibiotics compiling 54.3% followed by quinolones 7.5% of all prescribed antibiotics. Cefotaxime and ceftriaxone seem to be the preferred prescribed antibiotic for both surgical and medical wards [32].

2.3 InPEC-associated diarrheagenic infection and antibiotic resistance

Diarrhea is one of the major causes of serious issues among children in the developing world. More than 4 million children die annually from diarrhea in developing world. Diarrheagenic *E. coli* (DEC) is the most common cause of bacterial diarrhea in children worldwide and responsible for about 600,000 deaths per year [73, 74]. Diarrheagenic *E. coli* infection manifests as watery or bloody diarrhea accompanied by mild-to-severe dehydration. Beta-lactamases are a big problem when produced by DEC rendering the infection hard to be treated or untreatable. The arising of resistance toward extended-spectrum cephalosporins is most often due to hydrolyzing them by extended-spectrum β-lactamases (ESBLs) or due to AmpC. AmpC β-lactamases can prompt resistance to cephalothin, cefoxitin, cefazolin, most penicillins and beta-lactamase inhibitor-beta-lactam combinations. *Escherichia coli* isolates with CTX-M ESBLs are spreading multiresistance in the community and in hospitals [75, 76]. The resistance rate of Iraqi local diarrheagenic *E. coli* to different antibiotic classes is summarized in **Table 4**. The resistance genes are listed in **Table 5**.

The average resistance rate is as follows: cefotaxime (76.34%), ceftriaxone (79.87%), ciprofloxacin (26.3%), amikacin (31.21%), gentamycin (35.68%) and imipenem (8.18%).

The possible explanation to high level of resistance to this drug may be as a result of it being the most commonly available antibiotic used as a routine therapy among

Antibiotic	No. of isolate/study	Resistance %	Province	Reference
Cefotaxime	18	92	Najaf	[66]
	89, 39, 114	82.9, 89.7, 100	Babylon	[76–78]
	100, 145	71.4, 96.4	Wasit	[79, 80]
	51, 37	4, 54	Baghdad	[81, 82]
	30	96.7	Tikrit	[83]
Ceftriaxone	18	90	Najaf	[66]
	89, 39, 114	74.6, 79.5, 100	Babylon	[76–78]
	100	81	Wasit	[79]
	37	40.5	Baghdad	[82]
	163	76.66	Basra	[84]
	30	96.7	Tikrit	[83]
Ciprofloxacin	18	46	Najaf	[66]
	89, 39, 114	0.0, 15.8, 72.7	Babylon	[76–78]
	24, 51, 37	0.0, 8, 45.9	Baghdad	[81, 82, 85]
	145	25	Wasit	[80]
	30	23.3	Tikrit	[83]
Amikacin	18, 535	44, 0.0	Najaf	[66, 86]
	89, 39, 114	22.6, 12.8, 36.4	Babylon	[76–78]
	100, 145	7.1, 50	Wasit	[79, 85]
	24, 51, 37	16.6, 59, 67.5	Baghdad	[81, 82, 85]
	30	40	Tikrit	[83]
Gentamycin	18, 535	46, 9.1	Najaf	[66, 86]
	89, 39, 114	2.8, 51.3, 54.5	Babylon	[76–78]
	24, 51, 37	0.0, 16, 100	Baghdad	[81, 82, 85]
	145	57.14	Wasit	[80]
	30	20	Tikrit	[83]
Imipenem	18, 535	6, 0.0	Najaf	[66, 86]
	89, 114	9.5, 36.4	Babylon	[76–78]
	100, 145	0.0, 0.0	Wasit	[79, 80]
	37	13.5	Baghdad	[82]
	163	0.0	Basra	[84]

Table 4.
Distribution of antibiotic resistance among Iraqi local diarrheagenic E. coli.

Antibiotic class	Genes	Province	Reference
Beta-lactams	*blaTEM, blaCTX-M, blaSHV, blaOXA, AmpC*	Najaf	[86]

Table 5.
Antibiotic resistance genes among Iraqi local diarrheagenic E. coli.

gastrointestinal infections and people readily purchasing it across the counter for self-medication in last years. This could be a reflection of use and misuse of these antibiotics in the society. This finding is a result of the fact that outside the hospital environment the general population has easy access to various antibiotics from any pharmacy without prescription from a doctor [82].

2.4 ExPEC-associated vaginosis and antibiotic resistance

Bacterial vaginosis (BV) is the most common vaginal infections among women in reproductive age. It is a condition of vaginal flora imbalance, in which the typically plentiful H_2O_2-producing lactobacilli are scarce and other bacteria such as *E. coli* are abundant [87, 88]. Multi-drug resistant cervicovaginal *Escherichia coli* (CVEC) infections are a serious health problem. Bacteria use several strategies to avoid the effects of antimicrobial agents and have evolved a highly efficient means for clonal spread and for the dissemination of resistance traits [4]. Extended-spectrum β-lactamases (ESBLs) are capable of hydrolyzing broad-spectrum cephalosporins and monobactams. In addition, ESBL-producing organisms exhibit co-resistance to many other classes of antibiotics resulting in limitation of therapeutic options. Vaginal *E. coli* represents a real threat especially to neonates; however, little information is available regarding its antibiotic resistance [89, 90]. The resistance rate of Iraqi local cervicovaginal *E. coli* to different antibiotic classes is summarized in **Table 6**. The resistance genes are listed in **Table 7**.

The average resistance rate is as follows: cefotaxime (75%), ceftriaxone (47.5%), ciprofloxacin (29.4%), gentamycin (25.4%) and imipenem (7.8%).

2.5 ExPEC-associated otitis media, meningitis and cholecystitis infection and antibiotic resistance

Ear infection is a common clinical problem throughout the world and the major cause of preventable hearing loss in the developing world [92]. Its chronic form is a serious problem in all age groups with less chance of recovery. In certain cases,

Antibiotic	No. of isolate/study	Resistance %	Province	Reference
Cefotaxime	32	62.5	Babylon	[88]
	51	86.2	Wasit	[91]
Ceftriaxone	32	50	Babylon	[88]
	51	45	Wasit	[91]
Ciprofloxacin	51	29.4	Wasit	[91]
Gentamycin	51	25.4	Wasit	[91]
Imipenem	32	15.6	Babylon	[88]
	51	0.0	Wasit	[91]

Table 6.
Distribution of antibiotic resistance among Iraqi local cervicovaginal E. coli.

Antibiotic class	Genes	Province	Reference
Beta-lactamases	*blaCTX-M, blaSHV, blaOXA*	Wasit	[91]

Table 7.
Antibiotic resistance genes among Iraqi local cervicovaginal E. coli.

this condition can lead to serious life-threatening complications, such as hearing impairment, brain abscesses or meningitis, mostly in childhood and late in life [93]. *E. coli* is one of the major causative agents of ear infection. The burden and prevalence of ear infection are more intense in developing countries due to the poor living standard and hygienic conditions along with a lack of proper nutrition. Increased antimicrobial resistance is one of the greatest global public health challenges, which has been accelerated by overprescription of antibiotics worldwide. Infection with antibiotic-resistant bacteria may cause severe illness, increased mortality rates and an increased risk of complications and admission to hospital and longer stay. *E. coli* was the most prevalent multi-antibiotic-resistant pathogenic bacteria isolated from suspected patient ear discharges with otitis media [94–96].

Gram-negative bacillary meningitis continues to be an important cause of mortality and morbidity (15% and 50%, respectively) throughout the world despite advances in antimicrobial chemotherapy and supportive care [97]. *E. coli* is the most common Gram-negative bacillary organism causing meningitis. Recent reports of *E. coli* strains producing CTX-M-type or TEM-type extended-spectrum β-lactamases create a challenge. *E. coli* meningitis follows a high degree of bacteremia and invasion of the blood-brain barrier [21, 98].

Cholecystitis is most often caused by gall stones. Gall stones are one of the most common disorders of the gastrointestinal tract. Bacterial infection accounts for 50–85% of the disease's onset. *Escherichia coli* was the main biliary pathogenic microorganism [99]. It is strongly associated with retrograde bacterial infection and is an inflammatory disease that can be fatal if inappropriately treated [100]. The resistance rate of Iraqi local *E. coli* isolated from otitis media, meningitis and cholecystitis to different antibiotic classes is summarized in **Table 8**. The resistance genes are listed in **Table 9**.

The average resistance rate is as follows: cefotaxime (72.57%), ceftriaxone (68.39%), ciprofloxacin (8.5%), gentamycin (42.46%) and imipenem (0%).

2.6 ExPEC-associated wound infection and antibiotic resistance

A wound can represent a simple or a severe disorder to an organ (such as the skin) or a tissue and can spread to other tissues and anatomical structures (e.g., subcutaneous tissue, muscles, tendons, nerves, vessels and even to the bone). Among all human body (HB) organs, the skin is without doubt the most exposed to impairment and injury, scratches and burns. By damaging the epithelium and connective structures, the HB's capability to provide protection from the outer environment is weakened [109]. An improper repair process can cause severe damage, like the loss of skin, initiation of an infection, with consequent harms to the subjacent tissues and even systemic ones. The most common and inevitable impediment to wound healing is the installation of an infection [110].

Skin and soft tissue infections (SSTIs) are one of the most common infections in patients of all age groups. The most common causative agents are *Staphylococcus aureus* and aerobic streptococci. However, several reports associating the *Escherichia coli* with SSTI have been published: *E. coli* was found to be the causative agent of neonatal omphalitis, cellulitis localized to lower or upper limbs, necrotizing fasciitis,

Antibiotic	No. of isolate/study	Resistance %	Province	Reference
Cefotaxime	93	100	Anbar	[101]
	22	76	Najaf	[102]
	2	100	Baghdad	[103]
	7	14.29	Babylon	[104]
Ceftriaxone	93	90.57	Anbar	[101]
	22	76	Najaf	[102]
	2	50	Baghdad	[103]
	7	57	Babylon	[104]
Ciprofloxacin	10	20	Basra	[105]
	93	98.11	Anbar	[101]
	22	76	Najaf	[102]
	5	0.0	Tikrit	[106]
	7	14.29	Babylon	[104]
Levofloxacin	22	62	Najaf	[102]
Amikacin	10	20	Basra	[105]
	22	14	Najaf	[102]
	2	0.0	Baghdad	[103]
	5	0.0	Tikrit	[106]
Gentamycin	10	20	Basra	[105]
	4	100	Thi-Qar	[107]
	22	44.8	Najaf	[102]
	2	50	Baghdad	[103]
	7	0.0	Babylon	[104]
Imipenem	22	0	Najaf	[102]

Table 8.
Distribution of antibiotic resistance among Iraqi local E. coli *associated with otitis media, meningitis and cholecystitis.*

Antibiotic class	Genes	Province	Reference
Beta-lactamases	*acrAB*	Anbar	[101]
Quinolone	*gyrA, parC*	Anbar	[101]
Beta-lactamases	*blaCTX-M, blaSHV, blaOXA, blaTEM*	Najaf	[102]
Beta-lactamases	*blaCTX-M, blaSHV, blaOXA, blaTEM*	Al-Qadisiyah	[108]

Table 9.
Antibiotic resistance genes among Iraqi local E. coli *associated with otitis media, meningitis and cholecystitis.*

surgical site infections, infections after burn injuries and others [111, 112]. Cellulitis due to *Escherichia coli* is rare and usually secondary to a cutaneous portal of entry. Skin and soft tissue infections (SSTIs) secondary to *E. coli* bacteremia have been reported exclusively in immunodeficient patients. The resistance rate of Iraqi local *E. coli* isolated from wound infection to different antibiotic classes is summarized in **Table 10**. The resistance genes are listed in **Table 11**.

Antibiotic	No. of isolate/study	Resistance %	Province	Reference
Cefotaximes	7	85	Karbala	[113]
	165	100	Baghdad	[101]
	2	90	Anbar	[114]
	9	33.33	Karbala	[115]
	4	100	Baghdad	[116]
	16	88.09	Baghdad	[117]
	51	74.5	Najaf	[118]
Ceftriaxone	7	77	Karbala	[113]
	165	90.57	Baghdad	[101]
	2	90	Anbar	[114]
	4	100	Baghdad	[116]
	7	100	Erbil	[119]
	19	83.8	Erbil	[120]
	15	75	Erbil	[121]
	51	68.6	Najaf	[118]
Ciprofloxacin	7	42	Karbala	[113]
	165	98.11	Baghdad	[101]
	2	20	Anbar	[114]
	9	100	Karbala	[115]
	4	50	Baghdad	[116]
	7	85.8	Erbil	[119]
	16	73.8	Baghdad	[117]
	19	54.4	Erbil	[120]
	15	16.66	Erbil	[121]
	51	35.2	Najaf	[118]
	28	85.7	Diyala	[122]
Levofloxacin	2	20	Anbar	[114]
	4	0.0	Baghdad	[116]
	7	85.8	Erbil	[119]
	16	73.8	Baghdad	[117]
	15	17.7	Erbil	[121]
Amikacin	2	20	Anbar	[114]
	9	22.22	Karbala	[115]
	4	75	Baghdad	[116]
	7	14.2	Erbil	[119]
	19	1.9	Erbil	[120]
	15	18.33	Erbil	[121]
	51	39.2	Najaf	[118]

Antibiotic	No. of isolate/study	Resistance %	Province	Reference
Gentamycin	7	55	Karbala	[113]
	2	90	Anbar	[114]
	9	66.67	Karbala	[115]
	7	57.1	Erbil	[119]
	16	35.71	Baghdad	[117]
	19	40.6	Erbil	[120]
	15	25	Erbil	[121]
	51	78.4	Najaf	[118]
	28	64.3	Diyala	[122]
Imipenem	2	10	Anbar	[114]
	9	22.2	Karbala	[115]
	4	0.0	Baghdad	[116]
	7	0.0	Erbil	[119]
	16	4.76	Baghdad	[117]
	19	1.25	Erbil	[120]
	15	5	Erbil	[121]
	51	0.0	Najaf	[118]

Table 10.
Distribution of antibiotic resistance among Iraqi local E. coli *associated with wound infections.*

Antibiotic class	Genes	Province	Reference
Beta-lactamases	*blaNDM-1*	Baghdad	[123]
Carbapenem	*blaIMP*	Baghdad	[124]
Beta-lactamases	*blaNDM-1*	Basra	[125]
Beta-lactamases	*blaCTX-M*	Erbil	[121]
Beta-lactamases	*blaTEM, blaSHV, blaOXA51*	Babylon	[126]

Table 11.
Antibiotic resistance genes among Iraqi local E. coli *associated with wound infections.*

The average resistance rate is as follows: cefotaximes (81.56%), ceftriaxone (85.62%), ciprofloxacin (60.15%), levofloxacin (39.54%), amikacin (27.26%), gentamycin (56.97%) and imipenem (5.4%).

Author details

Hussein O.M. Al-Dahmoshi, Noor S.K. Al-Khafaji* and Mohammed H.O. Al-Allak
Biology Department, College of Science, University of Babylon, Hilla, Iraq

*Address all correspondence to: noorvaccine@gmail.com

References

[1] Gomes TA, Elias WP, Scaletsky IC, Guth BE, Rodrigues JF, Piazza RM, et al. Diarrheagenic *Escherichia coli*. Brazilian Journal of Microbiology. 2016;**47**:3-30. DOI: 10.1016/j.bjm.2016.10.015

[2] Foxman B. The epidemiology of urinary tract infection. Nature Reviews Urology. 2010;**7**(12):653. DOI: 10.1038/nrurol.2010.190

[3] Foxman B, Wu J, Farrer EC, Goldberg DE, Younger JG, Xi C. Early development of bacterial community diversity in emergently placed urinary catheters. BMC Research Notes. 2012;**5**(1):332. DOI: 10.1186/1756-0500-5-332

[4] Kaper JB, Nataro JP, Mobley HL. Pathogenic *Escherichia coli*. Nature Reviews Microbiology. 2004;**2**(2):123-140. DOI: 10.1038/nrmicro818

[5] Cabrera-Sosa L, Ochoa TJ. *Escherichia coli* diarrhea. In: Hunter's Tropical Medicine and Emerging Infectious Diseases. 10th ed. Elsevier. 2020. pp. 481-485. DOI: 10.1016/B978-0-323-55512-8.00046-6

[6] Zhang J, Xu Y, Ling X, Zhou Y, Lin Z, Huang Z, et al. Identification of diarrheagenic *Escherichia coli* by a new multiplex PCR assay and capillary electrophoresis. Molecular and Cellular Probes. 2020;**49**:101477. DOI: 10.1016/j.mcp.2019.101477

[7] Flores-Mireles AL, Walker JN, Caparon M, Hultgren SJ. Urinary tract infections: Epidemiology, mechanisms of infection and treatment options. Nature Reviews Microbiology. 2015;**13**(5):269-284. DOI: 10.1038/nrmicro3432

[8] Toval F, Köhler CD, Vogel U, Wagenlehner F, Mellmann A, Fruth A, et al. Characterization of *Escherichia coli* isolates from hospital inpatients or outpatients with urinary tract infection. Journal of Clinical Microbiology. 2014;**52**(2):407-418. DOI: 10.1128/JCM.02069-13

[9] Al-Hasnawy HH, Jodi MR, Hamza HJ. Molecular characterization and sequence analysis of plasmid-mediated quinolone resistance genes in extended-spectrum beta-lactamases producing uropathogenic *Escherichia coli* in Babylon Province, Iraq. Reviews in Medical Microbiology. 2018;**29**(3):129-135. DOI: 10.1097/MRM.0000000000000136

[10] Melsen WG, Rovers MM, Groenwold RH, Bergmans DC, Camus C, Bauer TT, et al. Attributable mortality of ventilator-associated pneumonia: A meta-analysis of individual patient data from randomised prevention studies. The Lancet Infectious Diseases. 2013;**13**(8):665-671. DOI: 10.1016/S1473-3099(13)70081-1

[11] Chastre J, Fagon JY. Ventilator-associated pneumonia. American Journal of Respiratory and Critical Care Medicine. 2002;**165**(7):867-903. DOI: 10.1164/ajrccm.165.7.2105078

[12] Russo TA, Johnson JR. Medical and economic impact of extraintestinal infections due to *Escherichia coli*: Focus on an increasingly important endemic problem. Microbes and Infection. 2003;**5**(5):449-456. DOI: 10.1016/S1286-4579(03)00049-2

[13] Fihman V, Messika J, Hajage D, Tournier V, Gaudry S, Magdoud F, et al. Five-year trends for ventilator-associated pneumonia: Correlation between microbiological findings and antimicrobial drug consumption. International Journal of Antimicrobial Agents. 2015;**46**(5):518-525. DOI: 10.1016/j.ijantimicag.2015.07.010

[14] Kollef MH, Ricard JD, Roux D, Francois B, Ischaki E, Rozgonyi Z, et al. A randomized trial of the amikacin

fosfomycin inhalation system for the adjunctive therapy of Gram-negative ventilator-associated pneumonia: IASIS Trial. Chest. 2017;**151**(6):1239-1246. DOI: 10.1016/j.chest.2016.11.026

[15] Atia A, Elyounsi N, Abired A, Wanis A, Ashour A. Antibiotic resistance pattern of bacteria isolated from patients with upper respiratory tract infections; a four year study in Tripoli city. Preprint. 20182018080435. DOI: 10.20944/preprints201808.0435.v1

[16] Zhang K, Schneider D, Biswas R, Espriella MG, Rao J, Baffoe-Bonnie A. 2603. Biofilm formation as a predictive marker of prognosis for *Escherichia coli* sepsis. InOpen Forum Infectious Diseases. 2019;**6**:S904-S905. DOI: 10.1093/ofid/ofz360.2281

[17] Gallardo L, Bauer M, Pannullo N, Michel LV. Determining the role of pal in *Escherichia coli* sepsis. The FASEB Journal. 2019;**33**(1_supplement): 648-643. DOI: 10.1096/fasebj.2019.33.1_supplement.648.3

[18] Pierpaoli E, Cirioni O, Simonetti O, Orlando F, Giacometti A, Lombardi P, et al. Potential application of berberine in the treatment of *Escherichia coli* sepsis. Natural Product Research. 2020;**31**:1-6. DOI: 10.1080/14786419.2020.1721729

[19] Xu X, Zhang L, Cai Y, Liu D, Shang Z, Ren Q, et al. Inhibitor discovery for the *E. coli* meningitis virulence factor IbeA from homology modeling and virtual screening. Journal of Computer-Aided Molecular Design. 2020;**34**(1):11-25. DOI: 10.1007/s10822-019-00250-8

[20] Kim M, Simon J, Mirza K, Swong K, Johans S, Riedy L, et al. Spinal Intradural *Escherichia coli* abscess masquerading as a neoplasm in a pediatric patient with history of neonatal *E. coli* meningitis: A case report and literature review. World Neurosurgery. 2019;**126**:619-623. DOI: 10.1016/j.wneu.2019.02.243

[21] Kasimahanti R, Satish SK, Anand M. Community-acquired *Escherichia coli* meningitis with ventriculitis in an adult—A rare case report. Journal of Intensive Care. 2018;**6**(1):1-5. DOI: 10.1186/s40560-018-0332-6

[22] Bichon A, Aubry C, Dubourg G, Drouet H, Lagier JC, Raoult D, et al. *Escherichia coli* spontaneous community-acquired meningitis in adults: A case report and literature review. International Journal of Infectious Diseases. 2018;**67**:70-74. DOI: 10.1016/j.ijid.2017.12.003

[23] Shrivastava SR, Shrivastava PS, Ramasamy J. World health organization releases global priority list of antibiotic-resistant bacteria to guide research, discovery, and development of new antibiotics. Journal of Medical Society. 2018;**32**:76-77. DOI: 10.4103/jms.jms_25_17

[24] Rodrigo-Troyano A, Sibila O. The respiratory threat posed by multidrug resistant gram-negative bacteria. Respirology. 2017;**22**(7):1288-1299. DOI: 10.1111/resp.13115

[25] Masters N, Wiegand A, Ahmed W, Katouli M. *Escherichia coli* virulence genes profile of surface waters as an indicator of water quality. Water Research. 2011;**45**(19):6321-6333. DOI: 10.1016/j.watres.2011.09.018

[26] Markowitz MA, Wood LN, Raz S, Miller LG, Haake DA, Kim JH. Lack of uniformity among United States recommendations for diagnosis and management of acute, uncomplicated cystitis. International Urogynecology Journal. 2019;**30**(7):1187-1194. DOI: 10.1007/s00192-018-3750-z

[27] Zude I, Leimbach A, Dobrindt U. Prevalence of autotransporters in

Escherichia coli: What is the impact of phylogeny and pathotype? International Journal of Medical Microbiology. 2014;**304**(3-4):243-256. DOI: 10.1016/j.ijmm.2013.10.006

[28] Behzadi P, Najafi A, Behzadi E, Ranjbar R. Microarray long oligo probe designing for *Escherichia coli*: An in-silico DNA marker extraction. Central European Journal of Urology. 2016;**69**(1):105. DOI: 10.1016/j.ijmm.2013.10.006

[29] Saeed NM. Detection of extended spectrum beta-lactamase gene production by *E. coli* isolated from human and broiler in Sulemania province/Iraq. Journal of Zankoy Sulaimani-Part A. 2014;**16**:2. DOI: 10.17656/jzs.10296

[30] Alkhudhairy MK, Alshammari MM. Extended spectrum β-lactamase-producing *Escherichia coli* isolated from pregnant women with asymptomatic UTI in Iraq. EurAsian Journal of BioSciences. 2019;**13**(2):1881-1889

[31] Alquraishi ZH, Alabbasy AJ, Alsadawi AA. Genotype and phenotype detection of *E. coli* isolated from children suffering from urinary tract infection. Journal of Global Pharmacy Technology. 2018;**10**(1):38-45

[32] Majeed HT, Aljanaby AA. Antibiotic susceptibility patterns and prevalence of some extended spectrum beta-lactamases genes in gram-negative bacteria isolated from patients infected with urinary tract infections in Al-Najaf City, Iraq. Avicenna Journal of Medical Biotechnology. 2019;**11**(2):192

[33] Al-Ouqaili MT. Molecular detection and sequencing of SHV gene encoding for extended-spectrum β-lactamases produced by multidrug resistance some of the gram-negative bacteria. International Journal of Green Pharmacy (IJGP). 2019;**12**(04):S910-S918. DOI: 10.22377/ijgp.v12i04.2274

[34] Al-Charrakh AH, AL-Tememy AZ. Phenotypic and genotypic characterization of IRS-producing *Escherichia coli* isolated from patients with UTI in Iraq. Medical Journal of Babylon. 2015;**12**(1):80-95

[35] Michael NS, Saadi AT. Detection of bla CTX-M, bla TEM-01 and bla SHV genes in multidrug resistant uropathogenic *E. coli* isolated from patients with recurrent urinary tract infections. International Journal of Medical Research & Health Sciences. 2018;**7**(9):81-89

[36] Al-Mayahie S, Al Kuriashy JJ. Distribution of ESBLs among *Escherichia coli* isolates from outpatients with recurrent UTIs and their antimicrobial resistance. The Journal of Infection in Developing Countries. 2016;**10**(06):575-583. DOI: 10.3855/jidc 6661

[37] Zirjawi AM, Hussein NH, Taha BM, Hussein JD. Molecular detection of Ctx-M-B-lactamases in *Pseudomonas aerogenosa* strains isolated from corneal infections. European Journal of Biomedical. 2017;**4**(4):70-74

[38] Promite S, Saha SK. Escherichia coli in respiratory tract infections: Evaluating antimicrobial resistance and prevalence of fimA, neuC and iutA virulence genes. Gene Reports. 2020;**18**:100576. DOI: 10.1016/j.genrep.2019.100576

[39] Alm'amoori KO, Hadi ZJ, Almohana AM. Molecular investigation of aminoglycoside modifying enzyme among aminoglycoside-resistant uropathogenic *Escherichia coli* isolates from Najaf Hospitals, Iraq. Indian Journal of Public Health Research & Development. 2019;**10**(10):2298-2303. DOI: 10.5958/0976-5506.2019.03199.1

[40] Alhamdany MH. Antibiotic susceptibility of bacteria isolated from patients with diabetes mellitus and

recurrent urinary tract infections in Babylon Province, Iraq. Medical Journal of Babylon. 2018;**15**(1):63-68. DOI: 10.4103/MJBL.MJBL_16_18

[41] Al-Hasnawy HH, Judi MR, Hamza HJ. The dissemination of multidrug resistance (MDR) and extensively drug resistant (XDR) among uropathogenic *E. coli* (UPEC) isolates from urinary tract infection patients in Babylon Province, Iraq. Baghdad Science Journal. 2019;**16** (4 Supplement):986-922. DOI: 10.21123/bsj.2019.16.4(Suppl.).0000

[42] AL-Sa'ady AT, Al-Mawla YH. Comparison of effects antibiotics and natural honey and extracts of plants on *Escherichia coli* growth isolated from different pathogenic cases. Journal of University of Babylon for Pure and Applied Sciences. 2019;**27**(3):420-434

[43] Abed ZH, Jarallah EM. Antibiotics susceptibility pattern of clinical and environmental *Escherichia coli* isolates from Babylon hospitals. In Journal of Physics: Conference Series. 2019;**1294**(6):062105. DOI: 0.1088/1742-6596/1294/6/062105

[44] Mohammed S, Ahmed M, Karem K. Incidence of multi-drug resistant *Escherichia coli* isolates from blood and urine in Kerbala, Iraq. Journal of Kerbala University. 2014;**12**(4):222-227

[45] AL-Samarraie MQ, Omar MK, Yaseen AH, Mahmood MI. The wide spread of the gene haeomolysin (Hly) and the adhesion factor (Sfa) in the *E. coli* isolated from UTI. Journal of Pharmaceutical Sciences and Research. 2019;**11**(4):1298-1303

[46] Mansoor IY, AL-Otraqchi KI, Saeed CH. Prevalence of urinary tract infections and antibiotics susceptibility pattern among infants and young children in Erbil city. Zanco Journal of Medical Sciences. 2015;**19**(1):915-922. DOI: 10.15218/zjms.2015.0012

[47] Polse R, Yousif S, Assafi M. Prevalence and antimicrobial susceptibility patterns of uropathogenic *E. coli* among people in Zakho, Iraq. International Journal of Research in Medical Sciences. 2016;**4**(4): 1219-1223. DOI: 10.18203/2320-6012.ijrms20160813

[48] Alsamarai AG, Ali S. Urinary Tract Infection in Female in Kirkuk City, Iraq: Causative Agents and Antibiogram. Vol. 5. WJPPS; 2016. pp. 261-273. DOI: 10.20959/wjpps20166-6807

[49] Assafi MS, Ibrahim NM, Hussein NR, Taha AA, Balatay AA. Urinary bacterial profile and antibiotic susceptibility pattern among patients with urinary tract infection in Duhok city, Kurdistan region, Iraq. International Journal of Pure and Applied Sciences and Technology. 2015;**30**(2):54

[50] Hussein NR, Daniel S, Salim K, Assafi MS. Urinary tract infections and antibiotic sensitivity patterns among women referred to Azadi Teaching Hospital, Duhok, Iraq. Avicenna Journal of Clinical Microbiology and Infection. 2017;**5**(2):27-30. DOI: 10.34172/ajcmi.2018.05

[51] Merza NS, Jubrael JM. The prevalence of virulence factors among uropathogenic *Escherichia coli* strains isolated from different hospitals in Kurdistan Region-Iraq. International Journal of Bioinformatics and Biomedical Engineering. 2015;**1**(4):338-343

[52] Hindi NK, Hasson SO, Hindi SK. Bacteriological study of urinary tract infections with antibiotics susceptibility to bacterial isolates among honeymoon women in Al Qassim Hospital, Babylon Province, Iraq. Biotechnology Journal International. 2013;**3**:332-340. DOI: 10.9734/BBJ/2013/3573

[53] Ibrahim AA. Identification of iha and kpsMT virulence genes in *Escherichia coli* isolates with urinary tract infection in Iraqi patients. Indian Journal of Natural Sciences. 2019;**52**(9):16675-16682

[54] Al-Jebouri MM, Mdish SA. Antibiotic resistance pattern of bacteria isolated from patients of urinary tract infections in Iraq. Open Journal of Urology. 2013;**3**(2):124-131. DOI: 10.4236/oju.2013.32024

[55] Lafi SA, Alkarboly AA, Lafi MS. Bacterial urinary tract infection in adults, hit district Anbar governorate, west of Iraq. Egyptian Academic Journal of Biological Sciences, G. Microbiology. 2012;**4**(1):21-26, 10.21608/EAJBSG.2012.16656

[56] Spellberg B, Guidos R, Gilbert D, Bradley J, Boucher HW, Scheld WM, et al. Infectious Diseases Society of America. The epidemic of antibiotic-resistant infections: A call to action for the medical community from the Infectious Diseases Society of America. Clinical Infectious Diseases. 2008;**46**(2):155-164. DOI: 10.1086/524891

[57] Bell BG, Schellevis F, Stobberingh E, Goossens H, Pringle M. A systematic review and meta-analysis of the effects of antibiotic consumption on antibiotic resistance. BMC Infectious Diseases. 2014;**14**(1):13. DOI: 10.1186/1471-2334-14-13

[58] Hersh AL, Shapiro DJ, Pavia AT, Shah SS. Antibiotic prescribing in ambulatory pediatrics in the United States. Pediatrics. 2011;**128**(6): 1053-1061. DOI: 10.1542/peds.2011-1337

[59] Lee GC, Reveles KR, Attridge RT, Lawson KA, Mansi IA, Lewis JS, et al. Outpatient antibiotic prescribing in the United States: 2000 to 2010. BMC Medicine. 2014;**12**(1):96. DOI: 10.1186/1741-7015-12-96

[60] Hersh AL, Jackson MA, Hicks LA. Committee on infectious diseases. Principles of judicious antibiotic prescribing for upper respiratory tract infections in pediatrics. Pediatrics. 2013;**132**(6):1146-1154. DOI: 10.1542/peds.2013-3260

[61] Fleming-Dutra KE, Hersh AL, Shapiro DJ, Bartoces M, Enns EA, File TM, et al. Prevalence of inappropriate antibiotic prescriptions among US ambulatory care visits, 2010-2011. Journal of the American Medical Association. 2016;**315**(17):1864-1873. DOI: 10.1001/jama.2016.4151

[62] Smieszek T, Pouwels KB, Dolk FC, Smith DR, Hopkins S, Sharland M, et al. Potential for reducing inappropriate antibiotic prescribing in English primary care. Journal of Antimicrobial Chemotherapy. 2018;**73**(suppl_2): ii36-ii43. DOI: 10.1093/jac/dkx500

[63] Gerber JS, Ross RK, Bryan M, Localio AR, Szymczak JE, Wasserman R, et al. Association of broad-vs narrow-spectrum antibiotics with treatment failure, adverse events, and quality of life in children with acute respiratory tract infections. Journal of the American Medical Association. 2017;**318**(23):2325-2336. DOI: 10.1001/jama.2017.18715

[64] Kreitmeyr K, von Both U, Pecar A, Borde JP, Mikolajczyk R, Huebner J. Pediatric antibiotic stewardship: Successful interventions to reduce broad-spectrum antibiotic use on general pediatric wards. Infection. 2017;**45**(4):493-504. DOI: 10.1007/s15010-017-1009-0

[65] Poole NM, Shapiro DJ, Fleming-Dutra KE, Hicks LA, Hersh AL, Kronman MP. Antibiotic prescribing for children in United States emergency departments: 2009-2014. Pediatrics. 2019;**143**(2): e20181056. DOI: 10.1542/peds.2018-1056

[66] Aljanaby AA, Alfaham QM. Phenotypic and molecular characterization of some virulence factors in multidrug resistance *Escherichia coli* isolated from different clinical infections in Iraq. American Journal of Biochemistry and Molecular Biology. 2017;7:65-78. DOI: 10.3923/ajbmb.2017.65.78

[67] Al-Delaimi MS, Nabeel M. Early onset neonatal sepsis: Bacteriological antimicrobial susceptibility study in Duhok Province, Iraq. Journal of University of Babylon for Pure and Applied Sciences. 2019;**27**(3):196-212

[68] ABDULRAHMAN IS, SAADI AT. Bacterial isolates and their antimicrobial resistance patterns in neonatal sepsis recorded at Hevi Teaching Hospital in Duhok City/Kurdistan Region of Iraq. Duhok Medical Journal. 2019;**13**(2):84-95

[69] Saadi AT, Garjees NA, Rasool AH. Antibiogram profile of septic meningitis among children in Duhok, Iraq. Saudi Medical Journal. 2017;**38**(5):517

[70] Al-mousawi MR. Bacterial profile and antibiogram of bacteremic children in Karbala city, Iraq. Karbala Journal of Pharmaceutical Sciences. 2016;**11**:131-139

[71] Raham TF, Abood AM. Bacterial profile and antimicrobial susceptibility in neonatal sepses, Al-Alwyia Pediatric Teaching Hospital in Baghdad. Al-Kindy College Medical Journal. 2017;**13**(2):21-25

[72] Ibrahim SA, Rahma S. Microbiological profile of neonatal septicemia. Iraqi Postgraduate Medical Journal. 2012;**11**:13-18

[73] Lozano R, Naghavi M, Foreman K, Lim S, Shibuya K, Aboyans V, et al. Global and regional mortality from 235 causes of death for 20 age groups in 1990 and 2010: A systematic analysis for the Global Burden of Disease Study 2010. The Lancet. 2012;**380**(9859):2095-2128. DOI: 10.1016/S0140-6736(12)61728-0

[74] Kotloff KL, Nataro JP, Blackwelder WC, Nasrin D, Farag TH, Panchalingam S, et al. Burden and aetiology of diarrhoeal disease in infants and young children in developing countries (the Global Enteric Multicenter Study, GEMS): A prospective, case-control study. The Lancet. 2013;**382**(9888):209-222. DOI: 10.1016/S0140-6736(13)60844-2

[75] Livermore DM, Woodford N. The β-lactamase threat in Enterobacteriaceae. Pseudomonas and Acinetobacter. Trends in Microbiology. 2006;**14**(9):413-420. DOI: 10.1016/j.tim.2006.07.008

[76] Al-Marzoqi AH, Aziz HW, Dulaimi TH. Genotype, phenotype and virulence genes markers in *Escherichia coli*: Molecular characterization and antimicrobial susceptibility associated with Diarrhoea among children in Babil province, Iraq. International Journal of Biochemistry, Bioinformatics a nd Biotechnology Studies. 2019;**4**(1):11-20

[77] Al-Dahmoshi HO, Al-Yassari AK, Al-Saad NF, Al-Dabagh NN, Al-Khafaji NS, Mahdi RK, et al. Occurrence of AmpC, MBL, CRE and ESBLs among diarrheagenic *Escherichia coli* recovered from Infantile Diarrhea, Iraq. Journal of Pharmaceutical and Biomedical Sciences. 2015;**5**:189-195

[78] Al-Saadi ZH, Tarish AH, Saeed EA. Phenotypic detection and antibiotics resistance pattern of local serotype of *E. coli* O157: H7 from children with acute diarrhea in Hilla city/Iraq. Journal of Pharmaceutical Sciences and Research. 2018;**10**(3):604-609

[79] Abdul-hussein ZK, Raheema RH, Inssaf AI. Molecular diagnosis of diarrheagenic *E. coli* infections among

the pediatric patients in Wasit Province, Iraq. Journal of Pure and Applied Microbiology. 2018;**12**(4):2229-2241

[80] Shamki JA, Al-Charrakh AH, Al-Khafaji JK. Detection of ESBLs in enteropathogenic *E. coli* (EPEC) isolates associated with infantile diarrhea in Kut City. Medical Journal of Babylon. 2012;**9**(2):403-412

[81] Khalil ZK. Isolation and identification of different diarrheagenic (DEC) *Escherichia coli* pathotypes from children under five years old in Baghdad. Iraqi Journalof Community Medicine. 2015;**28**(3):126-132

[82] Hassan JS. Antimicrobial resistance patterns of *Escherichia coli* O157: H7 isolated from stool sample of children. Iraqi Journal of Medical Sciences. 2015; **13**(3):259-264

[83] Jameel ZJ, Al-Assie AH, Badawy AS. Isolation and characterization of enteroaggregative *Escherichia coli* among the causes of bacterial diarrhea in children. Tikrit Journal of Pure Science. 2018;**20**(4):30-37

[84] Yaqoob MM, Mahdi KH, Al-Hmudi HA, Mohammed-Ali MN. Detection of rotavirus a and *Escherichia coli* from diarrhea cases in children and Coliphage characterization. International Journal of Current Microbiology and Applied Sciences. 2016;**5**(4):68-83. DOI: 10.20546/ijcmas.2016.504.011

[85] AL-Shuwaikh AM, Ibrahim IA, Shwaikh RM. Detection of *E. coli* and rotavirus in diarrhea among children under five years old. Iraqi Journal of Biotechnology. 2015;**14**(1):85-92

[86] Alsherees HA, Ali SN. A preliminary occurrence of extended-spectrum and AmpC Beta-lactamases in clinical isolates of enteropathogenic *Escherichia coli* in Najaf, Iraq. bioRxiv. 2019;**1**:512731. DOI: 10.1101/512731

[87] Hemalatha R, Ramalaxmi BA, Swetha E, Balakrishna N, Mastromarino P. Evaluation of vaginal pH for detection of bacterial vaginosis. The Indian Journal of Medical Research. 2013;**138**(3):354

[88] Al-Khaqani MM, Alwash MS, Al-Dahmoshi HO. Investigation of phylogroups and some virulence traits among cervico-vaginal *Escherichia coli* (CVEC) isolated for female in Hilla City, Iraq. Malaysian Journal of Microbiology. 2017;**13**(2):132-138

[89] Paniagua-Contreras GL, Monroy-Pérez E, Solis RR, Cerón AB, Cortés LR, Alonso NN, et al. O-serogroups of multi-drug resistant cervicovaginal *Escherichia coli* harboring a battery of virulence genes. Journal of Infection and Chemotherapy. 2019;**25**(7):494-497. DOI: 10.1016/j.jiac.2019.02.004

[90] Chen LF, Chopra T, Kaye KS. Pathogens resistant to antibacterial agents. Infectious Disease Clinics. 2009;**23**(4):817-845. DOI: 10.1016/j.idc.2009.06.002

[91] Al-Mayahie SM. Phenotypic and genotypic comparison of ESBL production by vaginal *Escherichia coli* isolates from pregnant and non-pregnant women. Annals of Clinical Microbiology and Antimicrobials. 2013;**12**(1):7. DOI: 10.1186/1476-0711-12-7

[92] Ullauri A, Smith A, Espinel M, Jimenez C, Salazar C, Castrillon R. WHO ear and hearing disorders survey: Ecuador national study 2008-2009. Conference Papers in Science. 2014;**2014**:13. DOI: 10.1155/2014/847526.847526

[93] Fauci AS, Kasper DL, Longo DL, et al. Harrison's Principles of Internal Medicine. 17th ed. New York, NY, USA: McGraw-Hill; 2008

[94] Yiengprugsawan V, Hogan A. Ear infection and its associated risk factors, comorbidity, and health service use in Australian children. International

Journal of Pediatrics. 2013;**2013**:7. DOI: 10.1155/2013/963132.963132

[95] Llor C, Bjerrum L. Antimicrobial resistance: Risk associated with antibiotic overuse and initiatives to reduce the problem. Therapeutic Advances in Drug Safety. 2014;**5**(6):229-241. DOI: 10.1177/2042098614554919

[96] Afolabi OA, Salaudeen AG, Ologe FE, Nwabuisi C, Nwawolo CC. Pattern of bacterial isolates in the middle ear discharge of patients with chronic suppurative otitis media in a tertiary hospital in North central Nigeria. African Health Sciences. 2012;**12**(3):362-367. DOI: 10.4314/ahs.v12i3.18

[97] Okike IO, Johnson AP, Henderson KL, Blackburn RM, Muller-Pebody B, Ladhani SN, et al. Incidence, etiology, and outcome of bacterial meningitis in infants aged< 90 days in the United Kingdom and Republic of Ireland: Prospective, enhanced, national population-based surveillance. Clinical Infectious Diseases. 2014;**59**(10):e150-e157. DOI: 10.1093/cid/ciu514

[98] Kim KS. Human meningitis-associated *Escherichia coli*. EcoSal Plus. 2016;7(1):1-25. DOI: 10.1128/ecosalplus.ESP-0015-2015

[99] Liu J, Yan Q, Luo F, Shang D, Wu D, Zhang H, et al. Acute cholecystitis associated with infection of Enterobacteriaceae from gut microbiota. Clinical Microbiology and Infection. 2015;**21**(9):851-8e1. DOI: 10.1016/j.cmi.2015.05.017

[100] Kujiraoka M, Kuroda M, Asai K, Sekizuka T, Kato K, Watanabe M, et al. Comprehensive diagnosis of bacterial infection associated with acute cholecystitis using metagenomic approach. Frontiers in Microbiology. 2017;**8**:685. DOI: 10.3389/fmicb.2017.00685

[101] Al-Fayyadh ZH, Turkie AM, Al-Mathkhury HJ. New mutations in GyrA gene of *Escherichia coli* isolated form Iraqi patients. Iraqi Journal of Science. 2017;**58**(2B):778-788. DOI: 10.24996.ijs.2017.58.2B.1

[102] Alfatlawi AS, Alsaidi MA, Almaliky NK, Alsherees HA. Molecular characterization of extended spectrum â-lactamases (ESBLs) producing *Escherichia coli* isolated from Cholecystitis. EurAsian Journal of BioSciences. 2019;**13**(1):113-120

[103] Al-Rawazq HS, Mohammed AK, Hussein AA. Etiology and antibiotic sensitivity for otitis media in a central pediatric teaching hospital. Iraqi Medical Journal. 2013;**59**(2):84-90

[104] Kumar A, Jayachandran L, Kumar S. Antimicrobial susceptibility pattern in chronic suppurative otitis media patient in a tertiary care hospital. Value in Health. 2016;**19**(7):A845-A846

[105] Alsaimary IE, Alabbasi AM, Najim JM. Impact of multi drugs resistant bacteria on the pathogenesis of chronic suppurative otitis media. African Journal of Microbiology Research. 2010;**4**(13):1373-1382

[106] Ibrahim II. Bacteriological study of chronic suppurative otitis media among patients attending Tikrit Teaching Hospital for the year 2013. The Medical Journal of Tikrit. 2015;**20**(2):15-28

[107] Neamah AS. Detection of bacterial pathogens causing a chronic suppurative otits media and study of antibiotic susceptibility in Iraqi patients. International Journal of Research in Pharmaceutical Sciences. 2019;**10**(3):2567-2571. DOI: 10.26452/ijrps.v10i3.1511

[108] Wajid AR, Alwan SK. Bacteriological and genetic study on *Escherichia coli* causing acute calculus cholecystitis for diabetes patients

in AL-Diwanyia City. International Journal. 2015;**3**(6):1374-1382

[109] van Koppen CJ, Hartmann RW. Advances in the treatment of chronic wounds: A patent review. Expert Opinion on Therapeutic Patents. 2015;**25**(8):931-937. DOI: 10.1517/13543776.2015.1045879

[110] Sorg H, Tilkorn DJ, Hager S, Hauser J, Mirastschijski U. Skin wound healing: An update on the current knowledge and concepts. European Surgical Research. 2017;**58**(1-2):81-94. DOI: 10.1159/000454919

[111] Moet GJ, Jones RN, Biedenbach DJ, Stilwell MG, Fritsche TR. Contemporary causes of skin and soft tissue infections in North America, Latin America, and Europe: Report from the SENTRY Antimicrobial Surveillance Program (1998-2004). Diagnostic Microbiology and Infectious Disease. 2007;**57**(1):7-13. DOI: 10.1016/j.diagmicrobio.2006.05.009

[112] Tourmousoglou CE, Yiannakopoulou EC, Kalapothaki V, Bramis J, Papadopoulos JS. Surgical-site infection surveillance in general surgery: A critical issue. Journal of Chemotherapy. 2008;**20**(3):312-318. DOI: 10.1179/joc.2008.20.3.312

[113] Al-Abbas AK. Aerobic bacteria isolation from post-caesarean surgical site and their antimicrobial sensitivity pattern in Karbala city, Iraq. Iraq Medical Journal. 2017;**1**(4):94-98

[114] Lafi SA, Al-Shamarry M, Ahmed MS, Ahmed WI. Bacterial profile of infected traumatic wound and the antibiogram of predominant bacterial isolates using Viteck automated system in Ramadi Teaching Hospital, Iraq. Egyptian Academic Journal of Biological Sciences, G. Microbiology. 2018;**10**(1):69-76

[115] Abboud ZH, Al-Ghanimi NH, Ahmed MM. An insight into bacterial profile and antimicrobial susceptibility of burn wound infections in Kerbala, Iraq. Karbala Journal of Medicine. 2014;**7**(2):2023-2032

[116] Alkaabi SA. Bacterial isolates and their antibiograms of burn wound infections in Burns Specialist Hospital in Baghdad. Baghdad Science Journal. 2013;**10**(2):331-340

[117] Ali MR, Al-Taai HR, Al-Nuaeyme HA, Khudhair AM. Molecular study of genetic diversity in Escherichia coli isolated from different clinical sources. Biochemical and Cellular Archives. 2018;**18**(2):2553-2560

[118] Aljanaby AA, Aljanaby IA. Prevalence of aerobic pathogenic bacteria isolated from patients with burn infection and their antimicrobial susceptibility patterns in Al-Najaf City. Iraq-a three-year cross-sectional study. F1000Research. 2018;**7**(1157):1157-10.12688/f1000research.15088.1

[119] Abdulqader HH, AT S. The distribution of pathogens, risk factors and their antimicrobial susceptibility patterns among post-surgical site infection in Rizgari Teaching Hospital in Erbil/Kurdistan Region/Iraq. Journal of Duhok University. 2019;**22**(1):1-0. DOI: 10.26682/sjuod.2019.22.1.1

[120] Ali FA, Merza EM, Aula TS. Antibiotic resistance among *Escherichia coli* isolated from different clinical samples in Erbil City. International Journal of Research Studies in Science, Engineering and Technology. 2017;**4**(10):12-21

[121] Ali FA. Distribution of CTX-M gene among *Escherichia coli* strains isolated from different clinical samples in Erbil City. Iraqi Journal of Biotechnology. 2018;**17**(1):78-90

[122] Al-Azawi ZH. Antimicrobial susceptibility patterns of aerobic

bacterial species of wound infections in Baquba General Teaching Hospital-Diyala. Diyala Journal of Medicine. 2013;**4**(1):94-100

[123] Hussein NH. Genotypic detection of carbapenem-resistant *Escherichia coli* producing NDM-1 gene for the first time in Baghdad/Iraq. Journal of Global Pharma Technology. 2017;**9**(9):106-111

[124] ASK A-K, Albaayit A, Ibraheem OS, Abdul-Ilah HH. Molecular investigation of metallo-β-lactamase encoding gene in nosocomial carbapenem-resistant Enterobacteriaceae in Iraqi Hospitals. The Eurasia Proceedings of Science, Technology, Engineering & Mathematics. 2018;**2**:239-243

[125] Abas IJ, Al-Hamdani MA. New Delhi metallo-B-lactamase 1 (Ndm-1) producing *Escherichia coli* in Basrah Hospitals, Iraq. European Journal of Biomedical. 2017;**4**(02):102-107

[126] Al-Hasnawy HH, Saleh RH, Hadi BH. Existence of a ESBL genes in *Escherichia coli* and *Acinetobacter baumannii* isolated from different clinical specimens. Journal of Pharmaceutical Sciences and Research. 2018;**10**(5):1112-1117